Introduction to the Electron Theory of Small Molecules

Introduction to the Electron Theory of Small Molecules

A. C. Hurley

CSIRO Division of Chemical Physics,
Clayton, Victoria,
Australia

1976

Academic Press

London New York San Francisco

A Subsidiary of Harcourt Brace Jovanovich, Publishers

ACADEMIC PRESS INC. (LONDON) LTD.
24/28 Oval Road
London NW1

United States Edition published by
ACADEMIC PRESS INC.
111 Fifth Avenue
New York, New York 10003

Library of Congress Catalog Card Number: 76-41143
ISBN: 0-12-362460-6

Printed in Great Britain at The Spottiswoode Ballantyne Press
by William Clowes & Sons Limited, London, Colchester and Beccles

Preface

Because of recent advances in theory and computational techniques, the wave-mechanical treatment of the electronic states of molecules has become directly relevant to many experimental studies in chemistry, physics and molecular biology. Already the interactions of small atoms, molecules and ions can be computed to high accuracy from first principles and it has become clear that the extension of this work to larger molecules is a much less formidable task than was feared earlier.

This introduction consists of the earlier and simpler chapters of what was originally conceived as a comprehensive one-volume work on the electron theory of small molecules. As such it is suitable not only for the specialist theoretical chemist aiming to devise novel theories and to implement them computationally, but for a wider audience wishing to understand the basic theory so that they may employ the new results in a more critical and useful way. My primary objective is to provide a "royal road" from basic quantum mechanics, as exemplified by Schrödinger's wave equation and elementary atomic structure to the various theories and techniques of calculation which today are yielding such detailed information on molecular interactions. With this object in mind the simplest available mathematical method is used at each stage in the development; for example, Hartree–Fock theory is developed in terms of Slater determinants rather than density matrices. When unavoidable complications are encountered, as in the explicit reduction of the Hartree–Fock equations for open shell states, the text is expanded to assist the mathematically unsophisticated reader.

The text features a thorough treatment of symmetry and of the electrostatic and virial theorems, which have furnished such valuable qualitative insights into molecular interactions, and an extensive account of the simple systems H_2^+ and H_2. These systems are not only of great intrinsic interest, providing as they do crucial tests of molecular quantum mechanics, but also furnish simple introductions to concepts such as correlating natural orbitals and multi-configurational self-consistent field functions which find their principal applications in larger systems.

The final chapters deal with determinantal wave-functions, valence bond methods, molecular orbitals and Hartree–Fock theory. The latter method and its variants are treated at considerable length with examples drawn from recent calculations, since computations of this type have assumed a dominant place in the current literature. Although inherent short-comings of the Hartree–Fock method, especially in treating dissociation and chemical reaction, are pointed out, the general problem of electron correlation in larger molecules is reserved for a companion volume "Electron Correlation in Small Molecules" which is included in the current series of monographs on Theoretical Chemistry, edited by Professor D. P. Craig and Professor R. McWeeny for Academic Press.

I would like to acknowledge the generous help of several friends and colleagues: Professors R. G. Parr, R. D. Brown, D. P. Craig and R. McWeeny who read and commented on early drafts; especially Dr. V. W. Maslen and Mr. P. R. Taylor who read the whole text and made many valuable suggestions; and Mrs. Eunice Day for her skilful and meticulous typing of a difficult manuscript.

I acknowledge with gratitude permission to reproduce diagrams which I have received from Dr. W. Kolos, Dr. A. C. Wahl and the American Institute of Physics.

Melbourne
August 1976

A. C. Hurley

Contents

1 Potential Energy Curves and Surfaces

1.1. The adiabatic approximation 1
1.2. The nuclear wave equation for diatomic molecules . . . 3
1.3. Empirical potential curves 7
1.4. Dunham's method 9
1.5. The Rydberg-Klein-Rees method 11
1.6. Atomic units 14
References 15

2 Variational Wave Functions

2.1. The minimum energy principle 17
2.2. The linear variation method 19
2.3. Hellmann-Feynman formulae 20
2.4. Hellmann-Feynman theorems 26
2.5. Variational calculations with linear constraints 33
(a) Constrained variational calculations with a finite basis 36
(b) Use of orthogonalized basis 39
References 41

3 The Hydrogen Molecule Ion

3.1. Separation of the wave equation in elliptical coordinates . 42
3.2. Some simple variational approximations 47
(a) The linear combination of atomic orbitals (LCAO) method 48
(b) The scaled LCAO function 51
(c) The floating, scaled LCAO function. 53
(d) The Dickinson function 54
(e) Simple functions in elliptical coordinates 54
3.3. One-centre expansions. 55
3.4. Angular momentum operators and spin functions 61
References 65

4 The Symmetry of Molecular Electronic States

4.1. Symmetry groups of molecules 66
4.2. Transformation of electronic wave functions 70
4.3. Symmetry properties of exact wave functions 72
4.4. Derivation of irreducible representations 76
4.5. Orthogonality relations 82
4.6. The reduction of representations 85
4.7. The use of symmetry in approximate calculations 89
4.8. The symmetry paradox 97
References 103

5 The Hydrogen Molecule

5.1. Separation of the spin variables 104
5.2. The Heitler-London theory of H_2 and its extensions . . . 106
(a) The scaled Heitler-London function (Wang function) . 110
(b) The floating Wang function 110
(c) The Weinbaum function 112
(d) The Coulson-Fischer form and orthogonalized atomic orbitals 114
5.3. The molecular orbital method 117
(a) The LCAO approximation 119
(b) Configuration interaction 122
(c) Relationship with VB theory 123
(d) Hartree-Fock orbitals 127
(e) Optimum double configurations; the (A, B) form . . 132
5.4. Extended orbital calculations 138
(a) In-out correlation; the σ-limit 138
(b) Angular correlation 140
5.5. The James-Coolidge method 144
(a) The adiabatic calculation 147
(b) The non-adiabatic calculation 152
5.6. The natural orbital expansion 153
5.7. Summary 165
References 169

6 The Determinantal Method

6.1. The antisymmetry principle 171
6.2. Determinantal wave functions 172
6.3. Matrix elements between Slater determinants 174
6.4. The determinantal solution of the electronic Schrödinger equation 181
6.5. Spin eigenfunctions 182
6.6. The branching diagram 186
6.7. Extended Rumer diagrams 189

6.8. Evaluation of matrix elements 192
6.9. Spin projection operators 194
References 197

7 Molecular Orbitals and the Hartree-Fock Method

7.1. Qualitative molecular orbital theory 198
(a) The zeroth approximation 198
(b) Orbital energies and correlation diagrams 203
(c) The symmetry of MO states 210
(d) States of diatomic molecules and ions 217
(e) Simple polyatomic molecules 221
7.2. Hartree-Fock theory 225
(a) The unrestricted theory 225
(b) The closed-shell case 237
(c) Restricted Hartree-Fock theory for the open-shell case 242
(d) Approximate Hartree-Fock orbitals by the expansion method 253
(e) Use of orthogonalized bases 262
(f) Numerical integration and accumulative accuracy . . 264
7.3. Localized molecular orbitals and chemical bonds 266
7.4. Results of some molecular orbital calculations 275
(a) Basis functions 276
(b) Near Hartree-Fock wave functions for N_2 and NH_3 . 282
(c) Extrapolated Hartree-Fock total energies 285
(d) Hartree-Fock binding energies 286
(e) Hartree-Fock enthalpies of reaction 291
(f) Simpler SCF calculations 295
(g) Hartree-Fock potential curves and surfaces 297
(h) Ionization and excitation energies 301
(i) Molecular properties and charge density function . . 303
References 307

Appendix 1. Orthonormalization

A1.1. The Schmidt process 310
A1.2. Symmetric orthogonalization 312
A1.3. Canonical orthogonalization 313
References 314

Appendix 2. Character Tables and Basis Functions for the Crystallographic Point Groups

A2.1. Symmetry elements 315
(a) Pure rotations 315
(b) Improper rotations 315

A2.2. Direct product groups 316
A2.3. Characters and basis functions 316

Author Index 321

Subject Index 325

CHAPTER 1

Potential Energy Curves and Surfaces

The primary task in the electronic theory of molecules is to construct approximate solutions of the time-independent Schrödinger equation

$$H\Psi = E\Psi. \tag{1.1}$$

The first step in this process, which we consider in this chapter, is the separation of equation (1.1) into two equations, one determining the motion of the electrons in the electrostatic field of the nuclei in fixed positions and the other determining the motion of the nuclei. The earliest systematic discussion of this adiabatic approximation was given by Born and Oppenheimer[1] who carried out a perturbation expansion in terms of a parameter $(m/M)^{\frac{1}{4}}$, where m is the electronic mass and M a typical nuclear mass. However, for our purposes, a different approach also due to Born[2] is more convenient.

1.1. The adiabatic approximation

We use r, p, r_i, p_i $(i = 1, \ldots)$ to denote the electronic coordinates and momenta: R, P, R_α, P_α $(\alpha = 1, \ldots)$ to denote the coordinates and momenta of the nuclei. The Hamiltonian H of equation (1.1) may then be written

$$H = T_e + T_N + V(r, R), \tag{1.2}$$

where T_e, T_N are the electronic and nuclear kinetic energy operators

$$T_e = \sum_i \frac{p_i^2}{2m} = -\frac{\hbar^2}{2m} \sum_i \nabla_i^2 \tag{1.3}$$

$$T_N = \sum_\alpha \frac{P_\alpha^2}{2M_\alpha} = -\frac{\hbar^2}{2} \sum_\alpha \frac{1}{M_\alpha} \nabla_\alpha^2 \tag{1.4}$$

and the potential $V(r, R)$ contains nuclear–nuclear, electron–nuclear and electron–electron terms

$$V(r, R) = V_{NN} + V_{Ne} + V_{ee} = \sum_{\alpha<\beta} \frac{Z_\alpha Z_\beta e^2}{|R_\alpha - R_\beta|} - \sum_{\alpha, i} \frac{Z_\alpha e^2}{|R_\alpha - r_i|} + \sum_{i<j} \frac{e^2}{|r_i - r_j|}. \tag{1.5}$$

The electronic Schrödinger equation

$$\{T_e + V(r, R)\} \psi(r, R) = E(R)\psi(r, R) \tag{1.6}$$

determines a set of eigenvalues $E_n(R)$ and eigenfunctions $\psi_n(r, R)$ which, as the notation indicates, depend parametrically on the nuclear coordinates. These eigenfunctions form a complete set for functions of the electronic coordinates and we expand the total molecular wave function in terms of them

$$\Psi(r, R) = \sum_n \phi_n(R)\psi_n(r, R). \tag{1.7}$$

Substitution of this expression into equation (1.1) leads to a set of coupled equations for the nuclear wave functions $\phi_n(R)$

$$\{T_N + U_n(R) - E\} \phi_n(R) = - \sum_{n' \neq n} C_{nn'}(R, P)\phi_{n'}(R). \tag{1.8}$$

The term $U_n(R)$ of equation (1.8) differs from the eigenvalue $E_n(R)$ of the electronic Schrödinger equation (1.6) by a correction term involving the nuclear kinetic energy operator T_N. Thus

$$U_n(R) = E_n(R) + \Delta_n(R) \tag{1.9}$$

with

$$\Delta_n(R) = \int \psi_n^*(r, R) T_N \psi_n(r, R) \, dr. \tag{1.10}$$

The derivation of equation (1.8) and the form of the coupling terms $C_{nn'}(R, P)$ are given by Born and Huang.[2] The adiabatic approximation is obtained by dropping the right-hand side of equation (1.8). We then have

$$\{T_N + U_n(R)\}\phi_n(R) = E\phi_n(R). \tag{1.11}$$

The total molecular wave function now appears as the product of an electronic factor determined by equation (1.6) and a nuclear factor determined by equation (1.11), that is, the expansion (1.7) reduces to a single term

$$\Psi(r, R) = \phi_n(R)\psi_n(r, R). \tag{1.12}$$

For well-separated electronic states this is an excellent approximation. The order of magnitude of the relative error introduced in the energy is given by

$$\frac{m}{M}\frac{\text{separation of vibrational levels}}{\text{separation of electronic levels}} \approx 10^{-7}. \tag{1.13}$$

However, when two or more electronic levels are degenerate or near-degenerate it may be necessary to include several terms in the expansion (1.7).

The correction term $\Delta_n(R)$ of equation (1.10) is small and is frequently omitted to give the fixed nuclei approximation

$$\{T_N + E_n(R)\}\phi_n(R) = E\phi_n(R). \tag{1.14}$$

Equation (1.14), which also appears in Born and Oppenheimer's analysis,[1] is the most frequently used form of the adiabatic approximation. The electronic eigenvalues $E_n(R)$ play the part of the potential energy for the nuclear motions. For a diatomic molecule $E_n(R)$ depends only on the separation of the nuclei and we obtain a set of potential energy curves, one for each electronic state. For polyatomic molecules several variables are needed to specify the nuclear positions and the electronic eigenvalues give potential energy surfaces.

However, there is no need to drop the correction term $\Delta_n(R)$. As we see from equation (1.10) this term is easy to determine from the electronic eigenfunctions. It makes a significant contribution to the potential energy curves for several states of the H_2 molecule which have been calculated to a very high accuracy (Chapter 5). In these applications $\Delta_n(R)$ appears in a slightly different form, since it is more convenient to separate out the motion of the centre of mass at the beginning of the calculation rather than work in a fixed coordinate system as we have done.

1.2. The nuclear wave equation for diatomic molecules

As we have seen the nuclear motion, in the adiabatic approximation, is determined by the Schrödinger equation (1.11). For diatomic molecules, the potential energy term $U_n(R)$ depends only on the nuclear separation, now denoted by R, and after separation of the motion of the centre of mass equation (1.11) reduces to

$$\left\{-\frac{\hbar^2}{2\mu}\nabla^2 + U_n(R)\right\}\Phi_n(R, \theta, \varphi) = E\Phi_n(R, \theta, \varphi). \tag{1.15}$$

Here μ, the reduced mass, is related to the masses of the two nuclei by

$$\mu = \frac{M_a M_b}{M_a + M_b}, \tag{1.16}$$

and we have introduced spherical polar coordinates R, θ, φ to express the wave function Φ_n for the relative motion of the two nuclei. In these coordinates equation (1.15) is separable. The resulting radial wave equation is

$$\left\{-\frac{\hbar^2}{2\mu}\frac{d^2}{dR^2} + U_n(R) + \frac{\hbar^2 J(J+1)}{2\mu R^2}\right\} f_{nJ}(R) = E f_{nJ}(R) \tag{1.17}$$

where J, the rotational quantum number is restricted to the integral values

$$J = 0, 1, 2, \ldots \tag{1.18}$$

by the angular wave equation.

Details of the reduction of equation (1.11) for a diatomic molecule to the radial wave equation (1.17) are given in several standard texts (e.g. Eyring, Walter and Kimball[(3)]).

The two most common types of potential energy curve are illustrated in Fig. 1.1. For a stable electronic state the potential energy curve exhibits a minimum at the equilibrium nuclear separation R_e and, for larger R values, rises smoothly to approach a dissociation limit equal to the energy of the appropriate state of the separated

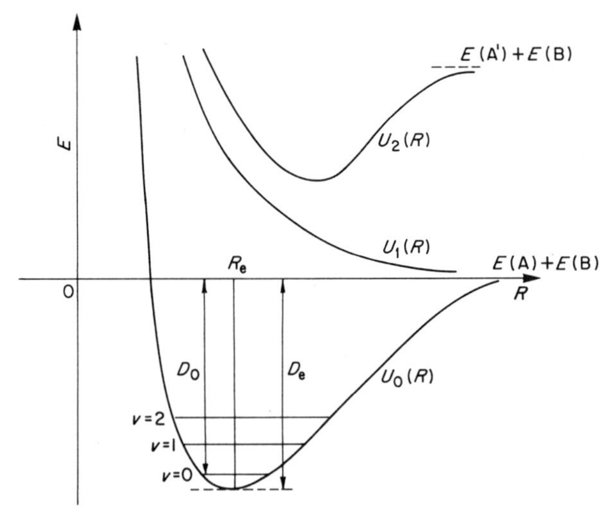

Fig. 1.1. Typical potential energy curves for a diatomic molecule AB. For the ground state curve $U_0(R)$, the first few vibrational levels and the quantities R_e, D_0 and D_e are shown.

atoms. For an unstable state the curve decreases monotonically with increasing R, that is, we have repulsion between the two atoms for all nuclear separations. In some cases more complex curves exhibiting maxima and subsidiary minima have been found. For small nuclear separations all the curves rise steeply to very large values because of the nuclear repulsion term in $U_n(R)$.

The dissociation energy D_0 for a stable electronic state is the difference between the energy of the molecular ground state, that is the lowest eigenvalue of equation (1.17) for the case $J = 0$, and the dissociation limit. The quantity D_e, defined as the depth of the minimum $U(R_e)$ below the energy of the separated atoms, differs from D_0 by the zero-point energy of the vibrational motion. In spectroscopic and thermochemical work D_0 is the more fundamental quantity, but D_e is useful for comparison with theoretical calculations which are often restricted to the single nuclear separation R_e. We shall refer to D_e as the binding energy.

For an unstable electronic state, equation (1.17) has only unquantized solutions with a continuous spectrum of eigenvalues. Such solutions are also obtained with a stable state for energies above the dissociation limit. However provided the potential well is deep enough, there will also be quantized discrete solutions lying below the continuous spectrum.

We may obtain a crude approximation to the lowest few of these discrete solutions by replacing the variable R in the centrifugal distortion term

$$\frac{\hbar^2 J(J+1)}{2\mu R^2}$$

by the constant R_e and retaining only the quadratic term in an expansion of $U_n(R)$ about the minimum. These approximations reduce (1.17) to the equation for a harmonic oscillator and the energy eigenvalues appear as the sum of electronic, vibrational and rotational contributions

$$E_{nvJ} = U_n(R_e) + E_{\text{vib}} + E_{\text{rot}}, \tag{1.19}$$

with

$$\frac{E_{\text{vib}}}{hc} = G(v) = \omega(v + \tfrac{1}{2}) \tag{1.20}$$

$$\frac{E_{\text{rot}}}{hc} = F(J) = BJ(J+1) \tag{1.21}$$

$$\omega = \frac{1}{2\pi c}\sqrt{\frac{U''(R_e)}{\mu}} \tag{1.22}$$

$$B = \frac{h}{8\pi^2 \mu c R_e^2}. \tag{1.23}$$

In equations (1.20) and (1.21) we have introduced a factor hc, so that the vibrational and rotational term values, $G(v)$ and $F(J)$, and the constants ω and B are expressed in wave number units (cm^{-1}). The vibrational quantum number v (equation (1.20)) takes the values 0, 1, 2, ...

We see from equations (1.22) and (1.23) that, in our present crude approximation, the spacings of the rotational and vibrational levels are determined by the location of the minimum in the potential curve (R_e) and the curvature at this point $U''(R_e)$. Conversely, if these spacings are determined from an analysis of molecular spectra we may infer these two properties of the potential curve.

Empirically it is found that the simple expressions (1.20) and (1.21) must be generalized to allow for several factors. The vibrational separations

$$\Delta G_{v+\frac{1}{2}} = G(v+1) - G(v) \tag{1.24}$$

are not constant but vary smoothly with v, usually decreasing and approaching zero at the dissociation limit. The rotational constant B also shows some dependence on both v and J.

The empirical vibrational and rotational term values are commonly expressed by generalizations of the simple equations (1.20) and (1.21). Thus

$$F(v, J) = G(v) + F_v(J), \tag{1.25}$$

with

$$G(v) = \omega_e(v+\tfrac{1}{2}) - \omega_e x_e(v+\tfrac{1}{2})^2 + \omega_e y_e(v+\tfrac{1}{2})^3 + \ldots \tag{1.26}$$

$$F_v(J) = B_v J(J+1) - D_v J^2(J+1)^2 + \ldots \tag{1.27}$$

$$B_v = B_e - \alpha_e(v+\tfrac{1}{2}) + \ldots \tag{1.28}$$

$$D_v = \mathrm{D}_e + \beta_e(v+\tfrac{1}{2}) + \ldots. \tag{1.29}$$

For a large number of electronic states of diatomic molecules accurate values of the constants ω_e, $\omega_e x_e$, B_e, α_e, D_e‡ . . . are known from the analysis of molecular spectra.[4, 5] The values of these constants contain much information about the shape of the potential energy curve. It is important to have methods of extracting this information for comparison with theoretical potential curves computed from the electronic wave equation (1.6).

For simple molecules, such as H_2 and H_2^+, very accurate potential curves have been computed from equation (1.6). In these cases the most satisfactory procedure is the numerical integration of equation (1.17) and a direct comparison of the theoretical and experimental term values. An example of this procedure is given in Chapter 5.

‡ The centrifugal distortion constant D_e has, of course, no relation to the binding energy D_e shown in Fig. 1.1.

For larger diatomic molecules less accurate theoretical potential curves can be computed, and it is convenient to reverse this procedure and determine an experimental potential curve from empirical values of the spectroscopic constants. As we shall see below (Section 1.5), this inverse problem is explicitly soluble to a high degree of accuracy, but first we consider a simpler procedure which is of adequate accuracy for many purposes.

1.3. Empirical potential curves

One way of obtaining the shape of a potential curve from empirical spectroscopic constants is to assume some simple functional form for $U_n(R)$ containing a number of variable parameters, calculate the vibrational and rotational term values and fix the variable parameters by comparison with the empirical term values. Two properties are required of the assumed functional form:

(i) It should be physically reasonable, that is, of the same general shape as the stable potential curves of Fig. 1.1.

(ii) It should permit an exact or sufficiently accurate approximate solution of the radial wave equation (1.17).

Quite a large variety of empirical potential curves have been derived in this way. Perhaps the most widely used and useful of them are the two forms of the Morse[6] curve and the generalization of this curve suggested by Hulburt and Hirschfelder.[7]

Morse assumes a potential curve of the form, in cm^{-1},

$$U(R) - U(R_e) = D(1 - e^{-\beta(R - R_e)})^2 \tag{1.30}$$

and shows that, for this form of $U(R)$, equation (1.17) is exactly soluble when $J = 0$. To a high degree of accuracy the vibrational term values for the potential (1.30) are given by[3, 6]

$$G(v) = \beta\sqrt{\frac{Dh}{2\pi^2 c\mu}}\,(v + \tfrac{1}{2}) - \frac{h\beta^2}{8\pi^2 \mu c}\,(v + \tfrac{1}{2})^2. \tag{1.31}$$

As ter Haar[8] points out, this simple form of the eigenvalues is obtained only if the boundary conditions for equation (1.17) are applied at $R = \pm\infty$, instead of at $R = 0$ and $R = \infty$ as they should be. However, since the potential (1.30) assumes very large positive values for $R \leqslant 0$, this approximate treatment of the boundary conditions introduces quite negligible errors in the eigenvalues.

The two forms of the Morse curve arise from different choices of the parameters D and β. If we equate the coefficients in equation (1.31) with the empirical ω_e and $\omega_e x_e$ we obtain

$$\beta = (8\pi^2 \mu c/h)^{\frac{1}{2}} (\omega_e x_e)^{\frac{1}{2}} \tag{1.32}$$

and

$$D = \frac{\omega_e^2}{4\omega_e x_e}, \tag{1.33}$$

whereas, if we fit only the first term in $G(v)$ and require that equation (1.30) gives the correct dissociation limit, we get

$$D = D_e \tag{1.34}$$

and

$$\beta = (2\pi^2 \mu c/D_e h)^{\frac{1}{2}} \omega_e. \tag{1.35}$$

In both cases the value of R_e is obtained from the rotational constant B_e of equation (1.28)

$$R_e = \left(\frac{h}{8\pi^2 \mu c B_e}\right)^{\frac{1}{2}}. \tag{1.36}$$

The Morse curve with parameters given by equations (1.32) and (1.33) is a good approximation for values of R near R_e but usually fails badly for large nuclear separations. For most states the parameters (1.34) and (1.35) provide a better overall fit, provided that an accurate value of D_e is available.

The generalization of the Morse function introduced by Hulburt and Hirschfelder, namely

$$U(R) - U(R_e) = D[(1 - e^{-\beta x})^2 + b\beta^3 x^3 e^{-2\beta x}(1 + a\beta x)] \tag{1.37}$$

where

$$x = R - R_e$$

contains five parameters R_e, D, β, a and b whose values are determined from the five spectroscopic constants D_e, B_e, ω_e, $\omega_e x_e$ and α_e. The fitting equations are (1.34), (1.35) and (1.36) above, and

$$b = 1 - \frac{1}{\beta R_e}\left(1 + \frac{\alpha_e \omega_e}{6B_e^2}\right) \tag{1.38}$$

$$a = 2 - \frac{1}{b}\left[\frac{7}{12} - \frac{1}{\beta^2 R_e^2}\left(\frac{5}{4} + \frac{5\alpha_e \omega_e}{12B_e^2} + \frac{5\alpha_e^2 \omega_e^2}{144B_e^4} - \frac{2\omega_e x_e}{3B_e}\right)\right]. \tag{1.39}$$

Hulburt and Hirschfelder derive these additional relationships from Dunham's analysis of the WKBJ phase integral (Section 1.4). Their extension of the Morse function is a particularly convenient one since the fitting of the parameters is quite simple and employs the most commonly available spectroscopic constants.

Many other empirical potential curves have been suggested by various authors. A review of this work, including comparisons with the more general methods of the next two sections, has been given by Steele, Lippincott and Vanderslice.[9]

1.4. Dunham's method[10]

Dunham's analysis of the term values of a rotating vibrator is based on the quasi-classical or WKBJ approximation (see e.g. Landau and Lifshitz[11]). In this approximation the eigenvalues for the one-dimensional motion of a particle in a potential well are given by the phase integral condition

$$\oint p \, \mathrm{d}x = 2\int_{x_1}^{x_2} p \, \mathrm{d}x = h(n + \tfrac{1}{2}) \tag{1.40}$$

where p is the momentum of the particle and x_1 and x_2 are the classical turning points.

If we introduce the variable

$$\xi = (R - R_{\mathrm{e}})/R_{\mathrm{e}} \tag{1.41}$$

and express all energy quantities in wave number units, the phase integral condition for the radial wave equation (1.17) becomes

$$\oint (F - U_J)^{\frac{1}{2}} \, \mathrm{d}\xi = 2\pi B_{\mathrm{e}}^{\frac{1}{2}}(v + \tfrac{1}{2}). \tag{1.42}$$

Here

$$B_{\mathrm{e}} = \frac{h}{8\pi^2 \mu R_{\mathrm{e}}^2 c} \tag{1.43}$$

and the effective potential U_J is the sum of two terms

$$U = \frac{U_n(R) - U_n(R_{\mathrm{e}})}{hc} \tag{1.44}$$

and the centrifugal term

$$U_r = \frac{1}{hc} \frac{\hbar^2 J(J+1)}{2\mu R^2} = \frac{B_{\mathrm{e}} J(J+1)}{(1+\xi)^2}. \tag{1.45}$$

Both these contributions are expanded as power series in the variable ξ

$$U = a_0 \xi^2 (1 + a_1 \xi + a_2 \xi^2 + \ldots) \tag{1.46}$$

$$U_{\mathrm{r}} = B_{\mathrm{e}} J(J+1)(1 - 2\xi + 3\xi^2 - 4\xi^3 + \ldots). \tag{1.47}$$

Using these expansions we may evaluate the phase integral in equation (1.42) as a power series in the term value F. Inserting the empirical expansion of F in terms of the spectroscopic constants (equation (1.25)) and comparing coefficients, we obtain expressions for the spectroscopic constants in terms of the coefficients in the expansion (1.46). The first few of these relations are as follows:

$$\begin{aligned}\omega_e &= (4B_e a_0)^{\frac{1}{2}} \\ -\alpha_e &= \frac{6B_e}{\omega_e}(1 + a_1) \\ -\omega_e x_e &= \frac{3B_e}{2}\left(a_2 - \frac{5a_1^2}{4}\right) \\ \mathrm{D}_e &= \frac{4B_e^3}{\omega_e^2}.\end{aligned} \tag{1.48}$$

Dunham derives relationships of this type for fifteen spectroscopic constants. He also goes beyond the first-order WKBJ approximation by including contributions from higher-order phase integrals in equation (1.42). For our purposes these additional terms are negligible.

Dunham's analysis provides a very simple and convenient method for computing spectroscopic constants from a theoretical potential curve. The parameters $a_0, a_1, a_2, \ldots$ are obtained by a least squares fit of the theoretical curve to the expression (1.46). The relations (1.48) then give theoretical spectroscopic constants for comparison with the empirical values.

However, the power series expansions (1.46) and (1.47) are not suitable for the inverse process of determining an "experimental" potential curve from known spectroscopic constants. This is because of their very poor convergence. It is obvious that the expansion (1.47) converges only for $|\xi| < 1$, that is for $R < 2R_e$, and in practice it is found that the useful range of convergence is much less than this.

One way of extending the range of convergence is to assume some specific functional form for the potential curve and to use Dunham's relations to fix the values of disposable parameters. The equations (1.38) and (1.39) which determine the parameters a and b of the Hulburt–Hirschfelder curve were obtained in this way.

The convergence is also greatly improved by inverting the expansion (1.46) (Sandeman[12]). The resulting expansions

$$\begin{aligned}\frac{R_1}{R_e} &= 1 - \left(\frac{U}{a_0}\right)^{\frac{1}{2}} + c_1\left(\frac{U}{a_0}\right) - c_2\left(\frac{U}{a_0}\right)^{\frac{3}{2}} + \ldots \\ \frac{R_2}{R_e} &= 1 + \left(\frac{U}{a_0}\right)^{\frac{1}{2}} + c_1\left(\frac{U}{a_0}\right) + c_2\left(\frac{U}{a_0}\right)^{\frac{3}{2}} + \ldots\end{aligned} \tag{1.49}$$

give the classical turning points R_1, R_2 in terms of the height U above the minimum of the potential curve.

However, it is possible to determine the classical turning points directly from the WKBJ phase integral condition without resorting to any expansions. This method, which grew out of work of Rydberg(13) and Klein(14) using the old quantum theory, and was greatly improved for practical applications by Rees,(15) is usually referred to as the Rydberg-Klein-Rees or RKR method. The derivations of Rydberg and Klein involve both vibrational and rotational conditions from the old quantum theory. However, it has been shown recently(16, 17) that the basic equations of the RKR method may be derived solely from the phase integral condition (1.40). We use this method here.

1.5. The Rydberg-Klein-Rees method

In discussing this method it is convenient to introduce the variables

$$I = h(v + \tfrac{1}{2}) \tag{1.50}$$

$$\kappa = \frac{\hbar^2}{2\mu} J(J + 1) \tag{1.51}$$

and to express our equations in energy units (ergs) rather than wave number units. We consider I and κ as continuously variable from zero upwards, the values for the quantized energy levels being obtained by inserting integral values for v and J in equations (1.50) and (1.51).

The effective potential curve is given in terms of the potential curve without rotation by

$$U_\kappa(R) = U(R) + \frac{\kappa}{R^2}. \tag{1.52}$$

The phase integral condition (1.40) becomes, in our present notation,

$$2\int_{R_1}^{R_2} \{2\mu(E - U_\kappa(R))\}^{\frac{1}{2}}\, dR = h(v + \tfrac{1}{2}) = I \tag{1.53}$$

and the equations (1.25)-(1.29) determine the total energy of the nuclear motion E as a function of I and κ. Thus

$$E(I, \kappa) = hcF(v, J). \tag{1.54}$$

Following Klein(14) we consider the quantity

$$S(V, \kappa) = (2\pi^2\mu)^{-\frac{1}{2}} \int_0^{I'} \{V - E(I, \kappa)\}^{\frac{1}{2}}\, dI \tag{1.55}$$

where the upper limit of integration is determined by the equation

$$E(I', \kappa) = V. \tag{1.56}$$

Differentiating the phase integral condition (1.53) we obtain

$$\frac{\mathrm{d}I}{\mathrm{d}E} = (2\mu)^{\frac{1}{2}} \int_{R_1}^{R_2} \{E - U_\kappa(R)\}^{-\frac{1}{2}} \,\mathrm{d}R. \tag{1.57}$$

We now use this equation to change the variable of integration in equation (1.55) to E, the integration limits being obtained from Fig. 1.2. We find

$$\begin{aligned} S &= (2\pi^2\mu)^{-\frac{1}{2}} \int_{E(0,\kappa)}^{V} (V - E)^{\frac{1}{2}} \frac{\mathrm{d}I}{\mathrm{d}E} \,\mathrm{d}E \\ &= \frac{1}{\pi} \int_{E(0,\kappa)}^{V} \mathrm{d}E \int_{R_1(E)}^{R_2(E)} \mathrm{d}R \left\{ \frac{V - E}{E - U_\kappa(R)} \right\}^{\frac{1}{2}} \\ &= \frac{1}{\pi} \int_{R_1(V)}^{R_2(V)} \mathrm{d}R \int_{U_\kappa(R)}^{V} \mathrm{d}E \left\{ \frac{V - E}{E - U_\kappa(R)} \right\}^{\frac{1}{2}}. \end{aligned} \tag{1.58}$$

The integration with respect to E in equation (1.58) is an elementary one with the value $(\pi/2)(V - U_\kappa(R))$, so that we end up with a very simple expression for S, namely

$$S(V, \kappa) = \tfrac{1}{2} \int_{R_1(V)}^{R_2(V)} (V - U_\kappa(R)) \,\mathrm{d}R. \tag{1.59}$$

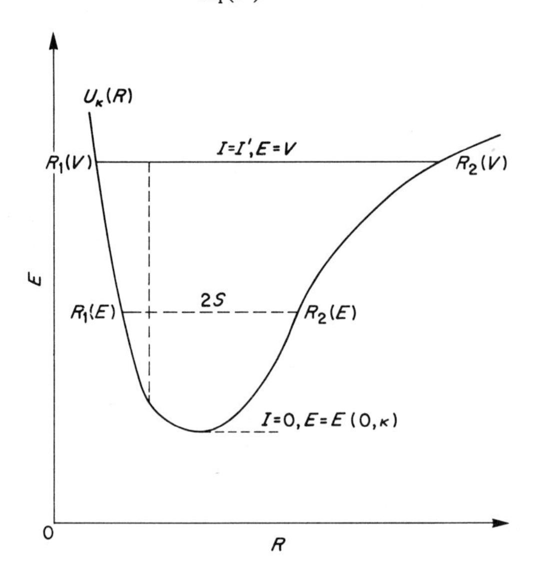

Fig. 1.2. Evaluation of Klein's function $S(V, \kappa)$.

We see, in fact, that S is just one half the area between the effective potential curve $U_\kappa(R)$ and the line of constant total energy $E = V$ (Fig. 1.2).

Differentiating equation (1.59) with respect to V and κ and noting that the integrand vanishes at both limits we obtain

$$f = \frac{\partial S}{\partial V} = \tfrac{1}{2}(R_2 - R_1) \tag{1.60}$$

$$g = -\frac{\partial S}{\partial \kappa} = \tfrac{1}{2}\left(\frac{1}{R_1} - \frac{1}{R_2}\right). \tag{1.61}$$

From these equations we may obtain explicit expressions for the turning points R_1 and R_2 in terms of f and g, namely

$$R_1(V) = \left(\frac{f}{g} + f^2\right)^{\frac{1}{2}} - f \tag{1.62}$$

$$R_2(V) = \left(\frac{f}{g} + f^2\right)^{\frac{1}{2}} + f. \tag{1.63}$$

Equations (1.54)-(1.56) and (1.60)-(1.63) are the basic equations of the RKR method. They lead directly from the vibrational and rotational term values to the ordinates of the potential curve at a given energy. In practical applications these equations are usually used to derive the potential curve without rotation, that is, the partial derivatives (1.60) and (1.61) are evaluated for $\kappa = 0$. The expressions which result for f and g are

$$f(V) = (8\pi^2\mu)^{-\frac{1}{2}} \int_0^{I'} \frac{\mathrm{d}I}{\sqrt{V - E(I, 0)}} \tag{1.64}$$

$$g(V) = \frac{(8\pi^2\mu c^2)^{\frac{1}{2}}}{h} \int_0^{I'} \frac{B_v(I)\,\mathrm{d}I}{\sqrt{V - E(I, 0)}} \tag{1.65}$$

where

$$E(I, 0) = hc\left[\omega_e\left(\frac{I}{h}\right) - \omega_e x_e\left(\frac{I}{h}\right)^2 + \ldots\right] \tag{1.66}$$

$$B_v(I) = B_e - \alpha_e\left(\frac{I}{h}\right) + \ldots \tag{1.67}$$

and the upper limit of integration is given by

$$E(I', 0) = V. \tag{1.68}$$

From equation (1.68) we see that both the integrands of equations (1.64) and (1.65) become infinite at the upper limit. For this reason the direct numerical integration of these equations and the graphical procedure of Rydberg and Klein are tedious and inaccurate, especially for the lower part of the curve.

A much more convenient technique, introduced by Rees[15] and extensively applied by Vanderslice and others[9] is to divide the range of integration $(0, I')$ into a number of sub-ranges and to fit the functions $E(I, 0)$ and $B_v(I)$ to different quadratic expressions in each sub-range. The functions f and g are then obtained simply and accurately by carrying out the integrations analytically.

A comparison of the potential curves obtained for the ground state of carbon monoxide by various methods is given in Table 1.1, which is taken from the work of Steele, Lippincott and Vanderslice.[9]

Table 1.1. Potential curves for the ground state of CO

R (Å)	RKR (eV)	Hulburt Hirschfelder	Morse (equations (1.34) and (1.35))
0.901	5.428	5.480	5.283
0.923	4.211	4.211	4.081
0.952	2.878	2.873	2.802
0.997	1.430	1.416	1.392
1.054	0.400	0.392	0.389
1.220	0.400	0.404	0.407
1.322	1.430	1.437	1.452
1.438	2.878	2.881	2.916
1.544	4.211	4.207	4.258
1.649	5.428	5.416	5.475

The differences among the values shown in Table 1.1 are typical of the results of Vanderslice *et al.*[9] for a large number of states and a variety of empirical potential curves. The better three-parameter functions differ by about 3% from the accurate RKR results, whereas the deviations of the five-parameter Hulburt–Hirschfelder function are not more than 1–2%.

1.6. Atomic units

In the remaining chapters we are chiefly concerned with the electronic Schrödinger equation. It is therefore very convenient to use a system of units in which the electronic mass m, the elementary charge e and $\hbar$ are reduced to unity. In this system the unit of length, the

Bohr, is the radius of the first Bohr orbit in hydrogen assuming an infinite nuclear mass

$$1 \text{ Bohr} = a_{\text{H}} = \frac{\hbar^2}{me^2} = 5.29167 \times 10^{-9} \text{ cm} \tag{1.69}$$

and the unit of energy, the Hartree, is twice the ionization potential of the hydrogen atom again assuming infinite nuclear mass

$$1 \text{ Hartree} = \frac{me^4}{\hbar^2} = 4.35942 \times 10^{-11} \text{ erg.} \tag{1.70}$$

Three other energy units are frequently used in spectroscopy and chemical physics: wave number units (cm^{-1}), electron volts (eV) and kcal mole^{-1}. The conversion factors relating these units are given in Table 1.2. We note that some authors use Rydberg units in place of Hartrees (1 Hartree = 2 Rydbergs) and that Hartrees and Bohrs are often listed simply as au (atomic units). The values given in equations (1.69), (1.70) and Table 1.2 are based on a recent compilation of fundamental constants.[18]

Table 1.2. Conversion factors for energy units

Unit	Hartree	eV	kcal mole^{-1}	cm^{-1}
1 Hartree	1	27.21070	627.51	219474.6
1 eV	3.675025×10^{-2}	1	23.061	8065.73
1 kcal mole^{-1}	1.59360×10^{-3}	4.3363×10^{-2}	1	349.76
1 cm^{-1}	4.556335×10^{-6}	1.23981×10^{-4}	2.8591×10^{-3}	1

If the coordinates are expressed in Bohrs and the energy in Hartrees, the electronic Schrödinger equation (1.6) may be written

$$\left(\sum_i -\tfrac{1}{2}\nabla_i^2 - \sum_i \sum_\alpha \frac{Z_\alpha}{r_{i\alpha}} + \sum_{i<j} \frac{1}{r_{ij}} + \sum_{\alpha<\beta} \frac{Z_\alpha Z_\beta}{R_{\alpha\beta}} \right) \Psi = E\Psi. \tag{1.71}$$

Here we have changed the notation somewhat, writing $r_{i\alpha} = |r_i - r_\alpha|$ etc. and using Ψ for the electronic wave function rather than the total wave function.

References

1. M. Born and J. R. Oppenheimer (1927), *Ann. Physik* **84**, 457.
2. M. Born and K. Huang (1954), "Dynamical Theory of Crystal Lattices", Oxford University Press, pp. 406, 407.
3. H. Eyring, J. Walter and G. E. Kimball (1944), "Quantum Chemistry", Wiley Interscience, New York.

4. G. Herzberg (1950), "Spectra of Diatomic Molecules". D. van Nostrand Company, Inc., Princeton, New Jersey.
5. P. G. Wilkinson (1961), *J. Mol. Spectroscopy* **6**, 1.
6. P. M. Morse (1929), *Phys. Rev.* **34**, 57.
7. H. M. Hulburt and J. O. Hirschfelder (1941), *J. Chem. Phys.* **9**, 61; (1961) erratum *J. Chem. Phys.* **35**, 1901.
8. D. ter Haar (1946), *Phys. Rev.* **70**, 222.
9. D. S. Steele, E. R. Lippincott and J. T. Vanderslice (1962), *Rev. Mod. Phys.* **34**, 239.
10. J. L. Dunham (1932), *Phys. Rev.* **41**, 721.
11. L. D. Landau and E. M. Lifshitz (1958), "Quantum Mechanics, Non-Relativistic Theory", Pergamon Press, London.
12. I. Sandeman (1940), *Proc. Roy. Soc. Edinburgh* **60**, 210.
13. R. Rydberg (1931), *Z. Physik.* **73**, 376; (1933) **80**, 514.
14. O. Klein (1932), *Z. Physik.* **76**, 226.
15. A. L. G. Rees (1947), *Proc. Phys. Soc.* **59**, 998.
16. W. R. Jarmain (1960), *Can. J. Phys.* **38**, 217.
17. A. C. Hurley (1962), *J. Chem. Phys.* **36**, 1117.
18. E. R. Cohen and J. W. M. DuMond (1965), *Rev. Mod. Phys.* **37**, 537.

CHAPTER 2

Variational Wave Functions

Most of the electronic wave functions used in discussions of the properties of atoms and small molecules are derived using some form of the variational method. Here we consider the basis of this method and some general theorems which are closely related to it.

2.1. The minimum energy principle

Let H be the electronic Hamiltonian of some molecular system. The eigenfunctions $\Psi_0, \Psi_1, \ldots$ of H may be chosen as a complete, orthonormal set, ordered in terms of the energy eigenvalues $E_0, E_1, \ldots$. Thus

$$H\Psi_i = E_i\Psi_i \tag{2.1}$$

with

$$E_i \leqslant E_{i+1} \tag{2.2}$$

$$\int \Psi_i^* \Psi_j \, d\tau = \delta_{ij}. \tag{2.3}$$

Here we have assumed, for simplicity, that the spectrum of H is entirely discrete. Although this is not the case for an isolated atom or molecule, we can always obtain a purely discrete spectrum by enclosing our system in a potential box. If this box is made sufficiently large its presence will have a negligible effect on the low-lying energy states which are our primary concern. Alternatively, contributions from the continuous spectrum may be included explicitly using the Dirac δ-function notation.[1]

Any approximate wave function Φ may be expanded in terms of the eigenfunctions Ψ_i, as follows:

$$\Phi = \sum_i c_i\Psi_i. \tag{2.4}$$

Consider now the expression

$$\Delta = \int \Phi^*(H - E_0)\Phi \, d\tau. \tag{2.5}$$

Substituting for Φ from equation (2.4) and using equation (2.1) we obtain

$$\Delta = \int \Big(\sum_i c_i^* \Psi_i^*\Big)(H - E_0)\Big(\sum_j c_j \Psi_j\Big) \mathrm{d}\tau \tag{2.6}$$

$$= \sum_i \sum_j c_i^* c_j (E_j - E_0) \int \Psi_i^* \Psi_j \, \mathrm{d}\tau \tag{2.7}$$

$$= \sum_i c_i^* c_i (E_i - E_0), \tag{2.8}$$

since the functions Ψ_i form an orthonormal set (equation (2.3)).

Now $c_i^* c_i \geqslant 0$ and, from equation (2.2), $E_i \geqslant E_0$. Hence

$$\Delta \geqslant 0, \tag{2.9}$$

that is, from equation (2.5),

$$\frac{\int \Phi^* H \Phi \, \mathrm{d}\tau}{\int \Phi^* \Phi \, \mathrm{d}\tau} \geqslant E_0. \tag{2.10}$$

The inequality (2.10) expresses the minimum energy principle: the expectation value of the Hamiltonian (or, more briefly, the energy) calculated with an approximate wave function is always greater than or equal to the true ground-state energy.

The variational method of calculating approximate wave functions is based on this principle. Some general form is assumed for the trial wave function Φ. This may be some analytic expression containing a number of variational parameters $\lambda_1, \lambda_2, \ldots$ or a more general form containing arbitrary functions. The energy

$$E = \frac{\int \Phi^* H \Phi \, \mathrm{d}\tau}{\int \Phi^* \Phi \, \mathrm{d}\tau} \tag{2.11}$$

is then evaluated and minimized with respect to variations of the parameters or the arbitrary functions. In the case of an analytic trial function the optimum values of the parameters, which minimize E, may be obtained from the equations.

$$\frac{\partial E}{\partial \lambda_i} = 0 \qquad (i = 1, 2, \ldots). \tag{2.12}$$

In this way we obtain both an estimate E_{min} of the ground-state energy and an approximation to the ground-state wave function. The estimate E_{min} provides an upper bound to the true ground-state energy. In general, the more appropriate and flexible the trial wave function the smaller will be the difference between E_{min} and E_0.

The electronic Hamiltonian for a molecule depends parametrically on the nuclear coordinates. Hence, to obtain an estimate of the potential energy curve (for a diatomic molecule) or surface (for a polyatomic molecule) the above process must be repeated for a number of values of the nuclear coordinates. In this way the estimate of the electronic energy and the optimum values of the parameters in the trial wave function are obtained as functions of the nuclear coordinates.

It is generally more difficult to use the variation method for the calculation of the energies and wave functions of excited states. If the trial wave function Φ is restricted to be orthogonal to the lower states

$$\int \Phi^* \Psi_i \, d\tau = 0 \qquad (i = 0, \ldots, n-1), \tag{2.13}$$

then, as is easily shown, the inequality (2.10) may be replaced by

$$\frac{\int \Phi^* H \Phi \, d\tau}{\int \Phi^* \Phi \, d\tau} \geqslant E_n, \tag{2.14}$$

giving a variational condition for determining E_n and Ψ_n. However, it is not possible to use equations (2.13) and (2.14) directly, since the accurate wave functions for the lower states are unknown. If approximate wave functions for the lower states are used in the conditions (2.13), the inequality (2.14) may be violated.

There are two important cases where this difficulty may be overcome. If Ψ_n is the lowest state of some particular symmetry type (cf. Chapter 4), then the conditions (2.13) are automatically satisfied and we may proceed as for the ground state. The other case arises in the linear variation method which we consider in the following section.

2.2. The linear variation method

For our trial function Φ we choose a linear expansion in terms of a known set of functions

$$\Phi = \sum_{i=1}^{n} C_i \Phi_i. \tag{2.15}$$

The expectation value of the Hamiltonian, calculated with the trial function Φ, is

$$E(\Phi) = \frac{\sum_i \sum_j C_i^* H_{ij} C_j}{\sum_i \sum_j C_i^* S_{ij} C_j}, \tag{2.16}$$

where

$$S_{ij} = \int \Phi_i^* \Phi_j \, d\tau \tag{2.17}$$

and

$$H_{ij} = \int \Phi_i^* H \Phi_j \, d\tau. \tag{2.18}$$

Minimization of $E(\Phi)$ with respect to the coefficients C_i leads to a set of linear equations

$$\sum_{j=1}^{n} (H_{ij} - ES_{ij})C_j = 0 \qquad (i = 1, \ldots, n). \tag{2.19}$$

This set of linear equations has a non-trivial solution only if E is one of the roots of the secular equation

$$\det \{H_{ij} - ES_{ij}\} = 0. \tag{2.20}$$

Since the matrices H_{ij} and S_{ij} are Hermitian, the n roots of equation (2.20) are all real. The lowest root provides an upper bound to the true ground state energy. Substituting this value for E in the equations (2.19), and solving for the C_j, we obtain the coefficients in the expansion (2.15) of the approximate ground-state wave function.

Furthermore, as MacDonald[2] has shown, the higher roots of the secular equation (2.20) provide upper bounds to the exact energies of the corresponding excited states. We may, therefore, use the linear variation method to construct approximate wave functions for excited states, without worrying about orthogonality with accurate wave functions for the lower states (equation (2.13)). We may even vary any non-linear parameters in the basis functions Φ_i to minimize the calculated energy of an excited state, provided we use the appropriate root of the secular equation.

2.3. Hellmann–Feynman formulae

The electronic energy of a molecule depends upon the values of certain parameters in the Hamiltonian, notably the nuclear coordinates and charges. A study of this dependence leads to a number of formulae which are closely related to the variational method. In the present section we state these formulae, and establish their validity for the case of exact wave functions. For simplicity we restrict the discussion to non-degenerate ground-state wave functions.

The first formula we consider applies to any quantum-mechanical system whose Hamiltonian H depends on some parameter α. Let Ψ be a normalized wave function and E the corresponding energy. Thus

$$E = \int \Psi^* H \Psi \, d\tau. \tag{2.21}$$

In equation (2.21) the three quantities E, H and Ψ all depend on the parameter α. The Hellmann–Feynman[3,4] formula expresses the derivative of E with respect to α in terms of $\partial H/\partial\alpha$, the terms involving the derivative of Ψ being dropped. Thus

$$\frac{dE}{d\alpha} = \int \Psi^* \frac{\partial H}{\partial\alpha} \Psi \, d\tau. \tag{2.22}$$

We regard the right-hand side of equation (2.22) as a *formula* (or procedure) for estimating the quantity on the left. The two sides of the equation are equal only for exact wave functions and for certain classes of approximate wave functions (cf. Section 2.4).

The proof that equation (2.22) holds for *exact* wave functions is very simple. Differentiating equation (2.21) we have

$$\frac{dE}{d\alpha} = \int \frac{\partial\Psi^*}{\partial\alpha} H\Psi \, d\tau + \int \Psi^* \frac{\partial H}{\partial\alpha} \Psi \, d\tau + \int \Psi^* H \frac{\partial\Psi}{\partial\alpha} \, d\tau. \tag{2.23}$$

The third term in equation (2.23) may be transformed using the Hermitian property of H. This gives

$$\int \Psi^* H \frac{\partial\Psi}{\partial\alpha} \, d\tau = \int \frac{\partial\Psi}{\partial\alpha} H^*\Psi^* \, d\tau. \tag{2.24}$$

Because Ψ is an exact wave function

$$H\Psi = E\Psi \tag{2.25}$$

and

$$H^*\Psi^* = E\Psi^*, \tag{2.26}$$

where we have used the fact that, since H is Hermitian, E is real.

Substituting equations (2.24), (2.25) and (2.26) into equation (2.23) we obtain

$$\begin{aligned}
\frac{dE}{d\alpha} &= \int \Psi^* \frac{\partial H}{\partial\alpha} \Psi \, d\tau + E\left(\int \frac{\partial\Psi^*}{\partial\alpha} \Psi \, d\tau + \int \Psi^* \frac{\partial\Psi}{\partial\alpha} \, d\tau\right) \\
&= \int \Psi^* \frac{\partial H}{\partial\alpha} \Psi \, d\tau + E \frac{\partial}{\partial\alpha}\left(\int \Psi^*\Psi \, d\tau\right) \\
&= \int \Psi^* \frac{\partial H}{\partial\alpha} \Psi \, d\tau,
\end{aligned}$$

since the wave function Ψ is normalized.

The implications of equation (2.22) naturally depend on the choice of the parameter α. Moreover, even when α has been chosen, the consequences of equation (2.22) depend upon the coordinate system which is used to express the wave function Ψ and the Hamiltonian H. This dependence arises from the definition of the partial derivative $\partial H/\partial\alpha$.

We consider two applications of equation (2.22), the electrostatic formula and the virial formula. For simplicity we derive the formulae for a diatomic molecule; the extension to polyatomic molecules is, in both cases, straightforward.

In both these applications the Hamiltonian is

$$H = T + V = \sum_i \left(-\tfrac{1}{2}\nabla_i^2\right) - \sum_i \left(\frac{Z_a}{r_{ai}} + \frac{Z_b}{r_{bi}}\right) + \sum_{i<j} \frac{1}{r_{ij}} + \frac{Z_a Z_b}{R} \tag{2.27}$$

and the nuclear separation R is chosen as the parameter α of equation (2.22).

For the derivation of the electrostatic formula, H and Ψ are expressed in Cartesian coordinates centred on nucleus a, the z-axis being directed towards nucleus b. In this coordinate system, the only quantities in H which depend on R are the distances r_{bi} of the electrons from nucleus b, and R itself. Consequently, we have from equations (2.22) and (2.27)

$$\begin{aligned}\frac{dE}{dR} &= \int \Psi^* \frac{\partial V}{\partial R} \Psi \, d\tau \\ &= \int \Psi^* \left(\sum_i \frac{Z_b}{r_{bi}^2} \frac{\partial r_{bi}}{\partial R} - \frac{Z_a Z_b}{R^2}\right) \Psi \, d\tau \\ &= \int \Psi^* \sum_i \frac{Z_b \cos\theta_{bi}}{r_{bi}^2} \Psi \, d\tau - \frac{Z_a Z_b}{R^2},\end{aligned} \tag{2.28}$$

where θ_{bi} is the angle between the z axis and the vector from electron i to nucleus b. The first term in equation (2.28), being the expectation value of a symmetric one-electron function, may be expressed as an integral over the charge density function ρ to give

$$\frac{dE}{dR} = Z_b \left\{ \int \frac{\rho \cos\theta_b}{r_b^2} \, dx \, dy \, dz - \frac{Z_a}{R^2} \right\}. \tag{2.29}$$

Equation (2.29) expresses the electrostatic formula, namely, the force on nucleus b, $F_{bz} = -(dE/dR)$, is that calculated by classical electrostatics from the quantum mechanical charge distribution ρ of the electrons and the point charge Z_a of nucleus a.

Similarly we have

$$F_{az} = \frac{dE}{dR} = Z_a \left\{ \int \frac{\rho \cos\theta_a}{r_a^2} \, dx \, dy \, dz - \frac{Z_b}{R^2} \right\}. \tag{2.30}$$

In order to derive the virial formula from equation (2.22), we express Ψ and H in terms of elliptical coordinates (μ, ν, φ) with foci at the nuclei a and b. In this coordinate system the operators T and V of equation (2.27) appear in the form

$$T = \frac{1}{R^2} \mathscr{T} \tag{2.31}$$

$$V = \frac{1}{R} \mathscr{V}, \tag{2.32}$$

where the operators $\mathscr{T}$ and $\mathscr{V}$ are functions of the elliptical coordinates μ_i, ν_i, φ_i of the electrons, and the derivative operators $\partial/\partial\mu_i$, $\partial/\partial\nu_i$, $\partial/\partial\varphi_i$ (see e.g. Eyring, Walter and Kimball[(5)]).

We now have from equations (2.27), (2.31) and (2.32)

$$\frac{\partial H}{\partial R} = -\frac{2}{R^3} \mathscr{T} - \frac{1}{R^2} \mathscr{V} = -\frac{1}{R}(2T + V), \tag{2.33}$$

so that equation (2.22) gives the result

$$\frac{\mathrm{d}E}{\mathrm{d}R} = -\frac{1}{R}(2\bar{T} + \bar{V}) \tag{2.34}$$

with

$$\bar{T} = \int \Psi^* T \Psi \, \mathrm{d}\tau, \qquad \bar{V} = \int \Psi^* V \Psi \, \mathrm{d}\tau. \tag{2.35}$$

Equation (2.34) is the virial formula for a diatomic molecule.[(6)] Since we also have the relation

$$E = \bar{T} + \bar{V}, \tag{2.36}$$

the virial formula leads to expressions for the mean kinetic and mean potential energies in terms of the total energy and its derivative. These are

$$\bar{T} = -E - R\frac{\mathrm{d}E}{\mathrm{d}R} \tag{2.37}$$

$$\bar{V} = 2E + R\frac{\mathrm{d}E}{\mathrm{d}R}. \tag{2.38}$$

In Chapter 1, various methods of determining E versus R curves from experimental band-spectroscopic data were considered. Given such a curve, the relations (2.37) and (2.38) enable us to derive curves showing the variation of the mean kinetic and mean potential energies with nuclear separation. For a stable electronic state, which dissociates into neutral atoms, these curves are of the general form shown in Fig. 2.1. At the equilibrium nuclear separation R_e, and at $R = \infty$, $\mathrm{d}E/\mathrm{d}R$ vanishes so that the relations (2.37) and (2.38) reduce to

$$\bar{T}(R_e) = -E(R_e), \qquad \bar{V}(R_e) = 2E(R_e) \tag{2.39}$$

$$\bar{T}(\infty) = -E(\infty), \qquad \bar{V}(\infty) = 2E(\infty). \tag{2.40}$$

Since the binding energy D_e is given by

$$D_e = E(\infty) - E(R_e), \tag{2.41}$$

we have, at equilibrium, the relations shown in Fig. 2.1

$$\bar{T}(R_e) - \bar{T}(\infty) = D_e \tag{2.42}$$

$$\bar{V}(R_e) - \bar{V}(\infty) = -2D_e. \tag{2.43}$$

If we integrate the differential equations (2.37) and (2.38) subject to the boundary conditions

$$\lim_{R \to \infty} R[E(R) - E(\infty)] = 0 \tag{2.44}$$

and

$$\lim_{R \to \infty} R^2 [E(R) - E(\infty)] = 0, \tag{2.45}$$

respectively, which are always satisfied if at least one of the dissociation products is a neutral atom, we obtain

$$\begin{aligned} E(R) - E(\infty) &= \frac{1}{R}\int_R^\infty [\bar{T}(R') - \bar{T}(\infty)]\, \mathrm{d}R' \\ &= -\frac{1}{R^2}\int_R^\infty R'[\bar{V}(R') - \bar{V}(\infty)]\, \mathrm{d}R'. \end{aligned} \tag{2.46}$$

The equations (2.46) imply that, for a stable electronic state $[E(R) < E(\infty)]$, the kinetic energy must initially fall and the potential energy must rise as the atoms are brought together from infinity (Fig. 2.1).

Consider now the change in the total electronic energy caused by a finite change in the value of some parameter, or parameters, appearing in the Hamiltonian. If we assume that the wave functions are real and normalized, the usual expression for this energy change is

$$E(\alpha_1) - E(\alpha_2) = \int \Psi(\alpha_1)H(\alpha_1)\Psi(\alpha_1)\, \mathrm{d}\tau - \int \Psi(\alpha_2)H(\alpha_2)\Psi(\alpha_2)\, \mathrm{d}\tau. \tag{2.47}$$

However, there is an alternative expression for the energy change which is in many ways simpler than (2.47). To derive this expression we start from the Schrödinger equations

$$H(\alpha_1)\Psi(\alpha_1) = E(\alpha_1)\Psi(\alpha_1) \tag{2.48}$$

$$H(\alpha_2)\Psi(\alpha_2) = E(\alpha_2)\Psi(\alpha_2). \tag{2.49}$$

If we again assume real, normalized wave functions, multiply equation (2.48) by $\Psi(\alpha_2)$ and integrate over all the electronic coordinates we obtain

$$\int \Psi(\alpha_2)H(\alpha_1)\Psi(\alpha_1)\, \mathrm{d}\tau = E(\alpha_1)S, \tag{2.50}$$

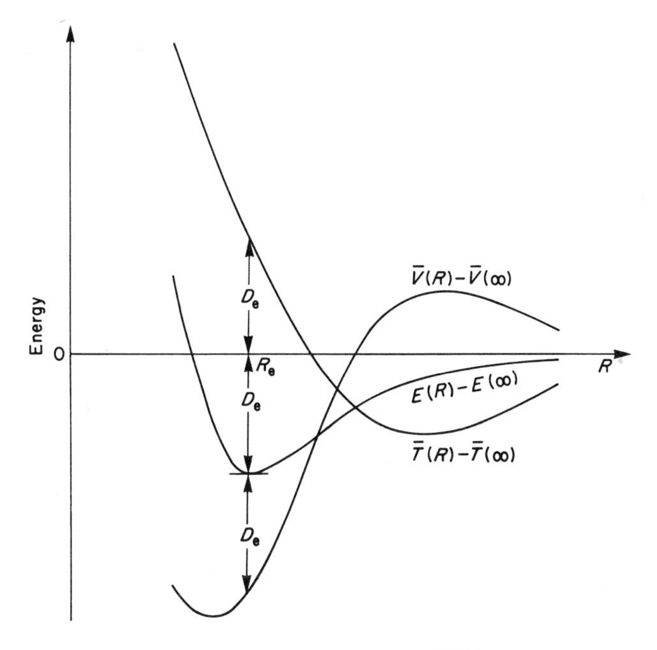

Fig. 2.1. Variation of mean kinetic and potential energies $\overline{T}$, $\overline{V}$ for a stable diatomic molecule which dissociates into neutral atoms or into a neutral atom and an ion.

where

$$S = \int \Psi(\alpha_1)\Psi(\alpha_2)\,\mathrm{d}\tau. \tag{2.51}$$

Similarly from equation (2.49) we obtain

$$\int \Psi(\alpha_1)H(\alpha_2)\Psi(\alpha_2)\,\mathrm{d}\tau = E(\alpha_2)S. \tag{2.52}$$

Now, since the operator $H(\alpha_1)$ is real and Hermitian, equation (2.50) may be written in the alternative form

$$\int \Psi(\alpha_1)H(\alpha_1)\Psi(\alpha_2)\,\mathrm{d}\tau = E(\alpha_1)S. \tag{2.53}$$

We now subtract equation (2.52) from equation (2.53) and divide by S to obtain

$$E(\alpha_1) - E(\alpha_2) = \frac{\int \Psi(\alpha_1)\{H(\alpha_1) - H(\alpha_2)\}\Psi(\alpha_2)\,\mathrm{d}\tau}{S} \cdot \tag{2.54}$$

Equation (2.54) is usually referred to as the integral Hellmann-Feynman formula or Parr's[7] formula. In the usual case where the parameters α specify the nuclear configuration, the expression (2.54) is much simpler to evaluate than the difference of expectation values (2.47). This is because the kinetic energy terms and the electron–electron repulsion terms cancel when the difference $H(\alpha_1) - H(\alpha_2)$ is taken. We are left with the change ΔV_{NN} in the nuclear repulsion term, which may be taken outside the integral, and the change ΔV_{Ne} in the electron–nuclear attractions, which is a symmetric one-electron operator of the form

$$\Delta V_{\mathrm{Ne}} = \sum_{i=1}^{N} V'(r_i). \qquad (2.55)$$

The expression (2.54) then reduces to

$$E(\alpha_1) - E(\alpha_2) = \Delta V_{\mathrm{NN}} + \int \rho_{\alpha_1\alpha_2}(r_1) V'(r_1)\, \mathrm{d}\tau_1, \qquad (2.56)$$

where the transition density $\rho_{\alpha_1\alpha_2}$ between the two nuclear configurations α_1 and α_2 is defined by integrating the product of the wave functions over all the electronic coordinates except those of electron 1, multiplying by the number of electrons N, and dividing by the overlap S, that is

$$\rho_{\alpha_1\alpha_2}(r_1) = \frac{N}{S} \int \Psi(\alpha_1)\Psi(\alpha_2)\, \mathrm{d}\tau_2 \ldots \mathrm{d}\tau_N. \qquad (2.57)$$

As in the case of the electrostatic formula (equations (2.29), (2.30)), the formula (2.56) has a simple classical interpretation. To calculate the energy change, first compute the nuclear–nuclear contribution. Then add the "electron-shielding" contribution, calculated by computing the energy of the charge distribution $\rho_{\alpha_1\alpha_2}$ in the nuclear frame α_1, in the nuclear frame α_2, and subtracting.

2.4. Hellmann–Feynman theorems

In the derivations of the electrostatic formula, the virial formula and Parr's formula given in the previous section, it was always assumed that the electronic wave functions were exact solutions of the Schrödinger equation. This does not necessarily prevent us from using the formulae to obtain approximate results from approximate wave functions. For example, any wave function which gives an accurate total electron charge density may be used to calculate forces on nuclei via the electrostatic formula. Nevertheless it is important to know which approximate wave functions will give the same results using one of the formulae as are obtained from the more usual method based on energy expectation values. In such cases we say that the approximate wave functions *satisfy* the electrostatic *theorem*, the virial *theorem* or Parr's *theorem*.

We consider first the Hellmann–Feynman theorem (equation (2.22)). Clearly the validity of this equation can be tested only if we have a well-defined approximate wave function for a continuous range of α values. We now denote this approximate wave function, assumed normalized, by $\Psi(\alpha)$ and define

$$E(\alpha) = \int \Psi^*(\alpha)H(\alpha)\Psi(\alpha)\,\mathrm{d}\tau. \tag{2.58}$$

In order to separate the change in E due to changes in H from those due to changes in Ψ we also consider the function

$$\mathscr{E}(\alpha, \lambda) = \int \Psi^*(\lambda)H(\alpha)\Psi(\lambda)\,\mathrm{d}\tau. \tag{2.59}$$

We then have

$$E(\alpha) = \mathscr{E}(\alpha, \alpha)$$

and

$$\frac{\mathrm{d}E}{\mathrm{d}\alpha} = \left.\frac{\partial\mathscr{E}(\alpha, \lambda)}{\partial\alpha}\right|_{\lambda=\alpha} + \left.\frac{\partial\mathscr{E}(\alpha, \lambda)}{\partial\lambda}\right|_{\lambda=\alpha} \tag{2.60}$$

where the notation indicates that α is substituted for λ after the differentiations have been performed.

From the definition (2.59) we have

$$\left.\frac{\partial\mathscr{E}(\alpha, \lambda)}{\partial\alpha}\right|_{\lambda=\alpha} = \int \Psi^*(\alpha)\frac{\partial H}{\partial\alpha}\Psi(\alpha)\,\mathrm{d}\tau, \tag{2.61}$$

so that equation (2.22) will hold for the approximate wave function $\Psi(\alpha)$, if and only if

$$\frac{\partial\mathscr{E}(\alpha, \lambda)}{\partial\lambda} = 0, \tag{2.62}$$

when the parameter λ is put equal to α.

Now equation (2.62) is just the condition used to determine the optimum value of the parameter λ in a variational calculation of an approximate wave function $\Psi(\lambda)$ for the Hamiltonian $H(\alpha)$. For the ground electronic state, the appropriate stationary point determined by equation (2.62) will be an absolute minimum. The condition for the validity of equation (2.22) may then be written

$$\mathscr{E}(\alpha, \alpha) < \mathscr{E}(\alpha, \lambda) \qquad \text{for all } \lambda \neq \alpha, \tag{2.63}$$

that is, of all the approximate wave functions $\Psi(\lambda)$ for the Hamiltonian $H(\alpha)$, $\Psi(\alpha)$ is the best by the criterion of the variation principle.

In general, equation (2.62) will be satisfied by some value of $\lambda \neq \alpha$, and will determine this optimum value of λ as a function,

$\lambda_0(\alpha)$, of α. We may then obtain, from the $\Psi(\alpha)$, wave functions $\Phi(\alpha)$ which do satisfy equation (2.22) simply by changing the association between the Hamiltonians and the wave functions. The procedure is as follows.

In place of the original association

$$H(\alpha) \leftrightarrow \Psi(\alpha), \tag{2.64}$$

we set up the association

$$H(\alpha) \leftrightarrow \Psi(\lambda_0(\alpha)) \equiv \Phi(\alpha). \tag{2.65}$$

Defining

$$E_0(\alpha) = \mathscr{E}(\alpha, \lambda_0(\alpha)) = \int \Phi^*(\alpha)H(\alpha)\Phi(\alpha)\, \mathrm{d}\tau, \tag{2.66}$$

we then have

$$\frac{\mathrm{d}E_0}{\mathrm{d}\alpha} = \left.\frac{\partial\mathscr{E}(\alpha,\lambda)}{\partial\alpha}\right|_{\lambda=\lambda_0(\alpha)} + \frac{\mathrm{d}\lambda_0}{\mathrm{d}\alpha}\left.\frac{\partial\mathscr{E}(\alpha,\lambda)}{\partial\lambda}\right|_{\lambda=\lambda_0(\alpha)}. \tag{2.67}$$

Here the second term vanishes from the definition of $\lambda_0(\alpha)$ (equation (2.62)) and we obtain

$$\frac{\mathrm{d}E_0}{\mathrm{d}\alpha} = \int \Phi^*(\alpha)\frac{\partial H}{\partial\alpha}\Phi(\alpha)\, \mathrm{d}\tau \tag{2.68}$$

as required.

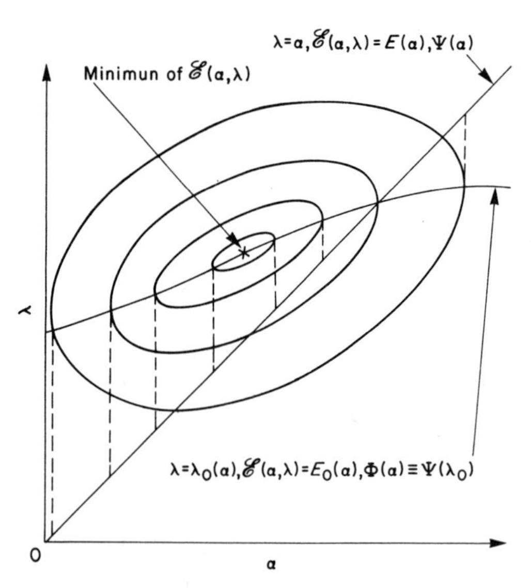

Fig. 2.2. Determination of the function $\lambda_0(\alpha)$ and the wave functions $\Phi(\alpha)$; contours of the function $\mathscr{E}(\alpha, \lambda)$ are shown for a typical case.

The contour lines of the function $\mathscr{E}(\alpha, \lambda)$, in a typical case, are shown in Fig. 2.2. This figure illustrates the determination of the function $\lambda_0(\alpha)$ and the relationship between the wave functions $\Psi(\alpha)$ and $\Phi(\alpha)$. In the example shown, the Hellmann–Feynman theorem is satisfied by the functions $\Psi(\alpha)$ at a single isolated value of α, whereas the functions $\Phi(\alpha)$ satisfy the theorem for all α values.

One can often determine on quite general grounds whether a set of approximate wave functions will satisfy the Hellmann–Feynman theorem. Suppose some class $\mathscr{C}$ of electronic wave functions is defined in any way that does not depend on the value of the parameter α and that, for each value of α, $\Psi(\alpha)$ is chosen as that function from $\mathscr{C}$ which minimizes the expectation value of $H(\alpha)$. The condition (2.63) is then automatically satisfied and the Hellmann–Feynman theorem (equation (2.22)) follows. As we shall see in later chapters, accurate Hartree–Fock wave functions satisfy this criterion for all parameters α, the class $\mathscr{C}$ in this case being composed of all functions expressible as single Slater determinants.

On the other hand, analytic variational functions may well violate the condition (2.63), even if all the usual variational parameters are determined by minimizing the energy. This will happen when the class $\mathscr{C}$ of functions, from which $\Psi(\alpha)$ is selected variationally, depends explicitly on α, that is, when we have different *classes* of functions for different values of α. Whether or not this is the case may depend, not only on α, but also on the coordinate system used in evaluating $\partial H/\partial\alpha$. In any particular case it is easy to see whether or not $\mathscr{C}$ depends on α.

For example, consider the simple wave function

$$\Psi(R) = N[e^{-k(R)r_a} + e^{-k(R)r_b}] \tag{2.69}$$

for the ground state of the H_2^+ ion. In equation (2.69), r_a, r_b are the distances of the electron from the two nuclei a and b, N is a normalization constant and $k(R)$ is a variable parameter (scale factor) determined, as a function of R, by minimizing the energy.

Let us see whether the function (2.69) will satisfy the virial theorem and the electrostatic theorem for the forces on the nuclei a and b. As we saw in Section 2.3, in all three cases the nuclear separation R is chosen as the parameter α but different coordinate systems are used.

In the case of the virial theorem, we must use the elliptical coordinates of equations (2.31) and (2.32). In this coordinate system

$$r_a = \frac{R}{2}(\mu + \nu), \qquad r_b = \frac{R}{2}(\mu - \nu) \tag{2.70}$$

and the wave function (2.69) becomes

$$\Psi(R) = N[\exp(-\tfrac{1}{2}Rk(R)(\mu + \nu)) + \exp(-\tfrac{1}{2}Rk(R)(\mu - \nu))]. \tag{2.71}$$

The class $\mathscr{C} \equiv \{\Psi\}_R$ of functions, from which $\Psi(R)$ is selected variationally, is

$$\{\Psi\}_R = \{N[\exp(-\tfrac{1}{2}Rk(\mu+\nu)) + \exp(-\tfrac{1}{2}Rk(\mu-\nu))]; \quad -\infty < k < \infty\}. \tag{2.72}$$

Clearly $\{\Psi\}_R$ is independent of the value of R, so that the condition (2.63) is satisfied and the function (2.69) satisfies the virial theorem (equation (2.34)).

Consider now the electrostatic theorem for the force on nucleus b (equation (2.29)). In deriving this equation, $\partial H/\partial R$ was evaluated in Cartesian coordinates centred on nucleus a with the x-axis directed towards nucleus b. In this coordinate system

$$r_{\rm a} = (x^2 + y^2 + z^2)^{\frac{1}{2}}, \qquad r_{\rm b} = [x^2 + y^2 + (z - R)^2]^{\frac{1}{2}} \tag{2.73}$$

and the class $\mathscr{C} \equiv \{\Psi\}_R$ becomes

$$\{\Psi\}_R = \{N(\exp(-k(x^2 + y^2 + z^2)^{\frac{1}{2}}) + \exp(-k(x^2 + y^2 + (z - R)^2)^{\frac{1}{2}})); \quad -\infty < k < \infty\}. \tag{2.74}$$

In this case the *class* of functions $\{\Psi\}_R$ depends explicitly on the value of R, so that the condition (2.63) is not satisfied and equation (2.29) does not hold.

To obtain a wave function which *does* satisfy the condition (2.63), the class $\{\Psi\}_R$ may be enlarged by replacing R by an additional variational parameter S, whose value, together with that of k, is determined by minimizing the energy. Clearly the enlarged *class*

$$\{\Psi\} = \{N(\exp(-k(x^2 + y^2 + z^2)^{\frac{1}{2}}) + \exp(-k(x^2 + y^2 + (z - S)^2)^{\frac{1}{2}})); \quad -\infty < k, S < \infty\} \tag{2.75}$$

is independent of the value of R, so that the condition (2.63) is automatically satisfied.

Alternatively, we may change the association between the Hamiltonians and the wave functions (Fig. 2.2 and equations (2.64)-(2.67) with α, λ replaced by R, S respectively). This leads to a wave function

$$\Phi(R) \equiv \Psi(S_0(R)) = N[\exp(-k(S_0)(x^2 + y^2 + z^2))^{\frac{1}{2}} + \exp(-k(S_0)(x^2 + y^2 + (z - S_0)^2))^{\frac{1}{2}}], \tag{2.76}$$

which also satisfies the condition (2.63), but in this case the parameter k will not have its optimum value.

In either case, in order to satisfy the electrostatic theorem for the force on nucleus b, the atomic orbital orbital $\exp(-kr_{\rm b})$ centred on nucleus b must be detached from the nucleus and allowed to take up an optimum position determined by minimizing the total electronic energy.

The arguments we have used in this simple case are quite general and may be used for any molecule and any variational wave function constructed from atomic orbitals. The main conclusions are:

(i) to satisfy the virial theorem it is sufficient to include a scale factor multiplying all lengths in the wave function and to determine this scale factor by minimizing the total electronic energy;

(ii) to satisfy the electrostatic theorem for the forces on *all* the nuclei it is sufficient to detach all the atomic orbitals from their nuclei and to minimize the total electronic energy with respect to their locations.

Condition (i) is easy to fulfil and most of the approximate wave functions which we consider in later chapters will satisfy the virial theorem either exactly or to good accuracy, the small deviations arising from a failure to fully optimize all variational parameters.

The procedure (ii) for satisfying the electrostatic theorem is, on the other hand, an awkward one and has been carried through in only a few calculations. A more generally useful procedure is to leave the atomic orbitals centred on the nuclei and to use the electrostatic theorem as a test of the accuracy of the wave function.

Wave functions which satisfy the Hellmann–Feynman theorem are referred to as floating functions.(8, 9) This name was suggested by the case of the electrostatic theorem for functions constructed from atomic orbitals but covers other cases as well. For example, accurate Hartree–Fock functions are floating functions for the variation of *any* parameter in the Hamiltonian and *any* choice of coordinate system.

Finally, let us seek conditions on approximate wave functions such that the integral Hellmann–Feynman theorem (equation (2.54)) will be satisfied. We consider the case of several parameters α continuously variable throughout some region A.

Suppose that an approximate wave function $\Psi(P)$ is obtained for each point P in A by applying the linear variational principle to some set Φ_i of real basis functions which is fixed, that is, the same basis functions are used for every point in A.

Then, if α_1 and α_2 are any two points in A, we have

$$E(\alpha_1) - E(\alpha_2) = \frac{\int \Psi(\alpha_1)[H(\alpha_1) - H(\alpha_2)]\,\Psi(\alpha_2)\,\mathrm{d}\tau}{S}, \qquad (2.77)$$

that is, the wave functions $\Psi(P)$ satisfy the integral Hellmann–Feynman theorem through the region A.

The proof of equation (2.77) is as follows. The approximate wave functions $\Psi(\alpha_1)$, $\Psi(\alpha_2)$ are given by

$$\Psi(\alpha_1) = \sum_i C_i(\alpha_1)\Phi_i, \qquad \Psi(\alpha_2) = \sum_i C_i(\alpha_2)\Phi_i \tag{2.78}$$

where

$$\sum_j [H(\alpha_1)]_{ij}C_j(\alpha_1) = E(\alpha_1) \sum_j S_{ij}C_j(\alpha_1) \tag{2.79}$$

$$\sum_j [H(\alpha_2)]_{ij}C_j(\alpha_2) = E(\alpha_2) \sum_j S_{ij}C_j(\alpha_2) \tag{2.80}$$

with

$$S_{ij} = \int \Phi_i\Phi_j \, d\tau, \qquad [H(\alpha_1)]_{ij} = \int \Phi_i H(\alpha_1)\Phi_j \, d\tau,$$

$$[H(\alpha_2)]_{ij} = \int \Phi_i H(\alpha_2)\Phi_j \, d\tau. \tag{2.81}$$

Multiplying equation (2.80) by $C_i(\alpha_1)$ and summing over i we obtain

$$\sum_i \sum_j C_i(\alpha_1)[H(\alpha_2)]_{ij}C_j(\alpha_2) = E(\alpha_2) \sum_i \sum_j C_i(\alpha_1)S_{ij}C_j(\alpha_2), \tag{2.82}$$

that is

$$\int \Psi(\alpha_1)H(\alpha_2)\Psi(\alpha_2) \, d\tau = SE(\alpha_2). \tag{2.83}$$

Similarly from equation (2.79), we get

$$\int \Psi(\alpha_1)H(\alpha_1)\Psi(\alpha_2) \, d\tau = SE(\alpha_1). \tag{2.84}$$

Subtracting (2.83) from (2.84) and dividing by S gives the required result (2.77).

The conditions we have established for the validity of the integral Hellmann–Feynman theorem are more restrictive than those for the ordinary theorem, as may be seen by rephrasing them as follows. A set of wavefunctions $\Psi(\alpha)$ will satisfy the integral Hellmann–Feynman theorem if they are all selected variationally from a single *linear* subspace of the space of all functions satisfying the appropriate boundary and smoothness conditions. Earlier, we saw that the ordinary Hellmann–Feynman theorem is satisfied if all the $\Psi(\alpha)$ are variationally selected from a single class of functions which may be non-linear.

In view of this relationship it is natural to refer to functions satisfying the integral Hellmann–Feynman theorem as super-floating. As the name implies, super-floating functions are always floating but the converse may not be true. For example, accurate Hartree–Fock functions are floating but not super-floating, since the class of functions involved (those expressible as single determinants) is clearly non-linear. A general procedure for constructing super-floating functions has been developed.[10] This procedure is a difficult one computationally, however, and has been carried through only for the ground state of the hydrogen atom.

2.5. Variational calculations with linear constraints

The electronic Schrödinger equation (2.1) may be regarded as the condition that the energy expectation value

$$E = \int \Psi^* H \Psi \, d\tau \tag{2.85}$$

should remain stationary for arbitrary variations $\delta\Psi$ of the many-electron wave function Ψ, which preserve the normalization condition

$$\int \Psi^* \Psi \, d\tau = 1. \tag{2.86}$$

For calculations on excited states and in self-consistent field theory it is sometimes necessary to impose additional linear constraints on the wave function and the admissible variations. Such a linear constraint may be represented by the orthogonality of Ψ to some fixed function Φ.

$$\int \Psi^* \Phi \, d\tau = \int \Phi^* \Psi \, d\tau = 0, \tag{2.87}$$

which we take to be normalized

$$\int \Phi^* \Phi \, d\tau = 1. \tag{2.88}$$

Incorporating the constraints (2.86) and (2.87) by means of the Lagrange multipliers E, ϵ and η we require that

$$\delta\left[\int \Psi^* H \Psi \, d\tau - E \int \Psi^* \Psi \, d\tau - \epsilon \int \Psi^* \Phi \, d\tau - \eta \int \Phi^* \Psi \, d\tau\right] = 0 \tag{2.89}$$

for arbitrary variations $\delta\Psi$ and suitably chosen values of the constants E, ϵ and η.

Using the Hermitian property of H the condition (2.89) may be reduced to the form

$$\int \delta\Psi^* (H\Psi - E\Psi - \epsilon\Phi) \, d\tau + \int \delta\Psi (H^*\Psi^* - E\Psi^* - \eta\Phi^*) \, d\tau = 0. \tag{2.90}$$

Since the variations $\delta\Psi^*$ and $\delta\Psi$ are arbitrary the condition (2.90) leads to two equations, namely

$$H\Psi = E\Psi + \epsilon\Phi \tag{2.91}$$

$$H^*\Psi^* = E\Psi^* + \eta\Phi^*. \tag{2.92}$$

Comparing equation (2.91) with the complex conjugate of (2.92) we see that the off-diagonal Lagrange multipliers are related by

$$\eta = \epsilon^*. \tag{2.93}$$

Once the condition (2.93) has been imposed equation (2.92) becomes equivalent to (2.91) and may be dropped.

Equation (2.91) together with the orthonormality conditions (2.86) and (2.87) provides the solution to our variational problem. The value of the constants E and ϵ may be found by multiplying (2.91) by Ψ^* and Φ^* respectively and integrating. Using equations (2.86), (2.87) and (2.88) we obtain

$$E = \int \Psi^* H \Psi \, d\tau, \qquad \epsilon = \int \Phi^* H \Psi \, d\tau. \tag{2.94}$$

Because of the second term on the right hand side the form of equation (2.91) is not that of an equation determining the eigenvalues and eigenfunctions of some Hermitian operator. However, it may be reduced to this standard form by introducing a suitable effective Hamiltonian. To achieve this end we define a projection operator $\mathscr{P}$, whose effect on an arbitrary function F is given by the equation

$$\mathscr{P}F = \Phi\left(\int \Phi^* F \, d\tau\right). \tag{2.95}$$

Regarding our functions as elements of an infinite-dimensional linear vector-space, we see from equation (2.95) that $\mathscr{P}$ has the effect of projecting F onto that component of F parallel to the fixed function Φ.

It is readily verified that the operator $\mathscr{P}$ is linear and Hermitian. Furthermore, since from equations (2.88) and (2.95)

$$\mathscr{P}\Phi = \Phi, \tag{2.96}$$

we see that $\mathscr{P}$ also satisfies the idempotency condition

$$\mathscr{P}^2 = \mathscr{P} \tag{2.97}$$

which is characteristic of projection operators. In the bra ket notation of Dirac[1] $\mathscr{P}$ has a natural expression as the product of ket and bra vectors in the opposite order to that involved in a scalar product

$$\mathscr{P} = |\,\Phi\,\rangle\langle\,\Phi\,|. \tag{2.98}$$

The defining equation (2.95) then appears as

$$\mathscr{P}|\,F\,\rangle = |\,\Phi\,\rangle\langle\,\Phi\,|\,F\,\rangle. \tag{2.99}$$

Using the definition (2.95) and the second of equations (2.94) equation (2.91) may be expressed as follows

$$H\Psi - \Phi\left(\int \Phi^* H\Psi \, d\tau\right) = H\Psi - \mathscr{P}H\Psi = E\Psi. \tag{2.100}$$

Equation (2.100) is still not in the standard form we are seeking, since, although the operators $\mathscr{P}$ and H are separately Hermitian, their product $\mathscr{P}H$ may not be. However, the symmetrized product $\mathscr{P}H + H\mathscr{P}$ of two Hermitian operators is necessarily Hermitian.(1) Furthermore, from equations (2.87) and (2.95) we have

$$H\mathscr{P}\Psi = H(\mathscr{P}\Psi) = 0. \tag{2.101}$$

Adding equations (2.100) and (2.101) we arrive at the equation

$$H_{\text{eff}}\Psi = [H - (\mathscr{P}H + H\mathscr{P})]\Psi = E\psi, \tag{2.102}$$

which is the standard form of the equation

$$H_{\text{eff}}\Psi_i = E_i\Psi_i \tag{2.103}$$

determining the eigenvalues E_i and eigenfunctions Ψ_i of the Hermitian operator

$$H_{\text{eff}} = H - (\mathscr{P}H + H\mathscr{P}). \tag{2.104}$$

Since the operator H_{eff} is Hermitian the eigenfunctions of equation (2.103) are automatically orthogonal, or in the case of degeneracy may be so chosen. This orthogonality of the eigenfunctions of H_{eff} includes the required orthogonality of the solutions to the fixed function Φ (equation (2.87)), since

$$\begin{aligned} H_{\text{eff}}\Phi &= H\Phi - \mathscr{P}H\Phi - H\mathscr{P}\Phi \\ &= H\Phi - \Phi\left(\int \Phi^* H\Phi \, \mathrm{d}\tau\right) - H\Phi \\ &= -\Phi\left(\int \Phi^* H\Phi \, \mathrm{d}\tau\right), \end{aligned} \tag{2.105}$$

that is, Φ itself is an eigenfunction of H_{eff} with the eigenvalue

$$E_0 = -\left(\int \Phi^* H\Phi \, \mathrm{d}\tau\right). \tag{2.106}$$

The introduction of the projection operator $\mathscr{P}$ has, therefore, enabled us to reduce equation (2.91) and the orthogonality constraint (2.87) to the single standard eigenvalue equation (2.103). In the context of self-consistent field theory this reduction is commonly referred to as "the elimination of the off-diagonal Lagrange multiplier ϵ".

The extension of this treatment to the case of several linear constraints, expressed as orthogonality to several fixed, orthonormal, functions Φ_α

$$\int \Psi^* \Phi_\alpha \, \mathrm{d}\tau = \int \Phi_\alpha^* \Psi \, \mathrm{d}\tau = 0 \tag{2.107}$$

is straightforward. For each constraint we have a projection operator $\mathscr{P}_\alpha$ defined as in equation (2.95)

$$\mathscr{P}_\alpha F = \Phi_\alpha\left(\int \Phi_\alpha^* F \, d\tau\right) \tag{2.108}$$

and the effective Hamiltonian is again given by equation (2.104), where now

$$\mathscr{P} = \sum_\alpha \mathscr{P}_\alpha. \tag{2.109}$$

In vector-space language, $\mathscr{P}$ of equation (2.109) projects onto that component of any function F which lies in the sub-space spanned by the orthonormal functions Φ_α.

(a) CONSTRAINED VARIATIONAL CALCULATIONS WITH A FINITE BASIS

Explicit calculations using the techniques of the previous section are usually carried out in terms of a finite set of basis functions of the type considered in Section 2.2. Denoting this set of functions by X_i ($i = 1, \ldots, n$), we obtain linear expansions for the functions Φ, Ψ and $\delta\Psi$.

Thus

$$\Phi = \sum_{i=1}^{n} X_i d_i, \tag{2.110}$$

$$\Psi = \sum_{i=1}^{n} X_i c_i, \tag{2.111}$$

$$\delta\Psi = \sum_{i=1}^{n} X_i \delta c_i. \tag{2.112}$$

Using these expansions it is a simple matter to reduce the integro-differential equations of the previous section to n-dimensional matrix equations. Thus substituting the expansions (2.110)–(2.112) into equation (2.90), carrying out the integrations and equating the coefficients of δc_i, δc_i^* to zero, we again find the condition (2.93) relating the off-diagonal Lagrange multipliers, whilst equation (2.91) is replaced by the linear algebraic equations

$$\sum_{j=1}^{n} H_{ij} c_j = E \sum_{j=1}^{n} S_{ij} c_j + \epsilon \sum_{j=1}^{n} S_{ij} d_j \qquad (i = 1 \ldots n) \tag{2.113}$$

where

$$S_{ij} = \int X_i^* X_j \, d\tau \tag{2.114}$$

and

$$H_{ij} = \int X_i^* H X_j \, d\tau. \qquad (2.115)$$

At this stage it is convenient to introduce an explicit vector notation. Bold-face lower-case letters **c**, **d**, etc. are used to denote column vectors ($(n \times 1)$ matrices), bold-face capitals **S**, **H**, etc. to denote $(n \times n)$ square matrices and a dagger (†) to denote the transposed complex conjugate (Hermitian conjugate) of a matrix. The Hermitian conjugate of a column vector is, of course, a row vector ($(1 \times n)$ matrix); for example

$$\mathbf{c}^\dagger = (c_1^*, c_2^*, \ldots c_n^*). \qquad (2.116)$$

Using this notation equation (2.113) becomes

$$\mathbf{Hc} = E\mathbf{Sc} + \epsilon\mathbf{Sd}, \qquad (2.117)$$

whilst the orthonormality conditions (2.86), (2.87) and (2.88) are expressed by

$$\mathbf{c}^\dagger \mathbf{Sc} = 1, \qquad (2.118)$$

$$\mathbf{c}^\dagger \mathbf{Sd} = \mathbf{d}^\dagger \mathbf{Sc} = 0, \qquad (2.119)$$

and

$$\mathbf{d}^\dagger \mathbf{Sd} = 1, \qquad (2.120)$$

respectively.

The elimination of the off-diagonal Lagrange multiplier ϵ from equation (2.117) is accomplished in much the same way as in the previous section, although the appearance of the formulae is somewhat different because of the role played by the overlap matrix **S**. Multiplying equation (2.117) on the left by $\mathbf{d}^\dagger$ and using equations (2.118)–(2.120) we obtain

$$\epsilon = \mathbf{d}^\dagger \mathbf{Hc}. \qquad (2.121)$$

Substituting this value of ϵ, equation (2.117) may be written

$$\mathbf{Hc} - \mathbf{Sd}(\mathbf{d}^\dagger \mathbf{Hc}) = E\mathbf{Sc}. \qquad (2.122)$$

The associative law for matrix multiplication may now be used to re-express the second term on the left-hand side of equation (2.122)

$$\mathbf{Sd}(\mathbf{d}^\dagger \mathbf{Hc}) = \mathbf{S} \cdot \mathbf{dd}^\dagger \cdot \mathbf{H} \cdot \mathbf{c} = \mathbf{SDHc}. \qquad (2.123)$$

Here we have introduced the $(n \times n)$ square matrix **D** as the product of the column matrix **d** and the row matrix $\mathbf{d}^\dagger$ (compare with (2.98)). Its matrix elements are simply

$$D_{ij} = (\mathbf{dd}^\dagger)_{ij} = d_i d_j^*. \qquad (2.124)$$

Substituting equation (2.123) into equation (2.122) we obtain

$$(\mathbf{H} - \mathbf{SDH})\mathbf{c} = E\mathbf{Sc}. \tag{2.125}$$

The matrices **S**, **H** and **D** appearing in equation (2.125) are all clearly Hermitian but the product **SDH** may not be. However, since

$$\mathbf{DSc} = \mathbf{dd}^\dagger\mathbf{Sc} = \mathbf{d}(\mathbf{d}^\dagger\mathbf{Sc}) = 0 \tag{2.126}$$

from equation (2.119), we may add the zero term $-\mathbf{HDSc}$ to equation (2.125) and hence obtain an Hermitian effective Hamiltonian matrix.

Thus we finally obtain

$$\mathbf{H}_{\text{eff}}\mathbf{c} = E\mathbf{Sc} \tag{2.127}$$

with

$$\mathbf{H}_{\text{eff}} = \mathbf{H} - \mathbf{SDH} - \mathbf{HDS} = \mathbf{H}_{\text{eff}}^\dagger. \tag{2.128}$$

Comparing these formulae with those of the preceding section we see that the role of the projection operator is played by the matrix

$$\mathbf{P} = \mathbf{SD}. \tag{2.129}$$

This matrix **P** is idempotent since

$$\begin{aligned} \mathbf{P}^2 &= \mathbf{SDSD} \\ &= \mathbf{Sd}(\mathbf{d}^\dagger\mathbf{Sd})\mathbf{d}^\dagger \\ &= \mathbf{Sdd}^\dagger \\ &= \mathbf{P}, \end{aligned} \tag{2.130}$$

but it is *not* Hermitian

$$\mathbf{P}^\dagger = \mathbf{D}^\dagger\mathbf{S}^\dagger = \mathbf{DS} \neq \mathbf{P}, \tag{2.131}$$

unless **D** and **S** happen to commute. This lack of Hermiticity arises from the use of a non-orthogonal set of basis functions. If an orthonormal basis is used the matrix **S** is replaced by the $(n \times n)$ unit matrix and **P**, like the operator $\mathscr{P}$ of the previous section, is Hermitian as well as idempotent.

Two non-degenerate eigenvectors $\mathbf{c}^{(1)}$ and $\mathbf{c}^{(2)}$ of equation (2.127) are easily shown to be orthogonal in the sense that

$$\mathbf{c}^{(1)\dagger}\mathbf{Sc}^{(2)} = 0, \tag{2.132}$$

which corresponds to the usual orthogonality relation

$$\int \Phi_1^*\Phi_2 \, d\tau = 0 \tag{2.133}$$

between the wavefunctions

$$\Phi_1 = \sum_{i=1}^{n} X_i c_i^{(1)} \tag{2.134}$$

and

$$\Phi_2 = \sum_{i=1}^{n} X_i c_i^{(2)}. \tag{2.135}$$

As in the previous section the relations (2.132) ensure that the restrictions (2.119) are automatically satisfied, since the vector $\mathbf{d}$ itself is a solution of (2.127) with $E = -\mathbf{d}^\dagger \mathbf{H}\mathbf{d}$. For we have

$$\begin{aligned} \mathbf{H}_{\text{eff}}\mathbf{d} &= \mathbf{Hd} - \mathbf{SDHd} - \mathbf{HDSd} \\ &= \mathbf{Hd} - \mathbf{Sd}(\mathbf{d}^\dagger \mathbf{Hd}) - \mathbf{Hd} \cdot \mathbf{d}^\dagger \mathbf{Sd} \\ &= \mathbf{Hd} - (\mathbf{d}^\dagger \mathbf{Hd})\mathbf{Sd} - \mathbf{Hd} \\ &= -(\mathbf{d}^\dagger \mathbf{Hd})\mathbf{Sd} \end{aligned} \tag{2.136}$$

as required.

Consequently, once the effective Hamiltonian matrix (2.128) has been constructed, the calculation proceeds exactly as in Section 2.2. The eigenvalues $E_i (i = 1, \ldots, n-1)$ are the $(n-1)$ roots of the determinantal equation

$$\det(\mathbf{H}_{\text{eff}} - E\mathbf{S}) = 0 \tag{2.137}$$

other than $E_0 = -\mathbf{d}^\dagger \mathbf{Hd}$, and the corresponding eigenvectors $\mathbf{c}^{(i)}$ $(i = 1, \ldots, n-1)$ are determined by the linear equations

$$(\mathbf{H}_{\text{eff}} - E_i\mathbf{S})\mathbf{c}^{(i)} = 0. \tag{2.138}$$

The extension to the case of several constraints of the form

$$\mathbf{c}^\dagger \mathbf{S}\mathbf{d}^{(\alpha)} = \mathbf{d}^{(\alpha)\dagger}\mathbf{S}\mathbf{c} = 0 \tag{2.139}$$

is again straightforward; the effective Hamiltonian matrix is given by (2.128) with

$$\mathbf{D} = \sum_{\alpha} \mathbf{D}_\alpha = \sum_{\alpha} \mathbf{d}^{(\alpha)}\mathbf{d}^{(\alpha)\dagger}. \tag{2.140}$$

(b) USE OF ORTHOGONALIZED BASIS

In the present case of a finite set of basis functions there is an alternative solution to the constrained variation problem, which is sometimes simpler than the above treatment using projection operators.

In this alternative approach the first step is to construct from the functions $X_i (i = 1, \ldots, n)$ a set of functions

$$X_i^{(0)} = X_i - \Phi\left(\int \Phi^* X_i \, d\tau\right) \qquad (i = 1, \ldots, n) \tag{2.141}$$

each of which is orthogonal to the fixed function(s) Φ (or Φ_α). Since *any* linear combination of the functions $X_i^{(0)}$ is orthogonal to

Φ, we may now proceed exactly as in Section 2.2 and obtain the eigenvalues and wave functions from the equations

$$\Psi = \sum_j X_j^{(0)} c_j' \tag{2.142}$$

$$\det \{H_{ij}^{(0)} - ES_{ij}^{(0)}\} = 0 \tag{2.143}$$

$$\sum_j (H_{ij}^{(0)} - ES_{ij}^{(0)}) c_j' = 0. \tag{2.144}$$

We note that, because of the relations (2.110), not all the functions $X_i^{(0)}$ can be linearly independent. In fact, starting from a set of n linearly independent X_i and imposing p constraints of the form (2.110), (2.107) we obtain a set $\{X_i^{(0)}\}$ containing $(n-p)$ linearly independent functions. This causes no difficulty in the solution of equations (2.142), (2.143) and (2.144). Any $(n-p)$ linearly independent functions $X_i^{(0)}$ may be chosen as a new basis. The sums in equations (2.142)–(2.144) are then over these $(n-p)$ functions and $H^{(0)}$ and $S^{(0)}$ become $(n-p) \times (n-p)$ square matrices.

Explicit expressions for the full $(n \times n)$ matrices $\mathbf{H}^{(0)}$ and $\mathbf{S}^{(0)}$ may be obtained by using the $X_i^{(0)}$ of equation (2.141) in place of the X_i in the definitions (2.114) and (2.115). In this way we find

$$\mathbf{S}^{(0)} = \mathbf{S} - \mathbf{SDS} \tag{2.145}$$

and

$$\mathbf{H}^{(0)} = \mathbf{H} - \mathbf{SDH} - \mathbf{HDS} + \mathbf{d}^\dagger \mathbf{Hd} \cdot \mathbf{SDS}. \tag{2.146}$$

The equivalence of the two methods of calculation which we have outlined is now readily established. Let us suppose that the vector $\mathbf{c}$ is a solution of equation (2.127) for a certain value of E. Defining the vector $\mathbf{c}'$ by

$$\mathbf{c} = \mathbf{c}' - \mathbf{DSc}' \tag{2.147}$$

we then have

$$\begin{aligned}
0 &= (\mathbf{H}_{\text{eff}} - E\mathbf{S})\mathbf{c} \\
&= (\mathbf{H} - \mathbf{SDH} - \mathbf{HDS} - E\mathbf{S})(\mathbf{c}' - \mathbf{DSc}') \\
&= (\mathbf{H} - \mathbf{SDH} - \mathbf{HDS} - E\mathbf{S} - \mathbf{HDS} + \mathbf{SDHDS} + \mathbf{HDSDS} + E\mathbf{SDS})\mathbf{c}' \\
&= (\mathbf{H} - \mathbf{SDH} - \mathbf{HDS} - E\mathbf{S} - \mathbf{HDS} + \mathbf{d}^\dagger\mathbf{Hd} \cdot \mathbf{SDS} + \mathbf{d}^\dagger\mathbf{Sd} \cdot \mathbf{HDS} + E\mathbf{SDS})\mathbf{c}' \\
&= [\mathbf{H} - \mathbf{SDH} - \mathbf{HDS} + \mathbf{d}^\dagger\mathbf{Hd} \cdot \mathbf{SDS} - E(\mathbf{S} - \mathbf{SDS})]\,\mathbf{c}',
\end{aligned} \tag{2.148}$$

that is, the vector $\mathbf{c}'$ is a solution of equation (2.144) for the given value of E. Furthermore, from the definition (2.141) of the $X^{(0)}$

basis, it is clear that the vectors c in the X basis and c′ in the $X^{(0)}$ basis represent the same electronic wave function Ψ.

The methods of the preceding two sections find their principal application in the self-consistent-field theory of open shell states (Chapter 7, Section 2, (c) and (d)) and in multi-configuration, self-consistent-field theory.

References

1. P. A. M. Dirac (1957), "Quantum Mechanics", Third Edition, Oxford University Press.
2. J. K. L. MacDonald (1933), *Phys. Rev.* **43**, 830.
3. H. Hellmann (1937), "Einführung in die Quantenchemie", Franz Deuticke, Leipzig and Vienna.
4. R. P. Feynman (1939), *Phys. Rev.* **56**, 340.
5. H. Eyring, J. Walter and G. E. Kimball (1944), "Quantum Chemistry", Wiley, New York.
6. J. C. Slater (1933), *J. Chem. Phys.* **1**, 687.
7. R. G. Parr (1964), *J. Chem. Phys.* **40**, 3726.
8. A. C. Hurley (1954), *Proc. Roy. Soc.* **A226**, 179.
9. S. T. Epstein, A. C. Hurley, R. E. Wyatt and R. G. Parr (1967), *J. Chem. Phys.* **47**, 1275.
10. A. C. Hurley (1967), *Int. J. Quantum Chem.* **1S**, 677.

CHAPTER 3

The Hydrogen Molecule Ion

We start our discussion of the electronic wave functions of molecules with the simplest case, the hydrogen molecule ion. In the adiabatic approximation we have the problem of a single electron moving in the electrostatic field of two protons in fixed positions. This problem is exactly soluble and the properties of the solutions, especially their symmetry, play a key role in the treatment of more complex diatomic molecules. We also consider various approximate treatments of the problem. Because of its simplicity the hydrogen molecule ion has been much used as a testing ground for approximate methods designed primarily for more complex molecules.

3.1. Separation of the wave equation in elliptical coordinates

For the hydrogen molecule ion the electronic Schrödinger equation (equation (1.71)) becomes

$$\left(-\tfrac{1}{2}\nabla^2 - \frac{1}{r_a} - \frac{1}{r_b} + \frac{1}{R}\right)\psi = E\psi, \tag{3.1}$$

where r_a, r_b are the distances of the electron from the nuclei a and b, and R is the nuclear separation.

We first express equation (3.1) in elliptical coordinates (μ, ν, φ) where

$$\mu = \frac{r_a + r_b}{R}, \quad 1 \leqslant \mu \leqslant \infty,$$

$$\nu = \frac{r_a - r_b}{R}, \quad -1 \leqslant \nu \leqslant 1,$$

and φ is the azimuthal angle about the internuclear axis, $0 \leqslant \varphi \leqslant 2\pi$.

In this coordinate system the kinetic energy operator $-\frac{1}{2}\nabla^2$ becomes

$$-\frac{2}{R^2(\mu^2-\nu^2)}\left\{\frac{\partial}{\partial\mu}\left[(\mu^2-1)\frac{\partial}{\partial\mu}\right]+\frac{\partial}{\partial\nu}\left[(1-\nu^2)\frac{\partial}{\partial\nu}\right]\right.$$
$$\left.+\left[\frac{\mu^2-\nu^2}{(\mu^2-1)(1-\nu^2)}\frac{\partial^2}{\partial\varphi^2}\right]\right\} \tag{3.2}$$

and the electronic Schrödinger equation (3.1) may be written

$$-\frac{2}{R^2(\mu^2-\nu^2)}\left\{\frac{\partial}{\partial\mu}\left[(\mu^2-1)\frac{\partial}{\partial\mu}\right]+\frac{\partial}{\partial\nu}\left[(1-\nu^2)\frac{\partial}{\partial\nu}\right]\right.$$
$$\left.+\left(\frac{1}{\mu^2-1}+\frac{1}{1-\nu^2}\right)\frac{\partial^2}{\partial\varphi^2}\right\}\psi(\mu,\nu,\varphi)-\frac{4\mu}{R(\mu^2-\nu^2)}\psi(\mu,\nu,\varphi)$$
$$=\left(E-\frac{1}{R}\right)\psi(\mu,\nu,\varphi). \tag{3.3}$$

Equation (3.3) is separable and, if we write

$$\psi(\mu,\nu,\varphi)=M(\mu)N(\nu)\Phi(\varphi), \tag{3.4}$$

we find that M, N and Φ must satisfy the ordinary differential equations

$$\frac{d^2\Phi}{d\varphi^2}=-\lambda^2\Phi \tag{3.5}$$

$$\frac{d}{d\mu}\left\{(\mu^2-1)\frac{dM}{d\mu}\right\}+\left(2R\mu+e\mu^2-\frac{\lambda^2}{\mu^2-1}+A\right)M=0 \tag{3.6}$$

$$\frac{d}{d\nu}\left\{(1-\nu^2)\frac{dN}{d\nu}\right\}+\left(-e\nu^2-\frac{\lambda^2}{1-\nu^2}-A\right)N=0 \tag{3.7}$$

where $e=(R^2/2)(E-(1/R))$ and A is a separation constant.

The φ equation (3.5) is of the same form as that for the hydrogen atom with the solutions

$$\Phi=e^{i\lambda\varphi} \qquad (\lambda=0,\pm 1,\pm 2,\ldots). \tag{3.8}$$

Since the operator for the component of the angular momentum of the electron about the internuclear axis is $(1/i)(\partial/\partial\varphi)$, this component of the angular momentum is quantized with the eigenvalue λ. By analogy with the $s, p, d, f\ldots$ notation for the one-electron states of atoms, the states of H_2^+ with $\lambda=0,\pm 1,\pm 2,\pm 3\ldots$ are referred to as $\sigma, \pi, \delta, \varphi\ldots$ states, respectively.

The solution of equations (3.6) and (3.7) is more difficult and involves expansions in series and the use of recurrence relations. The most complete treatment of these equations is that of Bates, Ledsham and Stewart[1] who obtained accurate energies and wave functions for ten low-lying states including the ground state. We shall not consider the details of this work but merely point out two properties of equations (3.6) and (3.7) which apply to the one-electron states of all homonuclear diatomic molecules.

Firstly, the parameter λ only enters these equations in the form λ^2. This implies that the energies of the states are independent of the sign of λ, so that all π, δ, φ . . . states are at least doubly degenerate. Secondly, equation (3.7) is even in the coordinate ν. This implies that the eigenfunctions are either even or odd functions of ν, that is

$$N(-\nu) = \pm N(\nu). \tag{3.9}$$

The usual classification of the states of homonuclear diatomic molecules is not based directly on the symmetry (3.9) but rather on symmetry with respect to inversion i through the nuclear centre. In elliptical coordinates i corresponds to the transformation

$$\mu \to \mu, \qquad \nu \to -\nu, \qquad \varphi \to \varphi + \pi. \tag{3.10}$$

From the form (3.8) of the φ eigenfunctions we see that the transformation $\varphi \to \varphi + \pi$ introduces a factor $(-1)^\lambda$ in addition to any sign change from the transformation $\nu \to -\nu$. One-electron states (*orbitals*) which are even under inversion are given a subscript g (from the German "gerade") those odd under inversion a subscript u (from "ungerade").

The states of H_2^+ may be given additional labels which show the states of the united atom (He^+) that are obtained from the molecular states as the nuclear separation R tends to zero. Combining the labels obtained from equation (3.8), the transformation (3.10) and the united atom states, the solutions of equations (3.5), (3.6) and (3.7) obtained by Bates *et al.*[1] are the molecular orbitals.

$$1s\sigma_g,\ 2s\sigma_g,\ 3s\sigma_g,\ 2p\sigma_u,\ 3p\sigma_u,\ 4p\sigma_u,\ 3d\sigma_g,\ 4f\sigma_u,\ 2p\pi_u,\ 3d\pi_g. \tag{3.11}$$

The variation in the energies of these states with the nuclear separation R is shown in Figs 3.1 and 3.2.

The curves for two lowest states, $1s\sigma_g$ and $2p\sigma_u$, are of particular interest, since they are typical of the behaviour of bonding and anti-bonding orbitals in molecular orbital (MO) theory. From Fig. 3.2, we see that, for large nuclear separations, the energy for both states is equal to the energy of a hydrogen atom in its ground $1s$ state. The wave functions must also go over into a linear combination of the ground states $1s_a$, $1s_b$ of an electron centred on nuclei a, b.

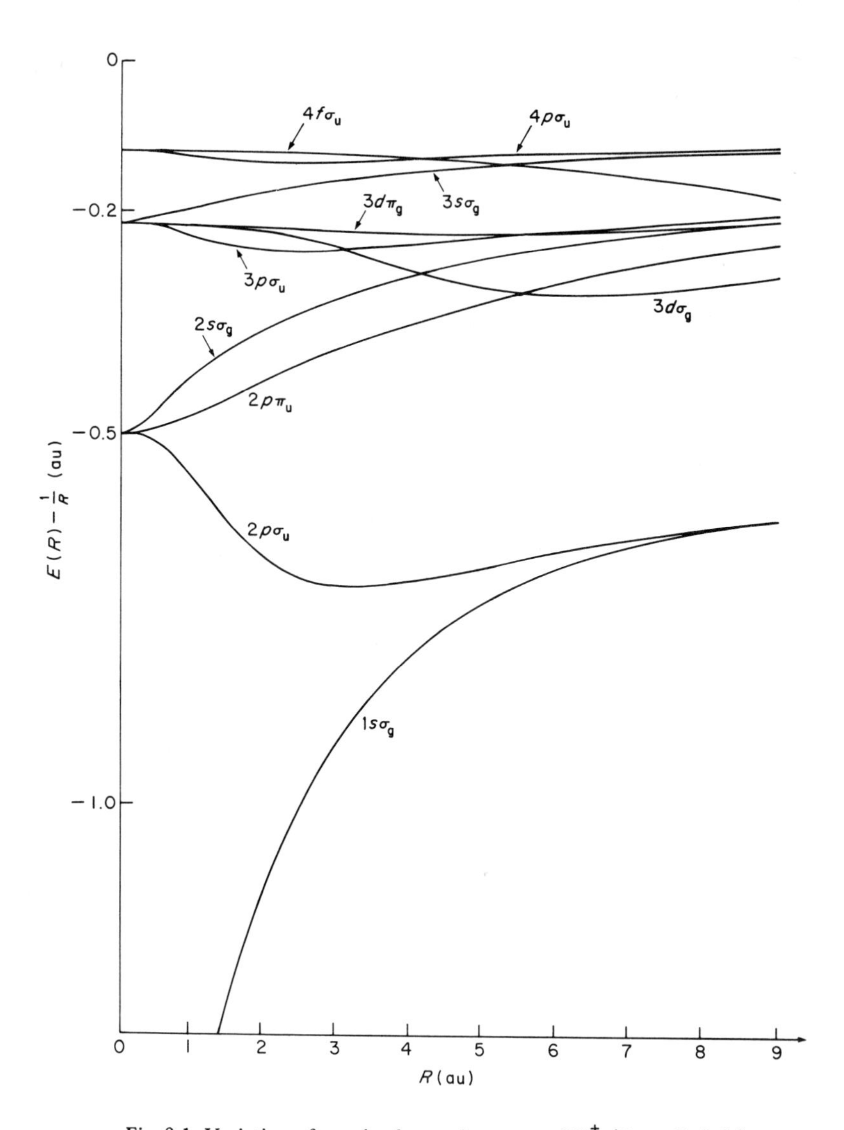

Fig. 3.1. Variation of purely electronic energy of H_2^+. (From Ref. 2.)

Taking account of the behaviour of the functions under inversion we obtain the correlations

$$1s\sigma_g \approx 1s_a + 1s_b \approx \sigma_g(1s) = \sigma(1s) \tag{3.12}$$

$$2p\sigma_u \approx 1s_a - 1s_b \approx \sigma_u(1s) = \sigma^*(1s). \tag{3.13}$$

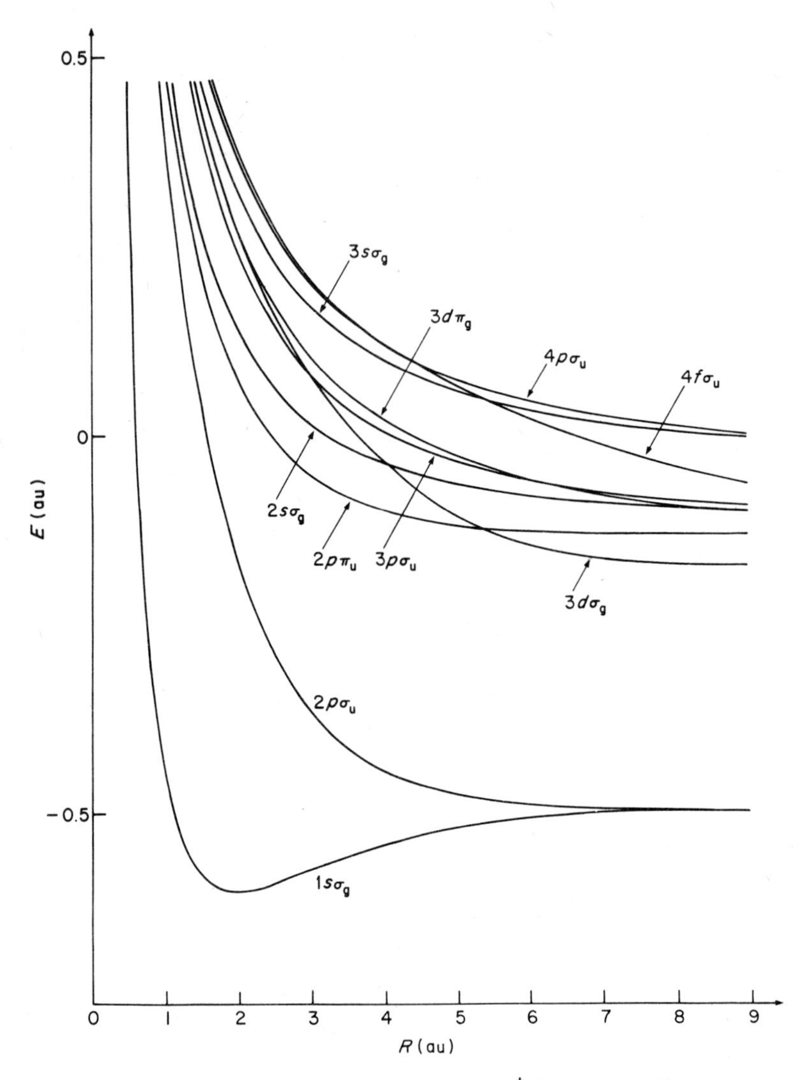

Fig. 3.2. Adiabatic potential curves for H_2^+. (From Ref. 2.)

The last two expressions for the molecular orbitals in equations (3.12) and (3.13) are the *separated-atom* designation. Here, to distinguish from the united-atom notation, the symmetry symbol is given first followed by the appropriate state of the separated atoms in parenthesis. An asterisk is used to distinguish *antibonding* orbitals, which are characterized by a nodal surface bisecting the internuclear axis, that is by antisymmetry under the transformation

$\nu \to -\nu$ of equation (3.9). Thus, for a homonuclear diatomic molecule, the bonding molecular orbitals are of types σ_g, π_u, δ_g . . . and the antibonding orbitals are of types σ_u, π_g, δ_u . . . .

It is easy to see in a qualitative way why the bonding orbital $\sigma_g(1s)$ leads to a stable potential energy curve, whereas the antibonding orbital $\sigma_u(1s)$ gives a repulsive curve (Fig. 3.2). As the two atoms approach each other the nodeless $\sigma_g(1s)$ orbital leads to a concentration of charge in the region between the two nuclei. This is a region of low potential energy for the negatively-charged electron. The nodeless form of the orbital also favours a relatively low value for the kinetic energy. These two factors combine to give a stable potential energy curve for the motion of the nuclei. Of course, as the nuclear separation changes, the accurate wave function for $\sigma_g(1s)$ automatically adjusts its scale in such a way as to satisfy the virial theorem (equation (2.34)). This leads to the more complicated behaviour of the potential and kinetic energies shown in Fig. 2.1.

For the antibonding $\sigma_u(1s)$ orbital the nodal plane between the two nuclei leads to a depletion of charge from the low potential energy region, and to a relatively high kinetic energy. Consequently we have repulsion between the two nuclei for all separations.

From a study of the behaviour of equations (3.5), (3.6) and (3.7) for large nuclear separation it is possible to derive separated atom correlations for the more highly excited states of H_2^+. However, these correlations for the higher states are affected by two features which are peculiar to this simple system. Firstly we have the degeneracy of all states of the hydrogen atom with the same principal quantum number. Secondly, as we see from Figs. 3.1 and 3.2, certain curves of the same symmetry type (e.g. $2s\sigma_g$, $3d\sigma_g$; $4f\sigma_u$, $4p\sigma_u$) cross each other as the nuclear separation varies. Neither of these complications arise for larger diatomic molecules where the separated atom correlations are most useful (see Chapter 7). The violation of the "non-crossing rule" for H_2^+ may be attributed to the separability of the wave equation,[(2)] which leads to symmetry transformations over and above the usual geometrical symmetries.[(3)] Separated atom correlations for all the states (3.11) of H_2^+ are given by Kotani, Ohno and Kayama.[(2)]

3.2. Some simple variational approximations

For molecules larger than H_2^+ it is no longer possible to solve the electronic Schrödinger equation accurately. Approximate methods must be used and in assessing their accuracy it is useful to compare the accurate results of the previous section with those obtained using simple approximate wave functions for H_2^+.

(a) THE LINEAR COMBINATION OF ATOMIC ORBITALS (LCAO) METHOD

We consider the two lowest states $\sigma_g(1s)$ and $\sigma_u(1s)$. In Section 3.1 we found that for large nuclear separations the wave functions for these states go over into the sum and difference of hydrogen $1s$ atomic orbitals centred on the nuclei a and b. In the LCAO approximation we use these same wave functions (apart from normalization factors) for all nuclear separations. Abbreviating the atomic orbitals $1s_a$, $1s_b$ by a, b respectively we have

$$\psi_g = N_g(a + b) \tag{3.14}$$

$$\psi_u = N_u(a - b). \tag{3.15}$$

If we use the normalized atomic orbitals

$$a = \pi^{-\frac{1}{2}} e^{-r_a}, \qquad b = \pi^{-\frac{1}{2}} e^{-r_b} \tag{3.16}$$

the normalization constants N_g and N_u are given by

$$\int \psi_g^2 \, dv = N_g^2(2 + 2S) = 1 \tag{3.17}$$

$$\int \psi_u^2 \, dv = N_u^2(2 - 2S) = 1 \tag{3.18}$$

where S is the overlap integral

$$S = \int ab \, dv. \tag{3.19}$$

Using the explicit forms (3.16) for the atomic orbitals the overlap integral is found to be

$$S = e^{-R}(1 + R + R^2/3). \tag{3.20}$$

The electronic Hamiltonian for H_2^+ is

$$H = -\tfrac{1}{2}\nabla^2 - \frac{1}{r_a} - \frac{1}{r_b} + \frac{1}{R}. \tag{3.21}$$

The energies corresponding to the wave functions (3.14) and (3.15) are found from the expectation values of H. Using equations (3.17) and (3.18) we find

$$E_g = \int \psi_g H \psi_g \, dv = (2 + 2S)^{-1} \int (a + b)H(a + b) \, dv \tag{3.22}$$

$$E_u = \int \psi_u H \psi_u \, dv = (2 - 2S)^{-1} \int (a - b)H(a - b) \, dv. \tag{3.23}$$

Equations (3.22) and (3.23) reduce to

$$E_{\mathrm{g}} = \frac{H_{aa} + H_{ab}}{1 + S} \tag{3.24}$$

$$E_{\mathrm{u}} = \frac{H_{aa} - H_{ab}}{1 - S} \tag{3.25}$$

where the matrix elements of the Hamiltonian are defined by

$$H_{aa} = \int aHa \, \mathrm{d}v = H_{bb} \tag{3.26}$$

$$H_{ab} = \int aHb \, \mathrm{d}v = H_{ba}. \tag{3.27}$$

These matrix elements are easily evaluated using the explicit forms (3.16) of the atomic orbitals and the fact that these orbitals are solutions of the Schrödinger equation for a hydrogen atom

$$\left(-\tfrac{1}{2}\nabla^2 - \frac{1}{r_{\mathrm{a}}}\right) a = -\tfrac{1}{2} a \tag{3.28}$$

$$\left(-\tfrac{1}{2}\nabla^2 - \frac{1}{r_{\mathrm{b}}}\right) b = -\tfrac{1}{2} b. \tag{3.29}$$

In this way we find

$$H_{aa} = -\tfrac{1}{2} + \frac{1}{R} - \int \frac{a^2}{r_{\mathrm{b}}} \, \mathrm{d}v \tag{3.30}$$

$$H_{ab} = \left(-\tfrac{1}{2} + \frac{1}{R}\right) S - \int \frac{ab}{r_{\mathrm{a}}} \, \mathrm{d}v \tag{3.31}$$

where

$$\int \frac{a^2}{r_{\mathrm{b}}} \, \mathrm{d}v = \frac{1}{R}\{1 - \mathrm{e}^{-2R}(1 + R)\} \tag{3.32}$$

$$\int \frac{ab}{r_{\mathrm{a}}} \, \mathrm{d}v = \mathrm{e}^{-R}(1 + R). \tag{3.33}$$

Substituting these results into equations (3.24) and (3.25) we end up with simple analytic expressions for the potential energy curves for the $\sigma_{\mathrm{g}}(1s)$ and $\sigma_{\mathrm{u}}(1s)$ states.

$$E_{\mathrm{g}}(R) = -\tfrac{1}{2} + \frac{1}{R} - \frac{(1/R)\{1 - \mathrm{e}^{-2R}(1 + R)\} + \mathrm{e}^{-R}(1 + R)}{1 + \mathrm{e}^{-R}(1 + R + \tfrac{1}{3}R^2)} \tag{3.34}$$

$$E_{\mathrm{u}}(R) = -\tfrac{1}{2} + \frac{1}{R} - \frac{(1/R)\{1 - \mathrm{e}^{-2R}(1 + R)\} - \mathrm{e}^{-R}(1 + R)}{1 - \mathrm{e}^{-R}(1 + R + \tfrac{1}{3}R^2)}. \tag{3.35}$$

These curves are shown in Fig. 3.3. Table 3.1 compares the binding energy D_e and the equilibrium nuclear separation R_e, obtained from the minimum in the $E_g(R)$ curve of equation (3.34), with the accurate values derived in Section 3.1. We see that the simple LCAO approximation gives results which are qualitatively correct; the binding energy is in error by 37%, the equilibrium nuclear separation by 25%.

For the bound state $\sigma_g(1s)$ it is of interest to compute the mean kinetic energy and mean potential energy separately and to compare their behaviour with that deduced in Chapter 2 from quite general considerations (Fig. 2.1). Proceeding as above we find

$$\bar{T}(R) = \tfrac{1}{2}\,\frac{1 + e^{-R}(1 + R - \tfrac{1}{3}R^2)}{1 + e^{-R}(1 + R + \tfrac{1}{3}R^2)} \tag{3.36}$$

$$\bar{V}(R) = \frac{1}{R} - \frac{1 + 2e^{-R}(1 + R) + (1/R)\{1 - e^{-2R}(1 + R)\}}{1 + e^{-R}(1 + R + \tfrac{1}{3}R^2)}. \tag{3.37}$$

From equation (3.36) we see immediately that $\bar{T}(0) = \bar{T}(\infty) = \tfrac{1}{2}$ and that $\bar{T}(R) < \tfrac{1}{2}$ for all intermediate values of R. This behaviour is quite different from that shown in Fig. 2.1. The virial theorem is not even approximately satisfied, so that the kinetic and potential energies given by the simple LCAO function must be quite wrong.

It is also a simple matter to use the electrostatic formula to calculate the forces on the nuclei for any nuclear separation. The charge density of the $\sigma_g(1s)$ state for the wave function (3.14) is

$$\rho = \frac{a^2 + b^2 + 2ab}{2(1 + S)}. \tag{3.38}$$

Substituting this charge density function into the electrostatic formula (equation (2.30)) we find for the forces on the nuclei

$$F_b(R) = -F_a(R) = \frac{1}{R^2} - [2(1 + S)]^{-1}\left[\int \frac{a^2}{r_a^2}\cos\theta_a\,dv + \int \frac{b^2}{r_a^2}\cos\theta_a\,dv + 2\int \frac{ab}{r_a^2}\cos\theta_a\,dv\right] \tag{3.39}$$

where θ_a is the angle between the internuclear axis and the vector from nucleus a to the electron.

Since the distribution a^2 is spherically symmetric the first integral in equation (3.39) vanishes. The other two integrals are easily evaluated in elliptical coordinates.[4] However, this force curve $F_a(R)$ calculated from the electrostatic formula is very inaccurate. We find,[5] indeed, that the attraction exerted on nucleus a by the electron cloud (the second term in equation (3.39)) is always less than the nuclear repulsion $1/R^2$, so that equation (3.39) would predict a purely repulsive potential energy curve for the $\sigma_g(1s)$ state.

Thus, although the potential energy curve (3.34) calculated from the LCAO function is qualitatively correct, the virial and electrostatic formulae give quite wrong results with this simple wave function. We next consider the improvements in the wave function which are needed to satisfy these two theorems.

(b) THE SCALED LCAO FUNCTION[6]

In Chapter 2 we saw that in order to satisfy the virial theorem we must introduce a scale factor α into the wave function and determine the value of α by minimizing the total energy. The normalized atomic orbitals (3.16) then become

$$a' = (\alpha^3/\pi)^{\frac{1}{2}} e^{-\alpha r_a}, \quad b' = (\alpha^3/\pi)^{\frac{1}{2}} e^{-\alpha r_b}, \tag{3.40}$$

and the scaled wave function for the $\sigma_g(1s)$ state, at a nuclear separation R, is related to the simple LCAO function by

$$\psi_g'(R;\mathrm{r}) = \alpha^{\frac{3}{2}}\psi_g(\alpha R;\alpha\mathrm{r}). \tag{3.41}$$

Since the kinetic energy operator is homogeneous of degree minus two in the coordinates and the potential energy operator is homogeneous of degree minus one, the mean kinetic energy $\bar{T}(R)$ and potential energy $\bar{V}(R)$ for the scaled wave function (3.41) are related to the corresponding quantities for the simple LCAO function by the equations

$$\bar{T}_\alpha(R) = \alpha^2\bar{T}(\alpha R), \qquad \bar{V}_\alpha(R) = \alpha\bar{V}(\alpha R). \tag{3.42}$$

For the total electronic energy of the scaled function we have

$$E_\alpha(R) = \alpha^2\bar{T}(\alpha R) + \alpha\bar{V}(\alpha R). \tag{3.43}$$

Equations (3.36), (3.37) and (3.43) determine the total electronic energy of the scaled wave function for all values of α and R. The optimum value of α is now found by minimizing $E_\alpha(R)$ for each value of R. Putting $t = \alpha R$ we have

$$0 = \left(\frac{\partial E}{\partial \alpha}\right)_R = 2\alpha\bar{T} + \bar{V} + R\left[\alpha^2\frac{\mathrm{d}\bar{T}}{\mathrm{d}t} + \alpha\frac{\mathrm{d}\bar{V}}{\mathrm{d}t}\right] \tag{3.44}$$

that is

$$\alpha = -\frac{\bar{V} + t\dfrac{\mathrm{d}\bar{V}}{\mathrm{d}t}}{2\bar{T} + t\dfrac{\mathrm{d}\bar{T}}{\mathrm{d}t}}. \tag{3.45}$$

Equation (3.45) determines the optimum value of the scale factor α as a function of t. The corresponding value of R is given by

$$R = \frac{t}{\alpha}. \tag{3.46}$$

Substitution of these values in equation (3.43) gives the results shown in Fig. 3.3 and Table 3.1. It is clear that the scaling procedure has removed a large part of the error obtained with the simple LCAO

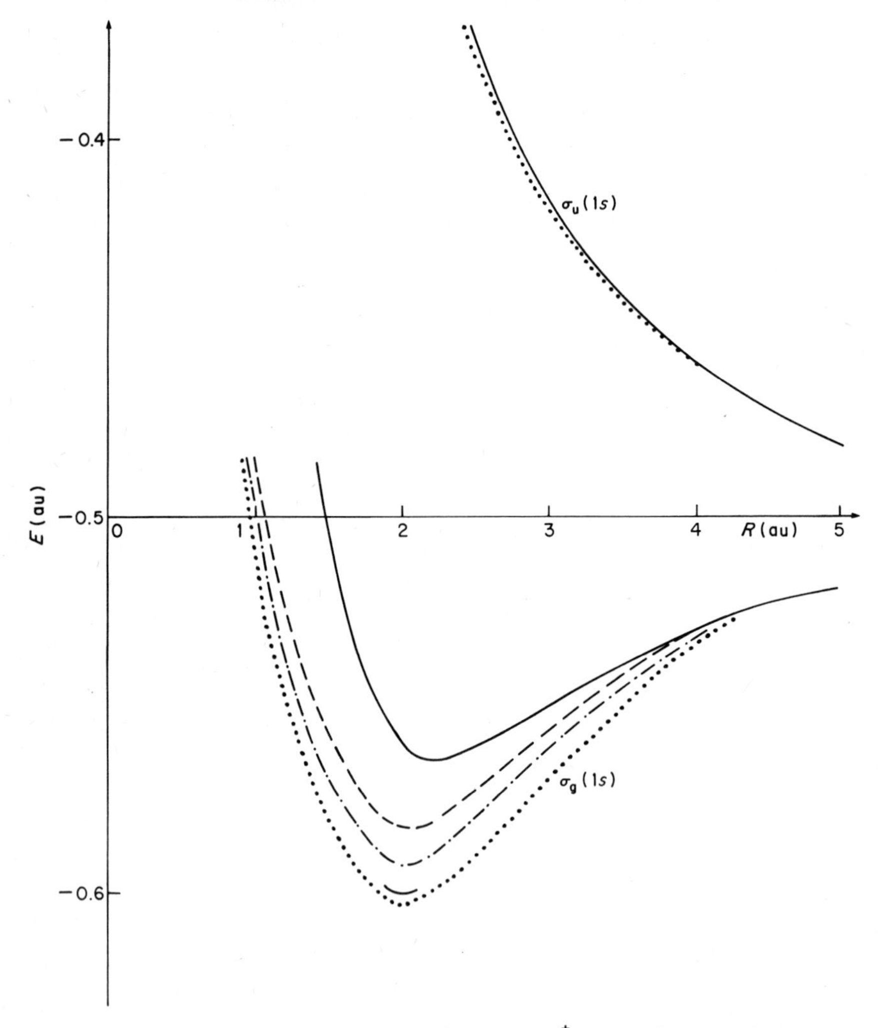

Fig. 3.3. Potential curves for $\sigma_g(1s)$ and $\sigma_u(1s)$ states of H_2^+. ——, LCAO calculation; – – – –, scaled LCAO calculation; –·–·–, floating, scaled LCAO calculation; –, James; •, Experiment.

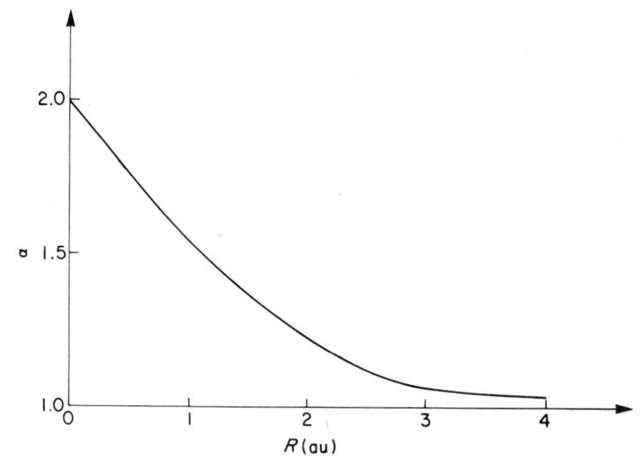

Fig. 3.4. Variation of scale factor α for the scaled LCAO calculation of the ground state of H_2^+.

function. The variation of the parameter α with R is shown in Fig. 3.4. We note that the scaled function becomes exact, not only for $R = \infty$, but also for $R = 0$, where $\alpha = 2$ and we have the exact wave function for the ground state of the united atom He^+.

(c) THE FLOATING, SCALED LCAO FUNCTION[7, 8]

Here we introduce additional flexibility into the wave function by detaching the atomic orbitals from the nuclei and allowing them to take up optimum positions. From the symmetry of the system it is clear that these optimum positions will lie on the internuclear axis and that their displacements from the nuclei will be equal and opposite (Fig. 3.5). The wave function now depends on two variable parameters, the scale factor α and the displacement x of the atomic orbitals from the nuclei. The total electronic energy is minimized with respect to variations of these two parameters. This ensures that

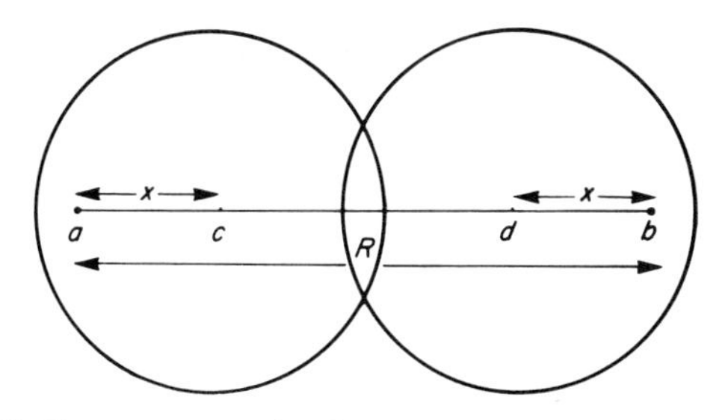

Fig. 3.5. The parameter x for the floating, scaled LCAO function.

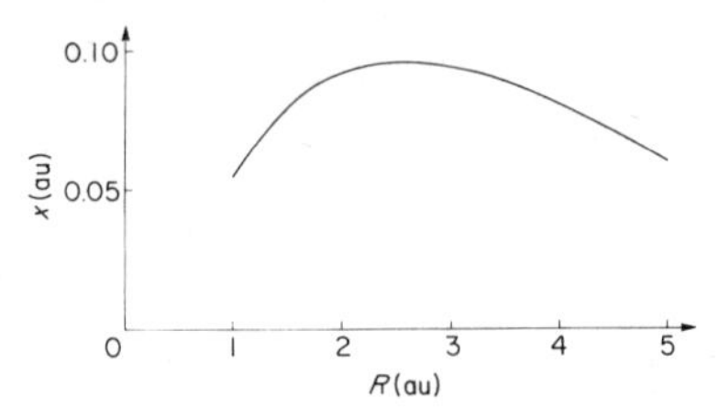

Fig. 3.6. Variation of x with nuclear separation for ground state of H_2^+.

both the virial theorem and the electrostatic theorem are satisfied (Chapter 2). The variation of the parameter x with R is shown in Fig. 3.6; the behaviour of α is similar to that for the scaled LCAO function (Fig. 3.4).

(d) THE DICKINSON FUNCTION[9]

The inward displacement of the atomic orbitals found with the floating LCAO function increases the value of the wave function in the region between the two nuclei. A similar effect may be achieved by introducing a contribution from $2p_\sigma$ atomic orbitals to obtain the hybridized LCAO function of Dickinson

$$\psi_g = N\{e^{-\alpha r_a} + e^{-\alpha r_b} + k(r_a \cos\theta_a\, e^{-\alpha' r_a} + r_b \cos\theta_b\, e^{-\alpha' r_b})\}. \quad (3.47)$$

For optimum values of the three variational parameters α, α' and k we obtain the equilibrium constants shown in Table 3.1. The binding energy obtained with the Dickinson function (2.65 eV) is considerably better than that obtained from the floating LCAO function (2.56 eV). The relatively poor energy of the floating function is associated with the incorrect location of the singularities in the wave function. The exact wave function and the various unfloated LCAO wave functions all have cusps at the nuclei. At these cusps the local kinetic energy becomes infinite and cancels, either exactly or approximately, the infinities in the local potential energy. For the floating function the infinities in the local kinetic and potential energies occur at different positions.

(e) SIMPLE FUNCTIONS IN ELLIPTICAL COORDINATES

There are two very simple and accurate variational wave functions for the ground state of H_2^+ which are quite different in form from the functions built up from atomic orbitals. These are the James[10] function

$$\psi_g = N\, e^{-\delta\mu}(1 + c\nu^2) \quad (3.48)$$

and the Guillemin and Zener[11] function

$$\psi_g = N\,e^{-\delta\mu} \cosh c\nu, \tag{3.49}$$

where μ, ν are the elliptical coordinates of Section 3.1.

For optimum values of the two variational parameters δ and c, both functions give a limiting energy within 0.005 eV of the accurate value. The Guillemin–Zener function retains this accuracy for all R values whereas the James function becomes inaccurate for large nuclear separations.

From the variational criterion of Section 2.4, it is clear that the functions (3.47), (3.48) and (3.49) all satisfy the virial theorem but not the electrostatic theorem.

Table 3.1. Calculated equilibrium constants for the ground state of H_2^+

Wave function	R_e (au)	D_e (eV)
(a) Simple LCAO	2.5	1.76
(b) Scaled LCAO	2.00	2.25
(c) Floating, scaled LCAO[a]	1.99	2.56
(d) Hybridized LCAO	2.00	2.65
(e) James, Guillemin and Zener	2.00	2.79
Accurate	2.00	2.79

[a] These values calculated by Shull and Ebbing[8] are more accurate than the earlier calculations of Hurley.[7]

3.3. One-centre expansions

There have been a number of treatments of the H_2^+ ion based on an expansion of the wave function about the mid point of the molecule. The object of these calculations is not so much to obtain accurate results for H_2^+ itself but to study the convergence of the expansions as a guide to their utility for larger molecules.

The simplest procedure is to use the wave functions of the united atom He^+. These are the well-known hydrogen-like orbitals of the form

$$\psi_{nlm} = R_{nl}(r) Y_{lm}(\theta, \varphi). \tag{3.50}$$

Here the axis of the spherical polar coordinates r, θ, φ is taken along the internuclear axis, Y_{lm} is a normalized surface harmonic

$$Y_{lm} = \frac{(-1)^{(m+|m|)/2}}{(4\pi)^{\frac{1}{2}}} \sqrt{\frac{(2l+1)(l-|m|)!}{(l+|m|)!}}\, P_l^{|m|}(\cos\theta)\, e^{im\varphi} \tag{3.51}$$

and R_{nl} is the radial wave function

$$R_{nl}(r) = \sqrt{\frac{4(n-l-1)!\,Z^3}{n^4\,[(n+l)!]^3}}\left(\frac{2Zr}{n}\right)^l L_{n+l}^{2l+1}\left(\frac{2Zr}{n}\right)\mathrm{e}^{-(Zr/n)}. \quad (3.52)$$

Matsen[12] has used these united atom functions to calculate the energies of the H_2^+ states $1s\sigma_g$, $2p\sigma_u$, $2p\pi_u$, $3d\pi_g$, $3d\delta_g$. The energy for these states is given by

$$E_{n,l,m} = \frac{Z^2}{2n^2} - 2\int \frac{\psi_{n,l,m}^2}{r_a}\,\mathrm{d}v. \quad (3.53)$$

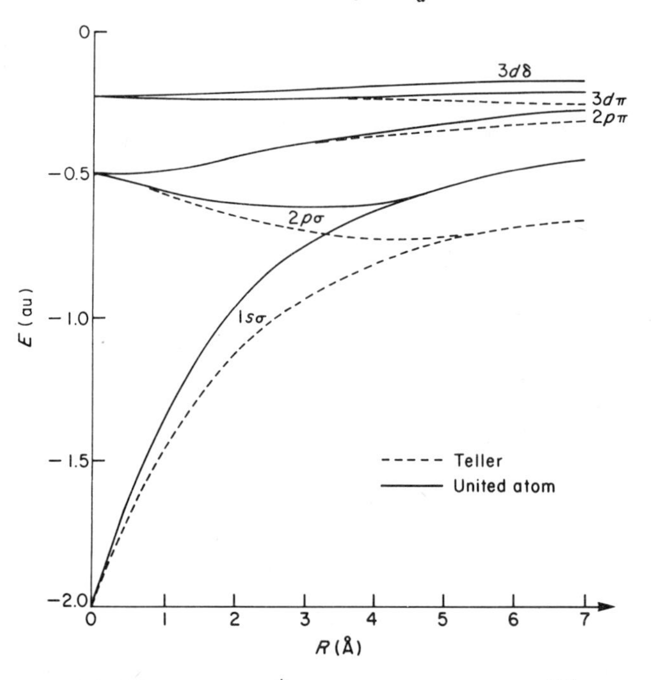

Fig. 3.7. Energy of H_2^+ states calculated by Matsen.[12]

For the united atom Z is equal to two. The use of this value leads to agreement with the accurate results of Bates *et al.*[1] and Teller[13] only for very small nuclear separations. Matsen found some improvement by varying Z to minimize the energy of each state. His results are shown in Fig. 3.7. The accuracy is higher for the excited states because the wave function extends far enough away from the two nuclei for them to appear united.

Other calculations by the one-centre method have aimed at convergence to the accurate result by including a large number of terms in the expansion of the wave function. The hydrogen-like functions (3.50) are not suitable for this purpose since they do not form a complete set of functions, that is, it is not possible to expand an arbitrary function of r, θ, φ in terms of them. To obtain a complete set the continuum wave functions corresponding to positive energy solutions of the Schrödinger equation would have to be included. The expansion of an arbitrary function would then involve a summation over the discrete functions (3.50) and an integration over the continuum functions. It is much more convenient to replace the hydrogen-like functions by a complete discrete set.

Howell and Shull[14] employ the set of associated Laguerre radial functions of order $2l + 2$

$$\Psi_{nl} = (2\eta)^{\frac{3}{2}}\{(n+l+1)!\}^{-\frac{1}{2}}\{(n-l-1)!\}^{\frac{1}{2}}(2\eta r)^{l} L_{n+l+1}^{2l+2}(2\eta r)\,\mathrm{e}^{-\eta r} \tag{3.54}$$

in place of the hydrogen-like radial functions (3.52). The functions (3.54), combined with the normalized spherical harmonics (3.51), give a complete orthonormal set of functions provided that the scale factor η has the same value for all Ψ_{nl} with a given value of l.

Houser, Lykos and Mehler[15], and Joy and Handler[16] use the generalized Slater orbitals

$$S_{nlm} = (2\zeta)^{n+\frac{1}{2}}[(2N)!]^{-\frac{1}{2}} r^{n-1}\,\mathrm{e}^{-\zeta r}\,Y_{lm}(\theta,\varphi). \tag{3.55}$$

These functions are normalized but not orthogonal. They again form a complete set. Indeed, if we put the orbital exponent $\zeta = \eta$, and suitably orthogonalize the radial parts of S_{nlm} for positive integral values of n we get back to the functions (3.54). However, in the calculations the ζ, n values for each function are varied independently in order to accelerate convergence. The use of non-integral values of n leads to appreciable improvement for short expansions but makes little difference if more than a few terms with the same l, m values are employed.

The results obtained for the electronic energy of the ground state at $R = 2$ au are shown in Table 3.2. Since this state is of symmetry σ_g, $m = 0$, and l is restricted to even values. The expansion of the wave function is then of the form

$$\psi(\sigma_g) = \sum_{t=0}^{\infty} f_{2t}(r) P_{2t}(\cos\theta), \tag{3.56}$$

that is, we have terms of the types $s, d_0, g_0 \ldots$.

Also shown in Table 3.2 are results obtained by Cohen and Coulson.[17] These authors curtail the expansion (3.56) to 1, 2, 3 or 4 terms, substitute in the electronic Schrödinger equation and solve the resulting coupled differential equations for the functions $f_{2t}(r)$ numerically.

Table 3.2. One-centre expansions for the ground state of H_2^+ (R = 2 au)

Term type	Electronic energy[a] using various radial functions			ϵ_l
	Laguerre	Generalized Slater	Numerical	
s	(6)[b] −1.01842	(4)[b] −1.01850	−1.01851	
s, d	(5) −1.08330	(4) −1.08367	—	0.06516
s, d, g	(4) −1.09563	(2) −1.09640	—	0.01273
s, d, g, i		(2) −1.09992	−1.09994	0.00354
s, d, g, i, k		(2) −1.10121		0.00127
		Exact	−1.10263	

[a] In atomic units. The nuclear repulsion $1/R$ is not included.
[b] The number of added terms of each symmetry type.

We see from Table 3.2 that the convergence of the analytic radial functions to the numerical values is quite rapid, especially for the generalized Slater functions. However, the overall convergence of the electronic energy to the accurate value of Bates *et al.* is rather slow. The final column of Table 3.2 gives ϵ_l, the total lowering of the energy brought about by all terms with a given value of l. As Joy and Handler(16) point out the asymptotic behaviour for large l appears to be

$$\epsilon_l \sim \frac{1}{l^4}. \tag{3.57}$$

On this basis we may estimate that some 25 terms in the expansion (3.56) are required to obtain an electronic energy approaching spectroscopic accuracy ($\pm$ 1 cm^{-1}).

The principal reason for this slow convergence lies in the singularities (cusps) of the accurate wave function at the nuclei. This difficulty is clearly apparent in Fig. 3.8 which shows the numerical values of various wave functions along the internuclear axis. The dashed curves are the results(15) of expansions using generalized Slater orbitals of the number and symmetry type indicated. The function $2s + 1d + 1g + 1i$ has converged quite well to the wave function of Cohen and Coulson, which contains the same symmetry types with numerical radial functions, but both wave functions are a poor fit to the accurate function of Bates *et al.* in the immediate vicinity of the nucleus at $Z = 1$.

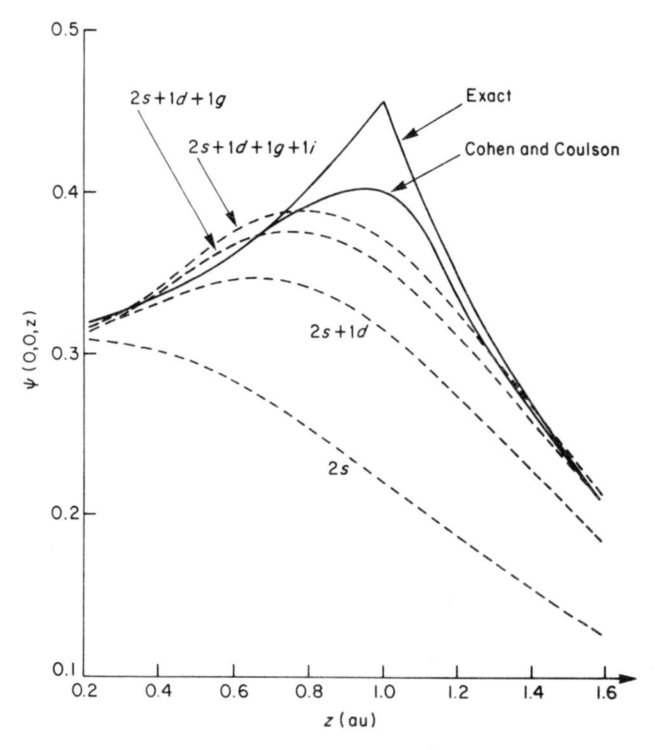

Fig. 3.8. Comparison of wave functions for H_2^+ ground state along the internuclear axis. (From Ref. 15.)

Cohen[18] has extended the method employing numerical radial functions to excited states of symmetry σ_u, π_u and π_g. For these states the expansion (3.56) is replaced by the following:

$$\psi(\sigma_u) = \sum_{t=0}^{\infty} f_{2t+1}(r) P_{2t+1}(\cos\theta) \tag{3.58}$$

$$\psi(\pi_u) = \sum_{t=0}^{\infty} f_{2t+1}(r) P_{2t+1}^1(\cos\theta)\, e^{i\varphi} \tag{3.59}$$

$$\psi(\pi_g) = \sum_{t=1}^{\infty} f_{2t}(r) P_{2t}^1(\cos\theta)\, e^{i\varphi}. \tag{3.60}$$

The electronic energies of several excited states using up to four terms in these expansions are shown in Table 3.3.

Table 3.3. One-centre expansions for some excited states of H_2^+ (R = 2 au)

State of united atom	Electronic energy (omitting nuclear repulsion) (au)				
	Single term	Two terms	Three terms	Four terms	Exact
$2p\sigma_u$	−0.61800	−0.65599	−0.66338	—	−0.66753
$3p\sigma_u$	−0.24592	−0.25316	−0.25460	—	−0.25541
$4p\sigma_u$	−0.13374	−0.13646	−0.13698	—	−0.13731
$4f\sigma_u$	—	−0.12659	−0.12664	—	−0.12664
$2p\pi_u$	—	—	—	−0.42876	−0.42877
$3d\pi_g$	—	—	—	−0.22669	−0.22670

It is clear from Table 3.3 that the convergence is much more rapid than for the ground state. The situation is similar to that for Matsen's simple united atom calculations. The wave functions for the excited states are much more spread out than for the ground state and are much smaller near the nuclei. Furthermore, for the states of π symmetry, the wave function is zero at the nuclei. For these reasons the poor fit of the expansions (3.58)-(3.60) to the singularities at the nuclei is much less important.

All the one-centre expansions we have considered satisfy the criteria established in Chapter 2 for the validity of the virial and electrostatic theorems provided that each variable parameter is fully optimized as a function of the nuclear separation. Furthermore the truncated expansions (3.56), (3.58)-(3.60) with numerical radial functions will satisfy the integral Hellmann–Feynman theorem. This follows from the fact that truncation of the expansions is a linear constraint independent of the nuclear separation.

One-centre expansions of electronic wave functions have been quite widely used for larger molecules, especially hydrides containing only one heavy atom (HF, OH, NH_3, CH_4 etc.). Such a calculation avoids the problem of computing the difficult two-, three- and four-centre integrals which arise from an LCAO expansion. Quite simple one-centre calculations can give very useful qualitative and semi-quantitative results. However, a comparison of Tables 3.1 and 3.2 suggests that it is foolish to try and extend these simple one-centre calculations to high accuracy by the brute-force method of including a large number of terms in the expansion. Thus the 14 term function of Table 3.2, which contains 27 variational parameters (14 of them non-linear) is comparable in accuracy with the very simple James function with 2 variational parameters (Table 3.1, equation (3.48)). If one wishes to obtain highly accurate wave functions by the one-centre method the convergence must be accelerated by some special treatment of the singularities at the nuclei.[22,23]

3.4. Angular momentum operators and spin functions

Although the electronic wave equation which we use contains no reference to electron spin, it is necessary, for many-electron systems, to include spin in the electronic wave function in order to satisfy the Pauli principle. This principle requires that the wave function for a many-electron system be antisymmetric under the simultaneous interchange of the space and spin coordinates of any two electrons.

The non-relativistic theory of electron spin is based on the general theory of angular momentum. Here we outline the principal results of this theory which we shall need, referring the reader elsewhere for proofs.[19-21]

The components J_x, J_y, J_z of any angular momentum operator $\mathbf{J}$ satisfy the commutation relations (in au, $\hbar = 1$)

$$\begin{aligned} [J_x, J_y] &= J_x J_y - J_y J_x = iJ_z \\ [J_y, J_z] &= iJ_x \\ [J_z, J_x &= iJ_y. \end{aligned} \tag{3.61}$$

From these commutation relations it follows that:

(i) The square of the total angular momentum operator

$$\mathbf{J}^2 = J_x^2 + J_y^2 + J_z^2 \tag{3.62}$$

commutes with J_z

$$[\mathbf{J}^2, J_z] = 0. \tag{3.63}$$

(ii) The eigenvalues of $\mathbf{J}^2$ are of the form $J(J+1)$ where J is integral or half-integral.

(iii) The eigenfunctions of $\mathbf{J}^2$ may be grouped into sets $\psi(J, M_J)$ ($M_J = -J, -J+1, \ldots, J-1, J$) of simultaneous eigenfunctions of $\mathbf{J}^2$ and J_z. Then

$$\mathbf{J}^2 \psi(J, M_J) = J(J+1)\psi(J, M_J) \tag{3.64}$$

$$J_z \psi(J, M_J) = M_J \psi(J, M_J). \tag{3.65}$$

If J is integral (half-integral) all the M_J are integral (half-integral).

(iv) All the functions $\psi(J, M_J)$ of a set may be obtained from one of them by repeated application of the step-up and step-down or shift operators

$$J_+ = J_x + iJ_y \tag{3.66}$$

$$J_- = J_x - iJ_y. \tag{3.67}$$

Thus

$$J_+\psi(J, M_J) = \sqrt{(J - M_J)(J + M_J + 1)}\,\psi(J, M_J + 1) \tag{3.68}$$

$$J_-\psi(J, M_J) = \sqrt{(J - M_J + 1)(J + M_J)}\,\psi(J, M_J - 1) \tag{3.69}$$

where, by convention, the positive sign is taken for all the square roots.

(v) There are convenient expressions for $\mathbf{J}^2$ in terms of J_+, J_- and J_z

$$\mathbf{J}^2 = J_+J_- + J_z^2 - J_z \tag{3.70}$$

$$\mathbf{J}^2 = J_-J_+ + J_z^2 + J_z. \tag{3.71}$$

The equations (3.65)-(3.69) and the commutation relations (3.61) determine the matrix elements of any function of the operators J_x, J_y and J_z between the functions $\psi(J, M_J)$ of a set.

An example of a set of functions satisfying these relationships is provided by the surface harmonics $Y_{lm}(\theta, \varphi)$ of equation (3.51). In this case $J = l$ (integral), $M_J = m$ and J_x, J_y, J_z are the orbital angular momentum operators

$$\begin{aligned} l_x &= -i\left(y\frac{\partial}{\partial z} - z\frac{\partial}{\partial y}\right) \\ l_y &= -i\left(z\frac{\partial}{\partial x} - x\frac{\partial}{\partial z}\right) \\ l_z &= -i\left(x\frac{\partial}{\partial y} - y\frac{\partial}{\partial x}\right). \end{aligned} \tag{3.72}$$

The phase factor $(-1)^{(m+|m|)/2}$ in the definition (3.51) of the surface harmonics is included to ensure positive signs in the equations (3.68) and (3.69).

Pauli's theory of electron spin is based on the case $J = s = \frac{1}{2}$. We then have just two functions in the set $\psi(s, m_s)$ which are denoted by α and β

$$\begin{aligned} \alpha &= \psi(\tfrac{1}{2}, \tfrac{1}{2}) \\ \beta &= \psi(\tfrac{1}{2}, -\tfrac{1}{2}). \end{aligned} \tag{3.73}$$

Writing s, s_x, s_y, s_z for the operators representing the electron spin and its components we have the relations

$$\begin{aligned} s^2\alpha &= \tfrac{1}{2}(\tfrac{1}{2}+1)\alpha = \tfrac{3}{4}\alpha \\ s^2\beta &= \tfrac{3}{4}\beta \end{aligned} \tag{3.74}$$

$$s_z\alpha = \tfrac{1}{2}\alpha, \qquad s_z\beta = -\tfrac{1}{2}\beta \tag{3.75}$$

$$\begin{aligned} s_+\alpha &= 0, \qquad s_+\beta = \alpha \\ s_-\alpha &= \beta, \qquad s_-\beta = 0. \end{aligned} \tag{3.76}$$

The components s_x, s_y, s_z of the spin of the electron may be represented by 2 × 2 matrices, the Pauli matrices, whose elements are determined by equations (3.75) and (3.76). Thus

$$s_x = \begin{bmatrix} 0 & \tfrac{1}{2} \\ \tfrac{1}{2} & 0 \end{bmatrix}, \qquad s_y = \begin{bmatrix} 0 & -\dfrac{i}{2} \\ \dfrac{i}{2} & 0 \end{bmatrix}, \qquad s_z = \begin{bmatrix} \tfrac{1}{2} & 0 \\ 0 & -\tfrac{1}{2} \end{bmatrix}. \tag{3.77}$$

The functions α and β are then represented by column matrices

$$\alpha = \begin{pmatrix} 1 \\ 0 \end{pmatrix}, \qquad \beta = \begin{pmatrix} 0 \\ 1 \end{pmatrix}. \tag{3.78}$$

Alternatively α and β may be regarded as functions of a spin variable ω, which takes on just two values, $\omega = \frac{1}{2}$ and $\omega = -\frac{1}{2}$. In this representation

$$\begin{aligned} \alpha(\tfrac{1}{2}) &= 1, \qquad \alpha(-\tfrac{1}{2}) = 0 \\ \beta(\tfrac{1}{2}) &= 0, \qquad \beta(-\tfrac{1}{2}) = 1 \end{aligned} \tag{3.79}$$

that is, in terms of the Kronecker δ symbol

$$\begin{aligned} \alpha(\omega) &= \delta_{\omega,\frac{1}{2}} \\ \beta(\omega) &= \delta_{\omega,-\frac{1}{2}}. \end{aligned} \tag{3.80}$$

It is easy to verify that these functions satisfy the orthonormality conditions

$$\begin{aligned} \sum_\omega \alpha^2(\omega) = \sum_\omega \beta^2(\omega) &= 1 \\ \sum_\omega \alpha(\omega)\beta(\omega) &= 0 \end{aligned} \tag{3.81}$$

where the summations extend over the two points $\omega = \frac{1}{2}$, $\omega = -\frac{1}{2}$.

To completely specify the state of a single electron we now need four coordinates, the three spatial coordinates of the position vector r and the spin coordinate ω. Each orbital $\varphi(\mathbf{r})$ leads to two spin orbitals

$$\begin{aligned}\psi_1(x) &= \varphi(\mathbf{r})\alpha(\omega)\\ \psi_2(x) &= \varphi(\mathbf{r})\beta(\omega)\end{aligned} \tag{3.82}$$

where x denotes the space and spin coordinates of the electron.

For a one-(many-) electron system the symbol $\int \mathrm{d}\tau$ is taken to imply integration over the position coordinates of the electron(s) and summation over the spin coordinate(s). Using this notation and equations (3.81) we have

$$\begin{aligned}\int \psi_1^*(x)\psi_1(x)\,\mathrm{d}\tau &= \int \varphi^*(\mathbf{r})\varphi(\mathbf{r})\,\mathrm{d}v \cdot \sum_\omega \alpha^2(\omega)\\ &= \int \varphi^*(\mathbf{r})\varphi(\mathbf{r})\,\mathrm{d}v,\\ \int \psi_2^*(x)\psi_2(x)\,\mathrm{d}\tau &= \int \varphi^*(\mathbf{r})\varphi(\mathbf{r})\,\mathrm{d}v \cdot \sum_\omega \beta^2(\omega)\\ &= \int \varphi^*(\mathbf{r})\varphi(\mathbf{r})\,\mathrm{d}v,\\ \int \psi_1^*(x)\psi_2(x)\,\mathrm{d}\tau &= \int \varphi^*(\mathbf{r})\varphi(\mathbf{r})\,\mathrm{d}v \cdot \sum_\omega \alpha(\omega)\beta(\omega) = 0,\end{aligned}$$

so that the spin orbitals (3.82) are orthonormal if $\varphi(\mathbf{r})$ is normalized.

For a one-electron system described by a spin-free Hamiltonian, such as in equation (3.1) for H_2^+, the only effect of the spin of the electron is to double the degeneracy of the energy levels. In Section 3.1 we found spatial solutions of the Schrödinger equation for H_2^+ of the symmetry types

$$\begin{array}{ll}\sigma_g, \sigma_u & \text{non-degenerate}\\ \pi_g, \pi_u, \delta_g, \delta_u, \ldots & \text{doubly degenerate.}\end{array} \tag{3.83}$$

Combining these spatial solutions with the spin functions α and β we obtain total electronic wave functions of the types

$$\begin{array}{ll}{}^2\Sigma_g^+, {}^2\Sigma_u^+ & \text{doubly degenerate}\\ {}^2\Pi_g, {}^2\Pi_u, {}^2\Delta_g, {}^2\Delta_u, \ldots & \text{four-fold degenerate.}\end{array} \tag{3.84}$$

In the standard notation used here, capital Greek letters are used to denote the spatial symmetry of the total electronic wave function, the prefix gives the spin degeneracy ($2S + 1 = 2$) and the superscript + indicates that the wave function is even under a reflection σ_v in a plane passing through the nuclear axis. All sigma states of a one-electron diatomic molecule are necessarily even under σ_v (this is why the superscript + is omitted from the σ orbitals in (3.83)) but, for more than one electron, states of symmetry Σ^- also arise.

References

1. D. R. Bates, K. Ledsham and A. L. Stewart (1953), *Phil. Trans. Roy. Soc.* **246**, 215.
2. M. Kotani, K. Ohno and K. Kayama (1961), "Quantum Mechanics of Electronic Structure of Simple Molecules", Handbuch der Physik, Vol. XXXVII/2, Springer-Verlag, Berlin.
3. S. P. Alliluev and A. V. Matveenko (1967), *Soviet Physics JETP* **24**, 1260 from *J. Exptl. Theoret. Phys. (USSR)* **51**, 1873 (1966).
4. C. A. Coulson (1941), *Proc. Cambridge Phil. Soc.* **38**, 210.
5. A. C. Hurley (1954), *Proc. Roy. Soc.* **A226**, 170.
6. B. N. Finkelstein and G. E. Horowitz (1928), *Z. Physik* **48**, 118, 448 .
7. A. C. Hurley (1954), *Proc. Roy. Soc.* **A226**, 179.
8. H. Shull and D. D. Ebbing (1958), *J. Chem. Phys.* **28**, 866.
9. B. N. Dickinson (1933), *J. Chem. Phys.* **1**, 317.
10. H. M. James (1935), *J. Chem. Phys.* **3**, 9.
11. V. Guillemin and C. Zener (1929), *Proc. Natl. Acad. Sci. U.S.* **15**, 314.
12. F. A. Matsen (1953), *J. Chem. Phys.* **21**, 928.
13. E. Teller (1930), *Z. Physik* **61**, 458.
14. K. M. Howell and H. Shull (1959), *J. Chem. Phys.* **30**, 627.
15. T. J. Houser, P. G. Lykos and E. L. Mehler (1963), *J. Chem. Phys.* **38**, 583.
16. H. W. Joy and G. S. Handler (1965), *J. Chem. Phys.* **42**, 3047.
17. M. Cohen and C. A. Coulson (1961), *Proc. Cambridge Phil. Soc.* **57**, 96.
18. M. Cohen (1962), *Proc. Cambridge Phil. Soc.* **58**, 130.
19. P. A. M. Dirac (1957), "Quantum Mechanics", Third Edition, Oxford University Press.
20. M. E. Rose (1957), "Elementary Theory of Angular Momentum", Wiley Interscience, New York.
21. E. U. Condon and G. H. Shortley (1935), "The Theory of Atomic Spectra", Cambridge University Press.
22. H. Conroy (1964), *J. Chem. Phys.* **41**, 1327.
23. A. C. Hurley (1976), "Electron Correlation in Small Molecules", Chapter 2, Section 20, Academic Press, London and New York.

CHAPTER 4

The Symmetry of Molecular Electronic States

In the previous chapter, we found that the electronic wave functions of H_2^+ could be characterized by their behaviour under spatial symmetry operations which either left the nuclei invariant or permuted the two equivalent nuclei. The systematic study of such symmetry properties involves the representation theory of the molecular symmetry group. The results of this theory are very useful since they enable us to deduce many qualitative properties of the electronic states such as degeneracies, selection rules and splitting under perturbations, without solving the Schrödinger equation. The theory may also be used to simplify calculation of approximate wave functions.

The brief account given here, which is largely based on that of Kotani, Ohno and Kayama,[1] emphasizes applications of the theory rather than the underlying mathematical structure; several key results are stated without proof, but it is shown how these results may be used in practical calculations. For a complete account, including proofs of the key theorems the reader is referred to the books by Wigner,[2] Hamermesh[3] and McWeeny.[4] McWeeny's treatment is particularly suited to chemical applications, but he does not consider the infinite groups $C_{\infty v}$ and $D_{\infty h}$. A recent discussion by Hall,[5] which emphasizes the group algebra rather than the matrix representations, is also recommended.

4.1. Symmetry groups of molecules

In the electronic wave equation (equation (1.71)) the nuclei appear simply as centres of force. The electronic Hamiltonian has the same spatial symmetry as the nuclear framework of the molecule. This spatial symmetry is described by the set of geometrical operations (rotations, reflections, inversion and rotary-reflections) which either leave the nuclei invariant or permute equivalent nuclei.

Consider a molecule, and suppose that all the nuclei are held fixed, usually at their equilibrium positions. Let $P, Q, \ldots$ denote symmetry operations allowed by this nuclear configuration. If we first apply P to the molecule, and then apply Q to the result of operation P, the result is equivalent to that of a single operation, which is called the product of P and Q and is written QP. Since the nuclear framework remains unchanged throughout, the product QP of two symmetry operations is a symmetry operation. The multiplication of operations is not necessarily commutative: $PQ \neq QP$ but is always associative $P(QR) = (PQ)R$. Every molecule allows the particular operation E (identity) which leaves all points unmoved: E satisfies $EP = PE = P$ for any operation P. If the operation P displaces configuration A to configuration B, then the operation which displaces B back to A is also a symmetry operation of the same molecule, and is denoted by P^{-1} (inverse of P). The equations $PP^{-1} = P^{-1}P = E$ holds for any P. These considerations show that the set of all symmetry operations of a molecule forms a *group, the symmetry group of the molecule.* The *order* of a group G is defined as the number of symmetry operations.

An operation P, of a symmetry group G, is said to be *conjugate* to another operation $Q(\in G)$, when there exists at least one operation $S \in G$ satisfying $P = SQS^{-1}$. Since $P = SQS^{-1}$ gives $Q = S^{-1}P(S^{-1})^{-1}$, and since S^{-1} also belongs to G, the relation of being conjugate is a mutual (reflexive) one. If P is conjugate to both Q and R, Q and R are mutually conjugate (transitivity) and the set of all operations which are conjugate to a particular operation of G is called a *class of conjugate operations*, or simply a *class* of G. The operations of G fall into a number of disjoint (non-overlapping) classes. A subset of the elements of a group G, which itself forms a group, is called a *sub-group* of G.

EXAMPLE 1. GROUP C_{3v}. TRIANGULAR PYRAMIDAL MOLECULES SUCH AS NH_3

We assume that H_3 forms a horizontal regular triangle and that the N nucleus is situated at a point above the centre of gravity of H_3. The choice of coordinate axes is indicated in Fig. 4.1.

The symmetry group consists of the following 6 operations:

E, the identity.

$C(2\pi/3)$, $C(-2\pi/3)$ rotations through angles $+2\pi/3$ and $-2\pi/3$ about the z-axis. The sign of the angle of rotation is determined by the right-hand screw rule.

σ_a, σ_b, σ_c reflections in vertical planes passing through the N nucleus and through the initial positions of the hydrogen nuclei a, b, c respectively.

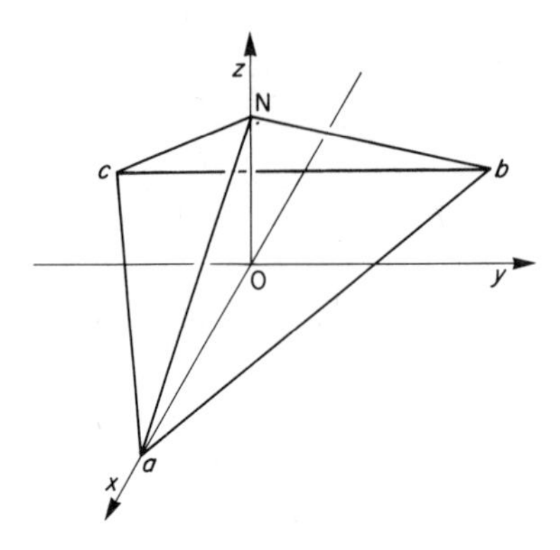

Fig. 4.1. Coordinate system for C_{3v} molecules (e.g. NH_3).

The products of all pairs of these symmetry operations are given in the multiplication Table 4.1. In this table the operations $C(2\pi/3)$ and $C(-2\pi/3)$ are abbreviated as C_+ and C_-.

Table 4.1. Multiplication table for C_{3v}. The entries give PQ

P \ Q	E	C_+	C_-	σ_a	σ_b	σ_c
E	E	C_+	C_-	σ_a	σ_b	σ_c
C_+	C_+	C_-	E	σ_c	σ_a	σ_b
C_-	C_-	E	C_+	σ_b	σ_c	σ_a
σ_a	σ_a	σ_b	σ_c	E	C_+	C_-
σ_b	σ_b	σ_c	σ_a	C_-	E	C_+
σ_c	σ_c	σ_a	σ_b	C_+	C_-	E

It is clear from Table 4.1 that the *proper* elements E, C_+, C_- form a *sub-group* (C_3) of C_{3v}. This sub-group C_3 is evidently commutative (*Abelian*).

Again using Table 4.1 we find that the operations of C_{3v} fall into 3 classes

$$E$$

$$\left\{C\left(\frac{2\pi}{3}\right),\quad C\left(-\frac{2\pi}{3}\right)\right\}$$

and

$$\{\sigma_a, \sigma_b, \sigma_c\}.$$

On the other hand for the Abelian group C_3, each element E, C_+, C_- forms a class by itself. From the definition of the relation of conjugacy it is clear that this is a general property of Abelian groups.

EXAMPLE 2. GROUP T_d. TETRAHEDRAL MOLECULES SUCH AS CH_4

We assume that the H_4 lie at the vertices of a regular tetrahedron with C at its centre. The symmetry group contains 24 operations which fall into 5 classes.

(i) E.

(ii) $C(2\pi/3)$ and $C(-2\pi/3)$ about each of four triple axes (CH_a, CH_b, CH_c, CH_d).

(iii) $C_x(\pi)$, $C_y(\pi)$, $C_z(\pi)$. Rotations of angle π about straight lines connecting mid-points of pairs of non-intersecting edges. There are 3 pairs and the 3 straight lines are mutually orthogonal. We choose these as the x, y, z axes of a Cartesian coordinate system, the vertex H_a lying in the positive octant.

(iv) σ_{ab}, σ_{ac}, σ_{ad}, σ_{bc}, σ_{bd}, σ_{cd}. Reflections in planes containing one edge and the centre of the tetrahedron.

(v) $S_x^+, S_x^-, S_y^+, S_y^-, S_z^+, S_z^-$. Rotations of $\pm\,\pi/2$ about the x axis followed by reflection in the yz plane. Similarly for the y, z axes.

The group T_d contains a number of sub-groups. For example, there is a sub-group C_{3v} for each of the four triple axes, and the identity E, together with the two-fold rotations $C_x(\pi)$, $C_y(\pi)$, $C_z(\pi)$, forms a sub-group D_2 of order 4.

EXAMPLE 3. GROUP $C_{\infty v}$. LINEAR ASYMMETRIC MOLECULES SUCH AS NO, HCN

We take the straight line passing through all nuclei as the z-axis. The group contains an infinite number of operations, which fall into an infinite number of classes, as shown in Table 4.2. Also shown are the results $(\mu', \nu', \varphi') = P(\mu, \nu, \varphi)$ of applying the group operations P to a general point with elliptical coordinates (μ, ν, φ).

Table 4.2. Operations and classes of $C_{\infty v}$

Class	Operations P falling in each class	$(\mu', \nu', \varphi') = P(\mu, \nu, \varphi)$
E	E	μ, ν, φ
$C\ (\pm\phi)$	$C(\phi), C(-\phi)$ (2 operations)	$\mu, \nu, \varphi \pm \phi$
$C\ (\pi)$	$C(\pi)$ (1 operation)	$\mu, \nu, \varphi + \pi$
$\sigma_v(\alpha)$	Reflection through a plane containing the z-axis, $\frac{\alpha}{2}$ is the azimuth of the normal to the plane (∞ operations)	$\mu, \nu, \alpha + \pi - \varphi$

EXAMPLE 4. GROUP $D_{\infty h}$. LINEAR SYMMETRIC MOLECULES SUCH AS H_2^+, N_2, OCO

We choose the Cartesian axes so that the z-axis coincides with the molecular axis and the origin is at the centre of symmetry. Again we have an infinite number of operations and an infinite number of classes (Table 4.3).

Table 4.3. Operations and classes of $D_{\infty h}$

Class	Operations P falling in each class	$(\mu', \nu', \varphi') = P(\mu, \nu, \varphi)$
E	E	μ, ν, φ
$C(\pm\phi)$	$C(\phi), C(-\phi)$	$\mu, \nu, \varphi \pm \phi$
$C(\pi)$	$C(\pi)$	$\mu, \nu, \varphi + \pi$
$C'_\alpha(\pi)$	Rotations through π about any line in the xy-plane, the azimuth of the line is $\frac{\alpha}{2}$	$\mu, -\nu, \alpha - \varphi$
i	Inversion through the origin	$\mu, -\nu, \varphi + \pi$
$iC(\pm\phi)$	$iC(\phi) = S(\pi + \phi), iC(-\phi) = S(\pi - \phi)$	$\mu, -\nu, \pi + \varphi \pm \phi$
$iC(\pi)$	$iC(\pi) = \sigma_h$, reflection in the xy-plane	$\mu, -\nu, \varphi$
$iC'_\alpha(\pi)$	$iC'_\alpha(\pi) = \sigma_v(\alpha)$	$\mu, \nu, \alpha + \pi - \varphi$

Notice the two different notations for the improper elements in the second column of Table 4.3. The rotary-reflection $S(\phi)$ is defined as $C(\phi)$ followed by the reflection σ_h in the xy-plane, and differs from the rotary-inversion $iC(\phi)$. Indeed we have the relation

$$iC(\phi) = S(\pi + \phi) \tag{4.1}$$

shown in Table 4.3. Usually the notation involving the inversion i is more convenient, since i commutes with all other operations.

4.2. Transformation of electronic wave functions

We now consider the transformations of electronic wave functions induced by the symmetry operations of a molecule. The symmetry operations themselves are regarded as point transformations in the space of the electronic position variables. The coordinate system remains fixed in space.

Consider an N-electron molecule with electronic spatial variables $\mathbf{r}_1, \mathbf{r}_2, \ldots \mathbf{r}_N$. A symmetry operation P of the molecule transforms the points (vectors) $\mathbf{r}_1, \ldots \mathbf{r}_N$ into the points

$$\begin{aligned} \mathbf{r}'_1 &= P\mathbf{r}_1, \\ &\ldots \\ \mathbf{r}'_N &= P\mathbf{r}_N. \end{aligned} \tag{4.2}$$

The relations (4.2) may be expressed in any convenient fixed coordinate system. For example, the transforms of a general point, expressed in elliptical coordinates (μ, ν, φ), under the operations of $C_{\infty v}$ and $D_{\infty h}$ are given in the final columns of Tables 4.2 and 4.3.

An operator O_P acting on any function $f(\mathbf{r}_1, \ldots, \mathbf{r}_N)$ is now defined by the identity

$$O_P f(P\mathbf{r}_1, \ldots, P\mathbf{r}_N) \equiv f(\mathbf{r}_1, \ldots, \mathbf{r}_N). \tag{4.3}$$

The operator O_P is linear

$$O_P(af + bg) = aO_P f + bO_P g \tag{4.4}$$

and preserves scalar products (overlap integrals)

$$(O_P f, O_P g) = (f, g) = \int f^* g \, \mathrm{d}\tau, \tag{4.5}$$

where f and g are any two functions of $\mathbf{r}_1, \ldots, \mathbf{r}_N$.

Substituting $P^{-1}\mathbf{r}_1, \ldots, P^{-1}\mathbf{r}_N$ for $\mathbf{r}_1, \ldots, \mathbf{r}_N$ in the identity (4.3) we obtain

$$O_P f(\mathbf{r}_1, \ldots, \mathbf{r}_N) \equiv f(P^{-1}\mathbf{r}_1, \ldots, P^{-1}\mathbf{r}_N). \tag{4.6}$$

This is a convenient form to use in an explicit determination of the transformed function $O_P f$.

Equation (4.6) may be regarded as an identity between two functions of the variables $\mathbf{r}_1, \ldots, \mathbf{r}_N$. Operating on both sides of this equation with O_Q, where Q is some other symmetry operation of the molecule, we obtain

$$\begin{aligned} O_Q O_P f(\mathbf{r}_1, \ldots, \mathbf{r}_N) &= f(P^{-1}(Q^{-1}\mathbf{r}_1), \ldots, P^{-1}(Q^{-1}\mathbf{r}_N)) \\ &= f((P^{-1}Q^{-1})\mathbf{r}_1, \ldots, (P^{-1}Q^{-1})\mathbf{r}_N) \\ &= f((QP)^{-1}\mathbf{r}_1, \ldots, (QP)^{-1}\mathbf{r}_N), \end{aligned} \tag{4.7}$$

so that, if $QP = R$, $O_Q O_P = O_R$. Thus the linear operators $O_P, O_Q, \ldots$ multiply in the same way as the symmetry operations $P, Q, \ldots$, that is

$$O_Q O_P = O_{QP}. \tag{4.8}$$

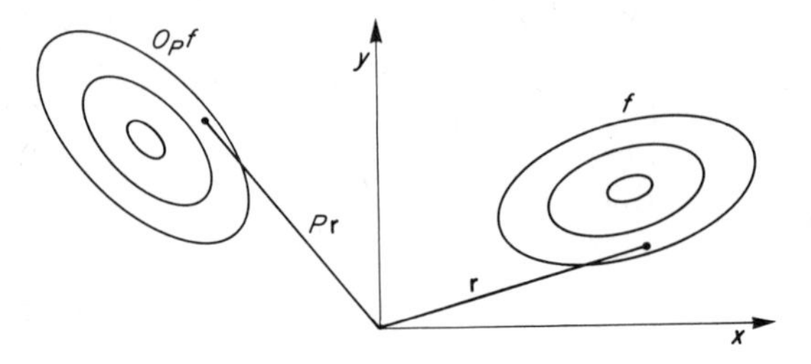

Fig. 4.2. Definition of the transformed function O_Pf.

The definition (4.3) of the transformed function O_Pf is illustrated in Fig. 4.2. Here f is a function of a single two-dimensional vector **r**, which is subject to the plane rotation $\mathbf{r}' = P\mathbf{r}$. We see that the pattern of values of f is rotated as a whole in the same way as the general vector **r**. This is just what we expect intuitively. Consider, for example, the symmetry operation $C(2\pi/3)$ for NH_3 (Example 1, Section 4.1), which permutes the hydrogen nuclei as follows: a → b, b → c, c → a. The corresponding linear operator $O_C(2\pi/3)$ effects the same permutation among s-type atomic orbitals centred on the nuclei:

$$s^{\mathrm{a}} \to s^{\mathrm{b}}, \qquad s^{\mathrm{b}} \to s^{\mathrm{c}}, \qquad s^{\mathrm{c}} \to s^{\mathrm{a}}.$$

4.3. Symmetry properties of exact wave functions

Suppose that $\Psi(\mathbf{r}_1, \ldots, \mathbf{r}_N)$ is an eigenfunction of the electronic Hamiltonian H belonging to the eigenvalue E

$$H\Psi = E\Psi. \tag{4.9}$$

Let P be any operation in the symmetry group G of the molecule. Then the linear operator O_P commutes with H

$$O_PH = HO_P, \tag{4.10}$$

and operating on equation (4.9) with O_P we obtain

$$O_PH\Psi = H(O_P\Psi) = E(O_P\Psi), \tag{4.11}$$

so that $O_P\Psi$ is also an eigenfunction of H belonging to the eigenvalue E. If the energy level E is non-degenerate this implies that $O_P\Psi$ is a constant multiple D of Ψ

$$O_P\Psi = D\Psi. \tag{4.12}$$

More generally, if the energy level E is f-fold degenerate, there will be f linearly independent solutions $\Psi_1, \ldots, \Psi_f$ of equation (4.9) and any eigenfunction belonging to E must be a linear combination of these f functions. Thus, for a degenerate energy level, equation (4.11) implies

$$O_P\Psi_i = \sum_{k=1}^{f} D_{ki}\Psi_k \qquad (i = 1, 2, \ldots, f). \tag{4.13}$$

In this way an $f \times f$ matrix $\mathbf{D}(P)$, with elements D_{ki}, is associated with each symmetry operation P.

Applying the linear operator O_Q to equation (4.13) we obtain

$$O_Q O_P \Psi_i = \sum_{k=1}^{f} D_{ki}(P) O_Q \Psi_k = \sum_{k=1}^{f} D_{ki}(P) \sum_{h=1}^{f} D_{hk}(Q)\Psi_h$$

$$= \sum_{h=1}^{f} \left\{ \sum_{k=1}^{f} D_{hk}(Q) D_{ki}(P) \right\} \Psi_h. \tag{4.14}$$

Now, if $QP = R$, then, from equation (4.8), $O_Q O_P = O_R$ and equation (4.14) must be the same as

$$O_R \Psi_i = \sum_{h=1}^{f} D_{hi}(R)\Psi_h. \tag{4.15}$$

Since the functions $\Psi_h (h = 1, \ldots, f)$ are linearly independent, we may equate the coefficients of Ψ_h on the right-hand sides of equations (4.14) and (4.15). Thus we obtain

$$D_{hi}(R) = \sum_{k=1}^{f} D_{hk}(Q) D_{ki}(P), \tag{4.16}$$

or in matrix notation

$$\mathbf{D}(R) = \mathbf{D}(Q)\mathbf{D}(P). \tag{4.17}$$

From equation (4.17) we see that the matrices $\mathbf{D}(P), \mathbf{D}(Q), \ldots$ associated with the symmetry operations multiply in the same way as the symmetry operations $P, Q, \ldots$ themselves. The matrices are said to form a *representation* of the symmetry group. We denote representations by the symbol Γ, and distinguish different representations by subscripts.

If the functions $\Psi_1, \Psi_2, \ldots, \Psi_f$ form an orthonormal set

$$(\Psi_i, \Psi_j) \equiv \int \Psi_i^* \Psi_j \, d\tau = \delta_{ij}, \tag{4.18}$$

then the transformed functions $O_P\Psi_1, O_P\Psi_2, \ldots, O_P\Psi_f$ are also orthonormal, since the linear operators O_P preserve scalar products (equation (4.5)). Thus we have

$$(O_P\Psi_i, O_P\Psi_j) = (\Psi_i \Psi_j) = \delta_{ij}. \tag{4.19}$$

Substituting equation (4.13) in equation (4.19) and using equation (4.18) we find

$$\sum_h D^*_{hi}(P) D_{hj}(P) = \delta_{ij} \tag{4.20}$$

or in matrix form

$$\mathbf{D}^\dagger(P)\mathbf{D}(P) = 1, \tag{4.21}$$

where † denotes the Hermitian conjugate, and 1 denotes the unit matrix. Equations (4.20) and (4.21) show that the matrices $\mathbf{D}(P)$ are *unitary*. In such a case the representation itself is said to be *unitary*.

In this way a representation of the symmetry group of the molecule is associated with each energy level. This representation completely specifies the spatial symmetry properties of the wave functions. We may use the representations generated by the wave functions to classify the different energy levels. To do this we need to know what different kinds of representation are possible for a given symmetry group. This involves the concepts of *equivalence* and *irreducibility* of representations.

If a set of matrices (with dimension $f > 1$)

$$\mathbf{D}(P), \mathbf{D}(Q), \ldots \tag{4.22}$$

forms a representation Γ of the symmetry group G, we can derive from Γ an infinite number of representations as follows. Let $\mathbf{T}$ be a constant f-dimensional square matrix, whose determinant does not vanish, and put

$$\bar{\mathbf{D}}(P) = \mathbf{T}^{-1}\mathbf{D}(P)\mathbf{T}, \bar{\mathbf{D}}(Q) = \mathbf{T}^{-1}\mathbf{D}(Q)\mathbf{T}, \ldots. \tag{4.23}$$

Then the set

$$\bar{\mathbf{D}}(P), \bar{\mathbf{D}}(Q), \ldots \tag{4.24}$$

also forms a representation of G, because $\bar{\mathbf{D}}(P)\bar{\mathbf{D}}(Q) = \bar{\mathbf{D}}(R)$ is an immediate consequence of $\mathbf{D}(P)\mathbf{D}(Q) = \mathbf{D}(R)$:

$$\begin{aligned}\bar{\mathbf{D}}(P)\bar{\mathbf{D}}(Q) &= \mathbf{T}^{-1}\mathbf{D}(P)\mathbf{T}\mathbf{T}^{-1}\mathbf{D}(Q)\mathbf{T} = \mathbf{T}^{-1}\mathbf{D}(P)\mathbf{D}(Q)\mathbf{T}\\ &= \mathbf{T}^{-1}\mathbf{D}(R)\mathbf{T} = \bar{\mathbf{D}}(R).\end{aligned} \tag{4.25}$$

Two representations, whose matrices are connected by a relation of the form (4.23) are said to be *equivalent*. Since $\mathbf{T}$ can be chosen in an infinite number of ways, we would have an infinite number of representations if we considered equivalent representations to be different.

However, in classifying the energy levels of a molecule, it is natural to regard equivalent representations as essentially the same. Thus, if the energy level E is degenerate ($f \geqslant 2$), the choice of f

linearly independent eigenfunctions $\Psi_1, \ldots, \Psi_f$ can be made in an infinite number of different ways. If $(\Psi_1, \ldots, \Psi_f)$ and $(\bar{\Psi}_1, \ldots, \bar{\Psi}_f)$ are two orthonormal bases, they are related by a unitary transformation.

$$\bar{\Psi}_i = \sum_{j=1}^{f} T_{ji}\Psi_j \qquad (i = 1, \ldots, f). \tag{4.26}$$

Applying the linear operator O_P to equation (4.26) we obtain

$$O_P\bar{\Psi}_i = \sum_{j=1}^{f} T_{ji}O_P\Psi_j = \sum_{k=1}^{f}\sum_{j=1}^{f} D_{kj}(P)T_{ji}\Psi_k. \tag{4.27}$$

On the other hand, we can write, using the representation matrix $\bar{\mathbf{D}}(P)$ generated by the new basis,

$$O_P\bar{\Psi}_i = \sum_{h=1}^{f} \bar{D}_{hi}(P)\bar{\Psi}_h = \sum_{k=1}^{f}\sum_{h=1}^{f} T_{kh}\bar{D}_{hi}(P)\Psi_k. \tag{4.28}$$

Equating coefficients of Ψ_k in the two expressions (4.27), (4.28) for $O_P\bar{\Psi}_i$ we obtain

$$\mathbf{D}(P)\mathbf{T} = \mathbf{T}\bar{\mathbf{D}}(P). \tag{4.29}$$

Since $\mathbf{T}$ is unitary its inverse $\mathbf{T}^{-1} = \mathbf{T}^{\dagger}$ exists, and equation (4.29) may be written as

$$\bar{\mathbf{D}}(P) = \mathbf{T}^{-1}\mathbf{D}(P)\mathbf{T}. \tag{4.30}$$

This result shows that the representation $\Gamma\{\mathbf{D}(P), \mathbf{D}(Q), \ldots\}$ is equivalent to the representation $\bar{\Gamma}\{\bar{\mathbf{D}}(P), \bar{\mathbf{D}}(Q), \ldots\}$. Thus, the representation associated with a definite energy level is changed into another one which is equivalent to the former. Hence, in order to consider the representation belonging to each energy level as definite without specifying the choice of basis functions, we should regard equivalent representations as essentially the same.

The second point we have to discuss is the irreducibility of a representation. If we take two representations $\Gamma_1\{\mathbf{D}_1(P), \ldots\}$ and $\Gamma_2\{\mathbf{D}_2(P), \ldots\}$ of dimensions f_1 and f_2 respectively, and build up the $(f_1 + f_2)$-dimensional matrices $\mathbf{D}(P), \mathbf{D}(Q), \ldots$ as shown in Fig. 4.3, we obtain a new representation $(\Gamma_1 \dot{+} \Gamma_2)\{\mathbf{D}(P), \mathbf{D}(Q) \ldots\}$ of dimension $f_1 + f_2 = f$ called the *sum* of Γ_1 and Γ_2.

Conversely, if all the matrices $\mathbf{D}(P), \mathbf{D}(Q), \ldots$ of a representation Γ can be brought into the form of Fig. 4.3 simultaneously by a suitable choice of basis functions, Γ is said to be *reducible* into the sum of Γ_1 and Γ_2. A representation for which such a simplification is not possible is said to be *irreducible.* If one of the component representations Γ_1 or Γ_2 of a reducible representation is, in turn,

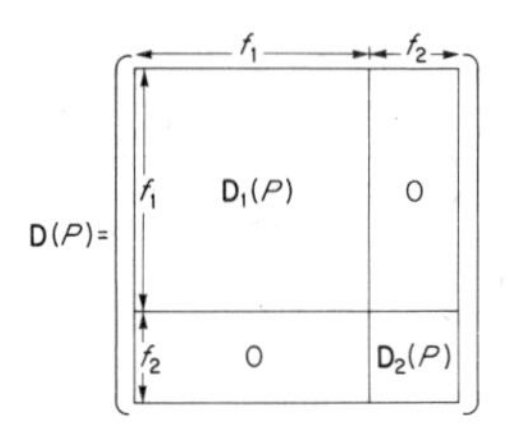

Fig. 4.3. Reduced form of a representation matrix **D**(P).

reducible, the process of reduction may be continued until eventually the original representation Γ appears as the sum of irreducible components. This complete reduction of Γ is unique apart from the order in which the irreducible representations appear and the possibility of equivalence transformations within each irreducible representation. Thus the irreducible representations provide building blocks in terms of which any representation may be expressed.

For a reducible representation of dimension f there exists a subset of functions $\Psi_1, \ldots, \Psi_{f_1}$ with $1 \leqslant f_1 < f$ which are transformed among themselves by all operators of the group, but this cannot occur for an irreducible representation. The representation Γ which belongs to an energy level E of a molecule is, in general, irreducible. If Γ were reducible, we could choose f basis functions $\Psi_1, \Psi_2, \ldots, \Psi_f$ in such a way as to put the representation matrices in reduced form. Then, functions of the sets $\{\Psi_1, \ldots, \Psi_{f_1}\}$ and $\{\Psi_{f_1+1}, \ldots, \Psi_f\}$ would transform separately, without mixing of the two sets. From the standpoint of the symmetry of the molecule there would be no reason for the energy values of these two sets of wave functions to be the same. If they happen to be the same we have a case of *accidental degeneracy*. Apart from accidental degeneracy, therefore, the representation of each energy level will be irreducible.

4.4. Derivation of irreducible representations

All molecular symmetry groups except $C_{\infty v}$ and $D_{\infty h}$ are finite groups, that is, groups containing a finite number (the order of the group) of elements (symmetry operations). The number of classes is, of course, also finite and the search for irreducible representations is limited by the following theorem valid for any finite group.

Theorem 1. The number of inequivalent, irreducible representations of a finite group is equal to the number of classes of conjugate elements.

For proofs of this theorem the reader is referred to the books by Wigner,[2] Hamermesh[3] and McWeeny[4] quoted in the introduction to this chapter.

Using Theorem 1 the irreducible representations of molecular symmetry groups may be built up by considering the transformation properties of simple functions. As examples we consider those groups specified in Section 4.1.

EXAMPLE 1. C_{3v}

Here we have 3 classes so there are three irreducible representations, which can be constructed as follows.

First, as for any group, we have the one-dimensional identity representation in which +1 is associated with every element. Another one-dimensional representation is obtained by associating +1 with all *proper* elements (pure rotations) and −1 with all *improper* elements (reflections and rotary-inversions or rotary-reflections). The third irreducible representation is a two-dimensional one, which can be built up using the pair of *basis functions* (x, y) (Fig. 4.1). For example, using equation (4.6) we obtain

$$O_C\left(\frac{2\pi}{3}\right) x = -\tfrac{1}{2}x + \frac{\sqrt{3}}{2}y \tag{4.31}$$

and

$$O_C\left(\frac{2\pi}{3}\right) y = -\frac{\sqrt{3}}{2}x - \tfrac{1}{2}y. \tag{4.32}$$

Notice, especially, that the transformations (4.31) and (4.32) of the basis functions x and y are *not* the same as the transformation

$$C\left(\frac{2\pi}{3}\right)\begin{pmatrix}x\\y\end{pmatrix} = \begin{bmatrix}-\tfrac{1}{2}, & -\dfrac{\sqrt{3}}{2}\\ +\dfrac{\sqrt{3}}{2}, & -\tfrac{1}{2}\end{bmatrix}\begin{pmatrix}x\\y\end{pmatrix} \tag{4.33}$$

of the Cartesian *components* x, y of a *vector* **r**, Indeed, one transformation is the inverse of the other (equation (4.6)). It is for this reason that we have used a notation which distinguishes clearly between transformations of *vectors* (P**r**) and transformations of *functions* ($O_P f(\mathbf{r})$). Expressed in matrix form, equations (4.31) and (4.32) become (cf. equation (4.13))

$$O_C\left(\frac{2\pi}{3}\right)(x, y) = (x, y)\begin{bmatrix}-\tfrac{1}{2}, & -\dfrac{\sqrt{3}}{2}\\ +\dfrac{\sqrt{3}}{2}, & -\tfrac{1}{2}\end{bmatrix}. \tag{4.34}$$

We see from equations (4.33) and (4.34) that the *matrices* expressing the transformations of vector components and functions *are* identical and both provide the same representation of the group.

Proceeding in this way for all the symmetry operations, we obtain the representation matrices shown in Table 4.4. The fact that the two dimensional representation Γ_3, E is, indeed, irreducible will be established below.

The labels A_1, A_2 and E for the representations in Table 4.4 are standard. In general, A denotes a one-dimensional representation even under rotation about the principal axis (or the body diagonals in the cubic groups), B is used for one-dimensional representations odd under rotation about the principal axis, and E, T are used for two- and three-dimensional representations respectively. The corresponding small letters a, b, t and e are used to indicate the symmetry of one-electron functions. Subscripts g and u are used for representations even and odd under inversion.

The last line of Table 4.4 gives the *character* of the two-dimensional representation E, defined as the *traces* (sums of diagonal elements) of the representation matrices. We see that the character has the same value for all elements in the same class, that is, it is a *class function.* This is a general result for any representation $\Gamma\{\mathbf{D}(P), \mathbf{D}(Q), \ldots\}$ of any group G. For, if P and Q are in the same class, then

$$Q = SPS^{-1}, \tag{4.35}$$

for some element S of G. Since the representation matrices multiply in the same way as the group elements, we must have

$$\mathbf{D}(Q) = \mathbf{D}(S)\mathbf{D}(P)[\mathbf{D}(S)]^{-1}. \tag{4.36}$$

Therefore, since the trace of a matrix is invariant under an equivalence transformation

$$\chi(Q) = \chi(P). \tag{4.37}$$

EXAMPLE 2. T_d

Here we have 5 classes (Section 4.1) so that we expect 5 irreducible representations. Two of these, A_1 and A_2, are obtained just as in the previous example. A third irreducible representation T_2 is obtained using the basis functions x, y, z, the coordinate axes being those specified in Section 1. By reversing the signs of the matrices of T_2 associated with improper elements we obtain a second three-dimensional representation T_1. Finally, the quadratic form $ax^2 + by^2 + cz^2$ $(a + b + c = 0)$ is invariant under all the operations of T_d, so that we obtain a two-dimensional representation E by choosing, for example, $2^{-\frac{1}{2}}(x^2 - y^2)$ and $6^{-\frac{1}{2}}(2z^2 - x^2 - y^2)$ as basis functions. Taking the

Table 4.4. Irreducible representations of C_{3v}

Representation	E	$C\left(\frac{2\pi}{3}\right)$	$C\left(-\frac{2\pi}{3}\right)$	σ_a	σ_b	σ_c
Γ_1, A_1	1	1	1	1	1	1
Γ_2, A_2	1	1	1	−1	−1	−1
Γ_3, E	$\begin{bmatrix} 1, & 0 \\ 0, & 1 \end{bmatrix}$	$\begin{bmatrix} -\frac{1}{2}, & -\frac{\sqrt{3}}{2} \\ \frac{\sqrt{3}}{2}, & -\frac{1}{2} \end{bmatrix}$	$\begin{bmatrix} -\frac{1}{2}, & \frac{\sqrt{3}}{2} \\ -\frac{\sqrt{3}}{2}, & -\frac{1}{2} \end{bmatrix}$	$\begin{bmatrix} 1, & 0 \\ 0, & -1 \end{bmatrix}$	$\begin{bmatrix} -\frac{1}{2}, & -\frac{\sqrt{3}}{2} \\ -\frac{\sqrt{3}}{2}, & \frac{1}{2} \end{bmatrix}$	$\begin{bmatrix} -\frac{1}{2}, & \frac{\sqrt{3}}{2} \\ \frac{\sqrt{3}}{2}, & \frac{1}{2} \end{bmatrix}$
$\chi(E)$	2	−1	−1	0	0	0

traces of the representation matrices we obtain the character table for T_d shown in Table 4.5. Also shown are the basis functions which generate each representation. These basis functions enable us to derive the full matrices for the two- and three-dimensional representations with a moderate amount of labour. These matrices are very useful in simplifying the calculation of approximate wave functions (Section 4.6).

Table 4.5. Character table for T_d

Representation	E	$8C\left(\frac{2\pi}{3}\right)$	$3C(\pi)$	6σ	$6S\left(\frac{\pi}{2}\right)$	Generating functions
A_1	1	1	1	1	1	1
A_2	1	1	1	−1	−1	$x^4(y^2 - z^2)$ $+ y^4(z^2 - x^2)$ $+ z^4(x^2 - y^2)$
E	2	−1	2	0	0	$2^{-\frac{1}{2}}(x^2 - y^2)$, $6^{-\frac{1}{2}}(2z^2 - x^2 - y^2)$
T_2	3	0	−1	1	−1	(x, y, z)
T_1	3	0	−1	−1	1	$x(y^2 - z^2)$, $y(z^2 - x^2)$, $z(x^2 - y^2)$

EXAMPLE 3. $C_{\infty v}$

The operations and classes of $C_{\infty v}$ are given in Table 4.2. Since, in this table, ϕ and α can take any value between 0 and π, we have an infinite number of operations and an infinite number of classes. There are also an infinite number of inequivalent irreducible representations, which can easily be generated using the coordinate transformations listed in the final column of Table 4.2. For the infinite group $C_{\infty v}$, it is not possible to use Theorem 1 to check that the list of representations is complete. However, a more general result (Theorem 3 below), applicable to both finite and infinite (continuous groups), can be used for this purpose.

Again we have the identity representation $A_1(\Sigma^+)$ and another one-dimensional representation $A_2(\Sigma^-)$ obtained by associating −1 with the reflections. The latter representation cannot occur for one-electron functions. The simplest basis function for Σ^- is the two-electron function $\sin(\varphi_1 - \varphi_2)$.

The remaining representations are obtained using eigenfunctions of the z component of orbital angular momentum as basis functions.

Thus from Table 4.2 and equations (4.6) and (4.13) we obtain

$$O_{C(\phi)}(\mathrm{e}^{-i\lambda\varphi}, \mathrm{e}^{i\lambda\varphi}) = (\mathrm{e}^{-i\lambda\varphi}, \mathrm{e}^{i\lambda\varphi}) \begin{bmatrix} \mathrm{e}^{i\lambda\phi}, & 0 \\ 0, & \mathrm{e}^{-i\lambda\phi} \end{bmatrix} \qquad (4.38)$$

and

$$O_{\sigma_v(\alpha)}(\mathrm{e}^{-i\lambda\varphi}, \mathrm{e}^{i\lambda\varphi}) = (\mathrm{e}^{-i\lambda\varphi}, \mathrm{e}^{i\lambda\varphi}) \begin{bmatrix} 0, & \mathrm{e}^{i\lambda(\alpha+\pi)} \\ \mathrm{e}^{-i\lambda(\alpha+\pi)}, & 0 \end{bmatrix}. \qquad (4.39)$$

Clearly the same transformations are obtained for a many-electron function with the z component of the *total* orbital angular momentum equal to λ.

Putting $\lambda = 1, 2, \ldots$ in equations (4.38) and (4.39) and taking the traces of the representation matrices we obtain Table 4.6. In this table we give two forms of the basis functions for the two-dimensional representations; the complex form of equations (4.38), (4.39) and the real form $(\cos \lambda\varphi, \sin \lambda\varphi)$. The latter basis functions yield *real* representation matrices, which are more convenient in some applications.

Table 4.6. Character table for $C_{\infty v}$

Representation	E	$2C(\phi)$	$\sigma_v(\alpha)$	Generating functions
A_1, Σ^+	1	1	1	1
A_2, Σ^-	1	1	−1	$\sin(\varphi_1 - \varphi_2)$
E_1, Π	2	$2\cos\phi$	0	$(\mathrm{e}^{-i\varphi}, \mathrm{e}^{i\varphi})$ $(\cos\varphi, \sin\varphi)$
E_2, Δ	2	$2\cos 2\phi$	0	$(\mathrm{e}^{-i2\varphi}, \mathrm{e}^{i2\varphi})$ $(\cos 2\varphi, \sin 2\varphi)$
⋮	⋮	⋮	⋮	⋮ ⋮
E_λ	2	$2\cos\lambda\phi$	0	$(\mathrm{e}^{-i\lambda\varphi}, \mathrm{e}^{i\lambda\varphi})$ $(\cos\lambda\varphi, \sin\lambda\varphi)$
⋮	⋮	⋮	⋮	⋮ ⋮

EXAMPLE 4. $D_{\infty h}$

The elements and classes of $D_{\infty h}$ are given in Table 4.3. To obtain the irreducible representations it is convenient to use rotary-inversions rather than rotary-reflections for the improper elements. We then obtain, from each irreducible representation of $C_{\infty v}$, a pair of irreducible representations by associating first +1 and then −1 with the inversion i. Since the representation matrices for $D_{\infty h}$ are, in this way, so simply related to those for $C_{\infty v}$, we do not list generating functions in Table 4.7.

Table 4.7. Character table for $D_{\infty h}$

Representation	E	$2C(\phi)$	$C'_\alpha(\pi)$	$iC(\phi) = S(\pi + \phi)$	$iC'_\alpha(\pi) = \sigma_v(\alpha)$
A_{1g}, Σ_g^+	1	1	1	1	1
A_{1u}, Σ_u^+	1	1	−1	−1	1
A_{2g}, Σ_g^-	1	1	−1	1	−1
A_{2u}, Σ_u^-	1	1	1	−1	−1
E_{1g}, Π_g	2	$2\cos\phi$	0	$2\cos\phi$	0
E_{1u}, Π_u	2	$2\cos\phi$	0	$-2\cos\phi$	0
E_{2g}, Δ_g	2	$2\cos 2\phi$	0	$2\cos 2\phi$	0
E_{2u}, Δ_u	2	$2\cos 2\phi$	0	$-2\cos 2\phi$	0
⋮	⋮	⋮	⋮	⋮	⋮
$E_{\lambda g}$	2	$2\cos\lambda\phi$	0	$2\cos\lambda\phi$	0
$E_{\lambda u}$	2	$2\cos\lambda\phi$	0	$-2\cos\lambda\phi$	0
⋮	⋮	⋮	⋮	⋮	⋮

Using the coordinate transformations listed in Table 4.3, it is easy to verify that the H_2^+ solutions discussed in Chapter 3 do, in fact, transform according to the appropriate irreducible representations of $D_{\infty h}$ given in Table 4.7.

4.5. Orthogonality relations

The matrix elements of the irreducible representations of a group satisfy orthogonality relations which are very useful in applications of the theory to quantum mechanics. For a finite group we have:

Theorem 2. Let

$$\begin{aligned} &\Gamma_1\,\{D^{(1)}_{hk}(P),\quad D^{(1)}_{hk}(Q),\ldots\}\ \textit{of dimension } f_1,\\ &\Gamma_2\,\{D^{(2)}_{hk}(P),\quad D^{(2)}_{hk}(Q),\ldots\}\ \textit{of dimension } f_2,\\ &\cdots\cdots\cdots\cdots\cdots\cdots\cdots \end{aligned} \tag{4.40}$$

be inequivalent, irreducible representations of a finite group G, of order g, and suppose that all these representations are chosen in unitary form. Then

$$\frac{1}{g}\sum_P D^{(\gamma)}_{hk}(P)^* D^{(\beta)}_{mn}(P) = \frac{1}{f_\gamma}\,\delta_{\gamma\beta}\,\delta_{hm}\,\delta_{kn}, \tag{4.41}$$

where the summation is over all elements P of G.

Theorem 2 is the key to applications of group representation theory to quantum mechanics. A complete proof of this theorem is somewhat complicated (see e.g. Wigner[2]) but, as we shall see, it is a relatively simple matter to *use* the result to simplify and clarify the quantum mechanical theory of the electronic structure of molecules. There is a corresponding theorem for a limited class of infinite groups, the *continuous compact* groups, in which the summation in equation (4.41) is replaced by integration over the *group manifold*. For the most general compact group, this theory of group integration is difficult, but the chemically important results for $C_{\infty v}$ and $D_{\infty h}$ are extremely simple. For these groups we have:

Theorem 2a. Let

$$\begin{aligned} &\Gamma_1 \{D_{hk}^{(1)}[C(\phi)], \quad D_{hk}^{(1)}[\sigma_v(\alpha)]\} \text{ of dimension } f_1, \\ &\Gamma_2 \{D_{hk}^{(2)}[C(\phi)], \quad D_{hk}^{(2)}[\sigma_v(\alpha)]\} \text{ of dimension } f_2, \\ &\vdots \end{aligned} \tag{4.40a}$$

be inequivalent, irreducible, unitary representations of the group $C_{\infty v}$. Then the orthogonality relations (4.41) become

$$\frac{1}{4\pi}\left\{\int_{-\pi}^{\pi} D_{hk}^{(\gamma)}[C(\phi)]^* D_{mn}^{(\beta)}[C(\phi)]\, d\phi + \int_{-\pi}^{\pi} D_{hk}^{(\gamma)}[\sigma_v(\alpha)]^* D_{mn}^{(\beta)}[\sigma_v(\alpha)]\, d\alpha\right\} = \frac{1}{f_\gamma}\delta_{\gamma\beta}\delta_{hm}\delta_{kn}. \tag{4.41a}$$

Theorem 2b. Let

$$\begin{aligned} &\Gamma_1\{D_{hk}^{(1)}[C(\phi)], D_{hk}^{(1)}[C'_\alpha(\pi)], D_{hk}^{(1)}[iC(\phi)], D_{hk}^{(1)}[iC'_\alpha(\pi)]\} \text{ of dimension } f_1, \\ &\Gamma_2\{D_{hk}^{(2)}[C(\phi)], D_{hk}^{(2)}[C'_\alpha(\pi)], D_{hk}^{(2)}[iC(\phi)], D_{hk}^{(2)}[iC'_\alpha(\pi)]\} \text{ of dimension } f_2, \\ &\vdots \end{aligned} \tag{4.40b}$$

be inequivalent, irreducible, unitary representations of the group $D_{\infty h}$. Then we have the orthogonality relations

$$\begin{aligned} \frac{1}{8\pi}\Bigg\{&\int_{-\pi}^{\pi} D_{hk}^{(\gamma)}[C(\phi)]^* D_{mn}^{(\beta)}[C(\phi)]\, d\phi + \int_{-\pi}^{\pi} D_{hk}^{(\gamma)}[C'_\alpha(\pi)]^* D_{mn}^{(\beta)}[C'_\alpha(\pi)]\, d\alpha \\ &+ \int_{-\pi}^{\pi} D_{hk}^{(\gamma)}[iC(\phi)]^* D_{mn}^{(\beta)}[iC(\phi)]\, d\phi + \int_{-\pi}^{\pi} D_{hk}^{(\gamma)}[iC'_\alpha(\pi)]^* D_{mn}^{(\beta)}[iC'_\alpha(\pi)]\, d\alpha\Bigg\} \\ &= \frac{1}{f_\gamma}\delta_{\gamma\beta}\delta_{hm}\delta_{kn}. \end{aligned} \tag{4.41b}$$

The very close analogy between the orthogonality relations for finite groups, and for the continuous groups $C_{\infty v}$, $D_{\infty h}$ is evident from equations (4.41), (4.41a) and (4.41b).

In the case of a finite group, the quantities $D_{hk}^{(\gamma)}(P), D_{hk}^{(\gamma)}(Q), \ldots$, for fixed values of γ, h and k, may be regarded as the components of a vector in a space of dimensionality g (the group manifold). The relations (4.41) then state that all the $f_1^2 + f_2^2 + \ldots$ vectors (4.40) are orthogonal. It follows that the number of these vectors cannot exceed the dimensionality g of the space. In fact, if *all* inequivalent irreducible representations are included in (4.40) we have

$$\sum_{\gamma} f_{\gamma}^2 = g. \tag{4.42}$$

Theorems 2, 2a and 2b may, therefore, be summed up in a single more general result.

Theorem 3. The matrix elements of all inequivalent, irreducible unitary representations of a finite (compact) group form a complete, *orthogonal set of vectors (functions) on the group manifold.*

The theorem in this form is equally applicable to finite groups and the continuous groups $C_{\infty v}$ and $D_{\infty h}$. The fact that Tables 4.6 and 4.7 include *all* representations of $C_{\infty v}$ and $D_{\infty h}$ respectively, is then a simple consequence of Fourier's expansion: any continuous function of period 2π may be expressed in terms of the functions $\cos \lambda\phi$, $\sin \lambda\phi$ ($\lambda = 0, 1, 2, \ldots$).

The *character* $\chi(P)$ of any representation $\Gamma\{\mathbf{D}(P), \mathbf{D}(Q), \ldots\}$ is given by the traces of the representation matrices. Thus

$$\chi(P) = \sum_{k} D_{kk}(P). \tag{4.43}$$

Now the matrices $\mathbf{D}(P)$ and $\mathbf{T}^{-1}\mathbf{D}(P)\mathbf{T}$ have the same trace. This has two consequences. Firstly, as we have already shown in the previous section (equation (4.37)), the character of a representation is a class function. Secondly the characters of two equivalent representations are the same.

Let us denote the characters of the irreducible representations (4.40) by $\chi_{\gamma}(P)$. Then, by putting $h = k$ and $m = n$ in equation (4.41) and summing over h and m, we obtain the orthogonality relations

$$\frac{1}{g} \sum_{P} \chi_{\gamma}^{*}(P) \chi_{\beta}(P) = \delta_{\gamma\beta}. \tag{4.44}$$

Proceeding in the same way from equations (4.41a) and (4.41b), for the continuous groups $C_{\infty v}$ and $D_{\infty h}$ we obtain

$$C_{\infty v}: \frac{1}{4\pi}\left\{\int_{-\pi}^{\pi} \chi_\gamma^*[C(\phi)]\chi_\beta[C(\phi)]\,d\phi + \int_{-\pi}^{\pi} \chi_\gamma^*[\sigma_v(\alpha)]\chi_\beta[\sigma_v(\alpha)]\,d\alpha\right\} = \delta_{\gamma\beta} \tag{4.44a}$$

$$D_{\infty h}: \frac{1}{8\pi}\left\{\int_{-\pi}^{\pi} \chi_\gamma^*[C(\phi)]\chi_\beta[C(\phi)]\,d\phi + \int_{-\pi}^{\pi} \chi_\gamma^*[C'_\alpha(\pi)]\chi_\beta[C'_\alpha(\pi)]\,d\alpha \right.$$
$$\left. + \int_{-\pi}^{\pi} \chi_\gamma^*[iC(\phi)]\chi_\beta[iC(\phi)]\,d\phi + \int_{-\pi}^{\pi} \chi_\gamma^*[iC'_\alpha(\pi)]\chi_\beta[iC'_\alpha(\pi)]\,d\alpha\right\} = \delta_{\gamma\beta} \tag{4.44b}$$

Since the character is a class function, the orthogonality relations (4.44) imply that the number of inequivalent, irreducible representations of a finite group cannot exceed the number of classes. As already stated in Section 4.4 (Theorem 1) these two numbers are, in fact, equal and we have the analogue of Theorem 3 for the characters.

Theorem 4. The characters of all inequivalent, irreducible representations of a group form a complete *orthogonal set of class functions.*

This result, like Theorem 3, applies unchanged to the continuous groups $C_{\infty v}$ and $D_{\infty h}$.

In Section 4.4 we have derived character tables for the groups C_{3v}, T_d, $C_{\infty v}$ and $D_{\infty h}$. These tables are a convenient means of classifying the energy levels of molecules according to their symmetry properties. Each energy level belongs to a particular row of the character table, that is, to a certain irreducible representation. In such a classification, no distinction is drawn between equivalent forms of the same irreducible representation. This is appropriate since, as we saw in Section 4.3, an equivalence transformation merely corresponds to taking new linear combinations of the independent eigenfunctions of a degenerate energy level. In Appendix 2 we list the character tables of most of the symmetry groups which arise in molecular studies. We also list basis functions, which may be used to generate the full matrix representations.

4.6. The reduction of representations

Consider now a reducible representation $\Gamma\{\mathbf{D}(P), \mathbf{D}(Q), \ldots\}$ of a group G. Then there exists a constant matrix $\mathbf{T}$, such that all the matrices

$$\bar{\mathbf{D}}(X) = \mathbf{T}^{-1}\mathbf{D}(X)\mathbf{T} \qquad (X = P, Q, \ldots) \tag{4.45}$$

take the fully reduced form shown in Fig. 4.4. This process is called the reduction of the representation Γ. If the irreducible representations which appear along the diagonal are equivalent to

$$\begin{aligned} &m_1 \text{ times } \mathbf{D}^{(1)}(P) \text{ of } \Gamma_1 \\ &m_2 \text{ times } \mathbf{D}^{(2)}(P) \text{ of } \Gamma_2 \end{aligned} \tag{4.46}$$

we say that the representation Γ contains Γ_1 m_1 times, Γ_2 m_2 times, . . ., and this may be written symbolically

$$\Gamma = m_1\Gamma_1 \dotplus m_2\Gamma_2 \dotplus \ldots. \tag{4.47}$$

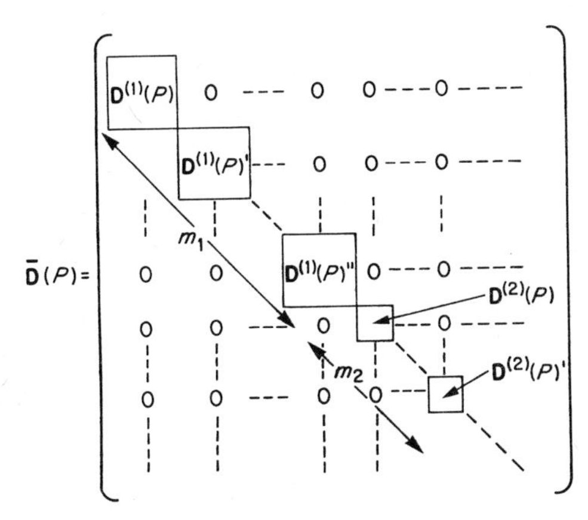

Fig. 4.4. Fully reduced form of a representation matrix **D**(*P*).

To specify **T** uniquely we may require that the equivalent representations in (4.45) all appear in some standard form. The m_γ matrices $\mathbf{D}^{(\gamma)}(P)$, $\mathbf{D}^{(\gamma)}(P)'$, . . . corresponding to the term $m_\gamma\Gamma_\gamma$ in equation (4.47) are then all identical rather than just equivalent.

If the characters of $\Gamma, \Gamma_1, \Gamma_2, \ldots$ are $\chi(P), \chi_1(P), \chi_2(P), \ldots$ respectively, then we obviously have

$$\chi(P) = m_1\chi_1(P) + m_2\chi_2(P) + \ldots. \tag{4.48}$$

Since the character of a representation is unchanged by the equivalence transformation $\mathbf{D}(X) \to \mathbf{T}^{-1}\mathbf{D}(X)\mathbf{T}$, the formula (4.48) may be used to find the number of times each irreducible representation occurs in the reduction of Γ. To determine $m_1, m_2, \ldots$ we need not actually reduce the representation, but simply write down the character and express it as a sum of irreducible characters appearing in the character table. We may, indeed, obtain an explicit expression for m_γ by multiplying equation (4.48) by $\chi_\gamma^*(P)$ and summing (or integrating) over all elements of G. Thus, using the orthogonality relations (4.44), we obtain, for a finite group

$$m_\gamma = \frac{1}{g} \sum_P \chi_\gamma^*(P)\chi(P). \tag{4.49}$$

The corresponding equations (4.49a), (4.49b) for the continuous groups $C_{\infty v}$, $D_{\infty h}$ are immediately obvious from equations (4.44a) and (4.44b). However, it is not usually necessary to resort to these equations; in most cases the reduction (4.48) can be written down simply by inspection of the characters. An application for which the orthogonality relations (4.44a) and (4.44b) *are* needed is given in Ref. 5a.

A useful test for the irreducibility of a representation is obtained by multiplying equation (4.48) by its complex conjugate and summing (or integrating) over all group elements. For a finite group, this leads to the equation

$$\frac{1}{g} \sum_P |\chi(P)|^2 = m_1^2 + m_2^2 + \ldots. \tag{4.50}$$

The modifications, (4.50a) and (4.50b), of this equation for the continuous groups $C_{\infty v}$, $D_{\infty h}$ are again obvious.

Since $m_1, m_2, \ldots$ are all non-negative integers, the right-hand side of equation (4.50) (or (4.50a), (4.50b)) will equal unity if and only if one of the m is unity and the remainder all vanish, that is, if and only if Γ is irreducible. Using this test we may immediately verify that all the representations listed in Tables 4.4–4.7 are, indeed, irreducible.

An important example of the reduction of representations is provided by the *direct product* representation. Suppose that $(\phi_1, \ldots, \phi_f)$ are basis functions for a representation

$$\Gamma_1\{\mathbf{D}^{(1)}(P), \mathbf{D}^{(1)}(Q), \ldots\}$$

and that $(\psi_1, \ldots, \psi_g)$ are basis functions for a representation

$$\Gamma_2\{\mathbf{D}^{(2)}(P), \mathbf{D}^{(2)}(Q), \ldots\}.$$

We then have

$$O_P \phi_k = \sum_h D_{hk}^{(1)}(P)\phi_h \tag{4.51}$$

and

$$O_P \psi_m = \sum_n D_{nm}^{(2)} \psi_n. \tag{4.52}$$

The *fg* products

$$\phi_k \psi_m \; (k = 1, \ldots, f; \; m = 1, \ldots, g) \tag{4.53}$$

form the basis for an *fg*-dimensional representation, which is called the direct-product representation and is denoted by $\Gamma_1 \times \Gamma_2$. Since we have

$$O_P \phi_k \psi_m = (O_P \phi_k)(O_P \psi_m) = \sum_h \sum_n D^{(1)}_{hk}(P) D^{(2)}_{nm} \phi_h \psi_n, \tag{4.54}$$

the character $\chi_{\Gamma_1 \times \Gamma_2}(P)$ of the direct product representation is given by

$$\chi_{\Gamma_1 \times \Gamma_2}(P) = \sum_k \sum_m D^{(1)}_{kk}(P) D^{(2)}_{mm}(P)$$

$$= \left(\sum_k D^{(1)}_{kk}(P) \right) \left(\sum_m D^{(2)}_{mm}(P) \right)$$

that is

$$\chi_{\Gamma_1 \times \Gamma_2}(P) = \chi_1(P) \chi_2(P). \tag{4.55}$$

That is, the character of the direct product representation is just the product of the characters of the original representations. Using this result and the character Tables 4.4–4.7 we derive Table 4.8 which gives the reduction of direct products of irreducible representations for the groups C_{3v}, T_d, $C_{\infty v}$ and $D_{\infty h}$.

Table 4.8. Reduction of direct products

C_{3v}

	A_1	A_2	E
A_1	A_1	A_2	E
A_2	A_2	A_1	E
E	E	E	$A_1 + A_2 + E$

T_d

	A_1	A_2	E	T_1	T_2
A_1	A_1	A_2	E	T_1	T_2
A_2	A_2	A_1	E	T_2	T_1
E	E	E	$A_1 + A_2 + E$	$T_1 + T_2$	$T_1 + T_2$
T_1	T_1	T_2	$T_1 + T_2$	$A_1 + E + T_1 + T_2$	$A_2 + E + T_1 + T_2$
T_2	T_2	T_1	$T_1 + T_2$	$A_2 + E + T_1 + T_2$	$A_1 + E + T_1 + T_2$

$C_{\infty v}$

	Σ^+	Σ^-	E_λ
Σ^+	Σ^+	Σ^-	E_λ
Σ^-	Σ^-	Σ^+	E_λ
E_λ	E_λ	E_λ	$\Sigma^+ + \Sigma^- + E_{2\lambda}$

($\Sigma^+ = A_1, \Sigma^- = A_2, E_\lambda = \Pi, \Delta, \Phi, \ldots$ for $\lambda = 1, 1, 3 \ldots$)

$D_{\infty h}$

	A_{1g}	A_{1u}	A_{2g}	A_{2u}	$E_{\lambda g}$	$E_{\lambda u}$
A_{1g}	A_{1g}	A_{1u}	A_{2g}	A_{2u}	$E_{\lambda g}$	$E_{\lambda u}$
A_{1u}	A_{1u}	A_{1g}	A_{2u}	A_{2g}	$E_{\lambda u}$	$E_{\lambda g}$
A_{2g}	A_{2g}	A_{2u}	A_{1g}	A_{1u}	$E_{\lambda g}$	$E_{\lambda u}$
A_{2u}	A_{2u}	A_{2g}	A_{1u}	A_{1g}	$E_{\lambda u}$	$E_{\lambda g}$
$E_{\lambda g}$	$E_{\lambda g}$	$E_{\lambda u}$	$E_{\lambda g}$	$E_{\lambda u}$	$A_{1g} + A_{2g} + E_{2\lambda g}$	$A_{1u} + A_{2u} + E_{2\lambda u}$
$E_{\lambda u}$	$E_{\lambda u}$	$E_{\lambda g}$	$E_{\lambda u}$	$E_{\lambda g}$	$A_{1u} + A_{2u} + E_{2\lambda u}$	$A_{1g} + A_{2g} + E_{2\lambda g}$

($A_{1g,u} = \Sigma^+_{g,u}; A_{2g,u} = \Sigma^-_{g,u}; E_{\lambda g,u} = \Pi_{g,u}; \Delta_{g,u}; \Phi_{g,u} \ldots$ for $\lambda = 1, 2, 3, \ldots$)

4.7. The use of symmetry in approximate calculations

In previous sections we have found that, for the symmetry classification of the electronic eigenfunctions of a molecule, it is appropriate to regard equivalent forms of an irreducible representation as essentially the same and to proceed via the character table. However, in calculating approximate wave functions we often need to form, from a given set of basis functions, explicit linear combinations with specified transformation properties. For this purpose it is very useful to have available, not just the character table of the appropriate symmetry group, but the explicit representation matrices in some standard basis. We may then use the following general result.

Theorem 5. Let

$$\Gamma_\alpha\{D^{(\alpha)}_{hk}(P), D^{(\alpha)}_{hk}(Q), \ldots\} \tag{4.56}$$

be a unitary, irreducible, matrix representation, of dimension f_α of a finite group G and let ψ be any function. Then the function

$$\phi^{(\alpha)}_{hk} = \sum_P D^{(\alpha)}_{hk}(P)^* O_P \psi, \tag{4.57}$$

transforms according to the equation

$$O_P \phi_{hk}^{(\alpha)} = \sum_{l=1}^{f_\alpha} D_{lh}^{(\alpha)}(P) \phi_{lk}^{(\alpha)} \tag{4.58}$$

for all elements P of G.

Equation (4.58) is expressed by saying that $\phi_{hk}^{(\alpha)}$ transforms as the hth partner in the representation α or, more briefly, belongs to the symmetry species (α, h). The f_α functions

$$(\phi_{1k}^{(\alpha)}, \phi_{2k}^{(\alpha)}, \ldots, \phi_{f_\alpha k}^{(\alpha)}), \tag{4.59}$$

obtained by letting h in equation (4.57) run down a column of the representation matrices, then form a set of basis functions for the representation Γ_α. By letting k vary, that is, by using different columns of the representation matrices, we obtain altogether f_α sets of basis functions like (4.59), each set having the same transformation properties. These f_α sets may not be linearly independent; some (or all) of the sets may vanish identically and others may be the same or differ only by a constant factor.

We now prove Theorem 5.

If we apply the linear operator O_Q to equation (4.57) and use equation (4.8) we obtain

$$O_Q \phi_{hk}^{(\alpha)} = \sum_P D_{hk}^{(\alpha)}(P)^* O_{QP} \psi. \tag{4.60}$$

Now as P runs through all elements of G so does $QP = R$ (say). Changing the summation variable in equation (4.60) to R, we have $P = Q^{-1}R$ and

$$O_Q \phi_{hk}^{(\alpha)} = \sum_R D_{hk}^{(\alpha)}(Q^{-1}R)^* O_R \psi. \tag{4.61}$$

But, since the matrices $\mathbf{D}^{(\alpha)}$ form a unitary representation of G, we have

$$D_{hk}^{(\alpha)}(Q^{-1}R) = \sum_l D_{hl}^{(\alpha)}(Q^{-1}) D_{lk}^{(\alpha)}(R) = \sum_l D_{lh}^{(\alpha)}(Q)^* D_{lk}^{(\alpha)}(R). \tag{4.62}$$

Taking the complex conjugate of equation (4.62) and substituting in equation (4.61) we obtain

$$O_Q \phi_{hk}^{(\alpha)} = \sum_l D_{lh}^{(\alpha)}(Q) \left(\sum_R D_{lk}^{(\alpha)}(R)^* O_R \psi \right) = \sum_l D_{lh}^{(\alpha)}(Q) \phi_{lk}^{(\alpha)}.$$

Since Q may be any element of G, this is the desired result (4.58).

Equation (4.57) provides a straightforward method of obtaining functions belonging to a definite symmetry species (α, h). The linear operators

$$\sum_P D_{hk}^{(\alpha)}(P)^* O_P \tag{4.63}$$

appearing in equation (4.57) are frequently referred to as *projection* operators; acting on any function, they project on to the component belonging to the symmetry species (α, h). On the other hand, some authors reserve the name projection operator for those operators $\mathscr{P}$ which satisfy the equation (cf. Section 2.5).

$$\mathscr{P}^2 = \mathscr{P}. \tag{4.64}$$

In this strict sense, only the normalized operators

$$N \sum_P D_{hh}^{(\alpha)}(P)^* O_P \tag{4.65}$$

constructed from the *diagonal* elements of the representation matrices are projection operators. The determination of the normalization constant N in equation (4.65), in order to satisfy the condition (4.64), is straightforward.

It is a simple matter to construct projection operators for the continuous groups $C_{\infty v}$ and $D_{\infty h}$ analogous to the operators (4.63). However, these operators are rarely needed. It is usually easier to carry out the analysis for these groups in two stages.

(i) The functions are first sorted out according to the value of the component of orbital angular momentum along the bond axis.

(ii) A further subdivision is obtained using projection operators for the *finite* groups

$$\mathscr{C}_{\infty v} = \{E, \sigma_v(0)\}, \tag{4.66}$$

$$\mathscr{D}_{\infty h} = \{E, C_2'(0), i, iC_2'(0) = \sigma_v(0)\}. \tag{4.67}$$

As an example of the use of equation (4.57) we construct symmetry orbitals for the molecule NH_3, using atomic orbitals of types s^N, p_x^N, p_y^N, p_z^N on the nitrogen atom s^a, p_x^a, p_y^a, p_z^a on hydrogen atom a etc. (cf. Fig. 4.1). The irreducible representations of the appropriate symmetry group C_{3v} are given in Table 4.4.

Clearly the nitrogen orbitals are already symmetry adapted; both s^N and p_z^N are invariant under all operations of C_{3v}, that is, are of symmetry species (1, 1) whereas p_x^N, p_y^N being proportional to the functions x, y used to generate the representation Γ_3 of Table 4.4 are of symmetry species (3, 1) and (3, 2) respectively.

It is evident that some of the hydrogen orbitals are equivalent in the sense that they are merely permuted by the symmetry operations; thus $O_{C(2\pi/3)}\, s^a = s^b$, etc. In order to simplify the calculations we choose local axes at each atom of an equivalent set (in this case the hydrogens) so that as many as possible of the basis orbitals become

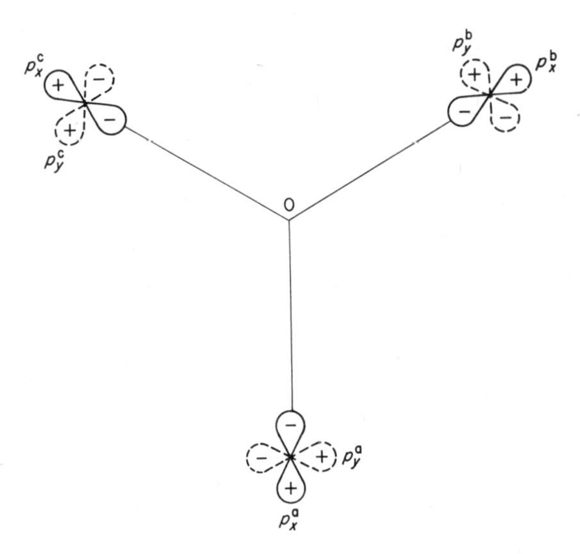

Fig. 4.5. Local coordinate axes for NH_3.

equivalent or, at most, suffer only a sign change under the symmetry operations. In this case it is expedient to choose the p_x and p_y orbitals of each hydrogen so that they lie in radial and tangential directions. This choice is illustrated in Fig. 4.5, where we are looking down from the apex of the pyramid of Fig. 4.1.

To obtain the symmetry orbitals we merely substitute one of the atomic orbitals s^a, p_x^a, p_y^a, p_z^a for ψ in equation (4.57) and use the appropriate matrix elements from Table 4.4. Thus from the orbital s^a we obtain successively

$\alpha = 1,$	$h = 1,$	$k = 1$	$2(s^a + s^b + s^c)$	species (1, 1)
$\alpha = 2,$	$h = 1,$	$k = 1$	0	
$\alpha = 3,$	$h = 1,$	$k = 1$	$2s^a - s^b - s^c$	species (3, 1)
$\alpha = 3,$	$h = 2,$	$k = 1$	$\sqrt{3}\,(s^b - s^c)$	species (3, 2)
$\alpha = 3,$	$h = 1,$	$k = 2$	0	
$\alpha = 3,$	$h = 2,$	$k = 2$	0	

Next we consider the p orbitals. Because we have chosen local axes the p_x, p_y and p_z types transform only among themselves without mixing. The p_z functions p_z^a, p_z^b, p_z^c and the p_x functions behave exactly like s^a, s^b, s^c under the operations of C_{3v} and therefore lead to a repetition of the above results. The p_y functions, however, are not merely permuted under the symmetry operations; their signs are reversed under each reflection. Thus from the function p_y^a we obtain the following symmetry orbitals.

$\alpha = 1,$	$h = 1,$	$k = 1$	0	
$\alpha = 2,$	$h = 1,$	$k = 1$	$2(p_y^a + p_y^b + p_y^c)$	species (2, 1)
$\alpha = 3,$	$h = 1,$	$k = 1$	0	
$\alpha = 3,$	$h = 2,$	$k = 1$	0	
$\alpha = 3,$	$h = 1,$	$k = 2$	$\sqrt{3}(p_y^c - p_y^b)$	species (3, 1)
$\alpha = 3,$	$h = 2,$	$k = 2$	$2p_y^a - p_y^b - p_y^c$	species (3, 2).

Collecting the results and normalizing each symmetry orbital so that the sum of the squares of the coefficients is unity we obtain Table 4.9.

We see then that equation (4.57) provides a completely straightforward way of analysing any function, or set of functions, into components each of which belongs to a definite symmetry species (α, h). To use this method we need the full matrices for the irreducible representations. For most of the finite groups which describe the spatial symmetry of molecules, these matrices can be generated using the standard basis functions listed in Appendix 2. McWeeny[4] lists full matrix representations for the 32 crystallographic point groups. These tables are adequate for almost all molecular symmetry groups.

If we put $k = h$ in equation (4.57) and sum over the f possible values of h we obtain

$$\sum_{h=1}^{f_\alpha} \phi_{hh}^{(\alpha)} = \sum_P \chi_\alpha^*(P) O_P \psi. \tag{4.68}$$

Equation (4.68) provides an alternative method of generating basis functions for the irreducible representation α which involves only the character of the representation.

Clearly for one-dimensional representations (4.57) and (4.68) are identical. For degenerate representations the method based on equation (4.68) is less powerful than that based on equation (4.57). We see that, for a general function ψ, equation (4.68) does *not* lead to a basis function for Γ_α but to a linear combination of f_α such functions. However, in simple cases, it is possible to choose ψ in such a way that the right hand side of equation (4.68) *does* yield a basis function $\phi_k^{(\alpha)}$ say. A sufficient condition for this is that the reducible representation generated by the basis functions $O_P \psi$, $O_Q \psi$ . . . should contain Γ_α only once.

Once $\phi_k^{(\alpha)}$ has been found the other partners may be obtained, either by operating on $\phi_k^{(\alpha)}$ with some of the linear operators $O_P, O_Q, \ldots$, or by taking other choices of ψ in equation (4.68).

Table 4.9. Symmetry orbitals for NH_3

Representation	Γ_1, A_1	Γ_2, A_2	Γ_3, E	
Species	(1, 1)	(2, 1)	(3, 1)	(3, 2)
	s^N	$3^{-\frac{1}{2}}(p_y^a + p_y^b + p_y^c)$	p_x^N	p_y^N
	p_z^N		$6^{-\frac{1}{2}}(2s^a - s^b - s^c)$	$2^{-\frac{1}{2}}(s^b - s^c)$
	$3^{-\frac{1}{2}}(s^a + s^b + s^c)$		$6^{-\frac{1}{2}}(2p_x^a - p_x^b - p_x^c)$	$2^{-\frac{1}{2}}(p_x^b - p_x^c)$
	$3^{-\frac{1}{2}}(p_x^a + p_x^b + p_x^c)$		$6^{-\frac{1}{2}}(2p_z^a - p_z^b - p_z^c)$	$2^{-\frac{1}{2}}(p_z^b - p_z^c)$
	$3^{-\frac{1}{2}}(p_z^a + p_z^b + p_z^c)$		$2^{-\frac{1}{2}}(p_y^c - p_y^b)$	$6^{-\frac{1}{2}}(2p_y^a - p_y^b - p_y^c)$

The bases obtained in this way are not orthogonal so that a final step of orthogonalization is necessary to express Γ_α in unitary form. If this procedure is used to derive several distinct sets of basis functions there is no guarantee that all the sets will generate identical matrices for Γ_α, the matrices generated by two sets may differ by an equivalence transformation. In other words the second symbol h determining the symmetry species (α, h) is not specified. In such a case we do not get all the simplifications which result from a transformation to a true symmetry adapted basis.

These points may be illustrated by using equation (4.68) to derive the symmetry orbitals of NH_3 belonging to the degenerate representation E of C_{3v}.

Substituting s^a for ψ in equation (4.68) and using the character $\chi(E)$ from Table 4.4 we obtain the function

$$\phi_1(E) = 2s^a - s^b - s^c. \tag{4.69}$$

Operating on $\phi_1(E)$ with $O_{C(2\pi/3)}$ or substituting s^b for ψ in equation (4.68), we obtain a second function

$$\phi_2(E) = 2s^b - s^c - s^a. \tag{4.70}$$

These two functions are not orthogonal. One way of obtaining orthogonal functions (the Schmidt method) is to replace $\phi_2(E)$ by

$$\phi_2'(E) = \phi_2(E) + k\phi_1(E) \tag{4.71}$$

and to choose k so that

$$\int \phi_2'(E)\phi_1(E)\, dv = \int \phi_2(E)\phi_1(E)\, dv + k \int \phi_1^2(E)\, dv = 0. \tag{4.72}$$

This condition is satisfied for

$$k = -\frac{\int \phi_2(E)\phi_1(E)\, dv}{\int \phi_1^2(E)\, dv} = \tfrac{1}{2}, \tag{4.73}$$

so that

$$\phi_2'(E) = \tfrac{3}{2}(s^b - s^c). \tag{4.74}$$

Normalizing the functions $\phi_1(E)$ and $\phi_2'(E)$ we obtain the same symmetry orbitals as from equation (4.57) (cf. Table 4.9). Applying exactly the same procedure to the functions ϕ_y^a and ϕ_y^b we end up with the pair of functions

$$\psi_1(E) = 6^{-\frac{1}{2}}(2p_y^a - p_y^b - p_y^c), \tag{4.75}$$

$$\psi_2(E) = 2^{-\frac{1}{2}}(p_y^b - p_y^c). \tag{4.76}$$

Comparing with the results of Table 4.9 we see that the two pairs of basis functions $(\psi_1(E), \psi_2(E))$ and $(\phi_1(E), \phi_2'(E))$ will generate *different* representation matrices for E. Thus, even when a standard pattern of substitution and orthogonalization is followed, equation (4.68) does *not* lead to a true symmetry adapted basis in which both the representation α and the partner h are specified.

The advantages of using a true symmetry adapted basis stem from the following general result, valid for compact continuous groups as well as for finite groups, although we shall prove it only for finite groups.

Theorem 6. Any operator F, such as the electronic Hamiltonian or the unit operator, which commutes with all the operators of a group G, has no off-diagonal matrix elements between two functions belonging to different symmetry species.

Thus, if the functions

$$\phi_h^{(\alpha)} \, (h = 1, \ldots, f_\alpha)$$

form a basis for the irreducible representation Γ_α, and if the functions

$$\psi_k^{(\beta)} \, (k = 1, \ldots, f_\beta)$$

form a basis for the irreducible representation Γ_β, then we have

$$\int \phi_h^{(\alpha)*} F \psi_k^{(\beta)} \, d\tau = 0 \tag{4.77}$$

unless $\alpha = \beta$ and $h = k$. Furthermore, if $\alpha = \beta$ and $h = k$ the integral (4.77) is independent of h, that is, the value of the integral is the same for all partners of the representation.

We now prove Theorem 6.

Firstly we express the integral as a scalar product and use equation (4.5) to obtain

$$\int \phi_h^{(\alpha)*} F \psi_k^{(\beta)} \, d\tau = (\phi_h^{(\alpha)}, F\psi_k^{(\beta)}) = (O_P \phi_h^{(\alpha)}, O_P F \psi_k^{(\beta)}), \tag{4.78}$$

where P is any element of group G. Therefore, since O_P commutes with F

$$\begin{aligned} \int \phi_h^{(\alpha)*} F \psi_k^{(\beta)} \, d\tau &= (O_P \phi_h^{(\alpha)}, F O_P \psi_k^{(\beta)}) \\ &= \sum_l \sum_m D_{lh}^{(\alpha)}(P)^* D_{mk}^{(\beta)}(P) (\phi_l^{(\alpha)}, F\psi_m^{(\beta)}). \end{aligned} \tag{4.79}$$

We now sum over all the g elements of G and use the orthogonality relations (4.41). This gives

$$\begin{aligned} g \int \phi_h^{(\alpha)*} F \psi_k^{(\beta)} \, d\tau &= \frac{g}{f_\alpha} \sum_l \sum_m \delta_{\alpha\beta} \delta_{lm} \delta_{hk} (\phi_l^{(\alpha)}, F\psi_m^{(\beta)}) \cdot \\ &= \frac{g}{f_\alpha} \delta_{\alpha\beta} \delta_{hk} \sum_l \int \phi_l^{(\alpha)*} F \psi_l^{(\beta)} \, d\tau. \end{aligned} \tag{4.80}$$

The properties of the integral (4.77) stated in Theorem 6 are obvious from equation (4.80). The factors $\delta_{\alpha\beta}$, δ_{hk} ensure that the integral vanishes unless $\alpha = \beta$ and $h = k$ and in this case the value of the integral is clearly independent of h.

A typical application of Theorem 6, and one of the most important, is the simplification of the secular equation in the linear variation method. The expansion of a general function Ψ in a symmetry adapted basis takes the form

$$\Psi = \sum_{\alpha} \sum_{h} \sum_{\mu} C_h^{(\alpha, \mu)} \phi_h^{(\alpha, \mu)} \tag{4.81}$$

where the third index μ is used to distinguish different basis functions of the same symmetry species (α, h).

Because there are no off-diagonal matrix elements between functions of different symmetry species, the secular equation arising from the expansion (4.81) will factorize into independent equations, one for each symmetry species. Furthermore, the equations for the *different* symmetry species

$$(\alpha, 1), (\alpha, 2), \ldots (\alpha, f_\alpha),$$

that is, for the different partners in the representation Γ_α, will all be the same. In other words, we have only one secular problem for each irreducible representation.

Corresponding to a particular root $E_1^{(\alpha)}$ of the secular equation for representation Γ_α we will have f_α orthogonal wave functions

$$\Psi_{1,h}^{(\alpha)} (h = 1, \ldots, f_\alpha)$$

with identical expansion coefficients $C_1^{(\alpha, \mu)}$. Thus

$$\Psi_{1,h}^{(\alpha)} = \sum_{\mu} C_1^{(\alpha, \mu)} \phi_h^{(\alpha, \mu)}. \tag{4.82}$$

These wave functions $\Psi_{1,h}^{(\alpha)}$ are clearly symmetry adapted.

Suppose, for example, that we are calculating molecular orbitals for NH_3 and are using some effective one-electron Hamiltonian (e.g. the Hartree–Fock Hamiltonian) which has the full symmetry of the nuclear framework. The transformation to the symmetry orbitals of Table 4.9 reduces the original 16 x 16 secular equation into one 1 x 1 equation for the A_2 orbitals, one 5 x 5 equation for the A_1 orbitals and two identical 5 x 5 equations for the E orbitals.

4.8. The symmetry paradox

It is necessary to close this chapter with a warning. In Section 4.3 we saw that the exact eigenfunctions of the electronic Schrödinger equation are necessarily symmetry adapted and in Section 4.7 we showed how to construct symmetry adapted functions from a set of arbitrary trial functions. However, it is *not* true that the result of a variational calculation, even on the ground electronic state of a molecule, is automatically symmetry adapted. Instead we have the following somewhat surprising situation called the symmetry paradox.

Let $\mathscr{S}$ be some symmetry condition which is satisfied by the exact eigenfunction Ψ of the Schrödinger equation for the ground electronic state of a molecule and let Φ be some variational trial wave function for this state. Then, in a variational calculation of the optimum form of Φ, the imposition of the symmetry condition $\mathscr{S}$ may have a catastrophic effect both on the calculated total electronic energy E and the optimum wave function Φ_0; if we relax the symmetry restraint $\mathscr{S}$ we may obtain a far better (lower) estimate of the total energy and a much better (asymmetrical) wave function Φ_{A}.

In principle, the symmetry paradox may always be resolved by applying the appropriate group theoretical projection operator to the asymmetrical wave function Φ_{A}. The best results are obtained if the variational parameters are revaried *after* projecting out the component of the trial wave function with the correct symmetry properties, but this procedure is sometimes difficult or even impracticable from the computational point of view.

The nature of the symmetry paradox and its resolution by projection operators is best brought out by some simple examples.

EXAMPLE 1. ONE-CENTRE, SCALED, FLOATING FUNCTION FOR THE GROUND STATE OF H_2^+

This example is so simple that we can deduce the qualitative behaviour of all parameters without carrying out any calculations. Nevertheless, the example illustrates all the important points which arise in more complicated situations.

The one-centre function is taken as the normalized Slater-type-orbital (STO)

$$\Phi = (\alpha^3/\pi)^{\frac{1}{2}}\, \mathrm{e}^{-\alpha r_{\mathrm{c}}} \tag{4.83}$$

where r_{c} is the distance of the electron from centre c and α is a variational scale factor.

Here the symmetry condition referred to in the above paradox is the evenness of the exact ground-state function Ψ under inversion through the centre of symmetry mid-way between the two protons a and b. That is

$$O_i \Psi = \Psi. \tag{4.84}$$

For the condition (4.84) to be satisfied by the trial function (4.83), the centre c must be located at the centre of symmetry for all values of the nuclear separation R. It is then easy to deduce the qualitative behaviour of the scale factor α and the electronic energy E as R varies from 0 to ∞. At $R = 0$, $\alpha = 2$ and (4.83) is the exact ground-state wave function of the united atom He^+ with $E = -2$ au. As R increases from 0 to ∞, α falls smoothly towards the asymptotic value *zero* and E rises smoothly to the asymptotic value *zero* (Fig. 4.6); that is, at $R = \infty$, the symmetry *restricted* variational function Φ leads to two bare protons and a free electron.

A much better wave function Φ_A for $R = \infty$ is immediately apparent, that in which the expansion centre c is located at one of the protons (say a) and the scale factor $\alpha = 1$; this gives the correct energy $E = -0.5$ au for dissociation into a ground-state hydrogen atom and a proton.

The qualitative behaviour of the parameter α, the energy E and the location of expansion centre c for an *unrestricted* variational calculation is now clear. For $R < R_c$, where R_c is some critical nuclear separation, the unrestricted wave function Φ' is, indeed, symmetry adapted with the expansion centre located at the centre of symmetry. As R increases beyond R_c, the central location of the expansion centre becomes a saddle point in the energy surface rather than the absolute minimum, and two true minima (with equal energies) develop, corresponding to a gradual drift of the expansion centre from the centre of symmetry towards one or other of the protons. As R varies from 0 to ∞, the scale factor α falls smoothly from 2 towards the asymptotic value 1 and E rises smoothly from -2 au at $R = 0$ to -0.5 au at $R = \infty$ (Fig. 4.6).

To obtain the *projected, unrestricted* function Φ'' we apply the group theoretical projection operator

$$\mathscr{P}_g = O_E + O_i = 1 + O_i, \tag{4.85}$$

obtained from the identity representation (A_g) of the group $I = \{E, i\}$, to (one of) the asymmetric function(s) Φ' corresponding to the true minima in the energy surface. Thus

$$\Phi'' = (1 + O_i)\Phi'. \tag{4.86}$$

Clearly for $R < R_c$ the operator $\mathscr{P}_g$ has no effect on Φ' (apart from normalization), but for $R > R_c$ we obtain the *two*-centre scaled, floating function for H_2^+ discussed in Chapter 3, Section 2(c), except that the values of the parameters α and x are not the optimum values.

Finally, by revarying the parameters α and x in Φ'' to minimize the energy, we find that the projected, revaried (or *extended*) function Φ''' is identical with the scaled, floating LCAO function of Section 3.2(c). The qualitative behaviour of the energy curves for the functions Φ, Φ', Φ'', Φ''' is shown in Fig. 4.6. Note that each step in the chain leading from Φ to Φ''' is accompanied by a decrease in

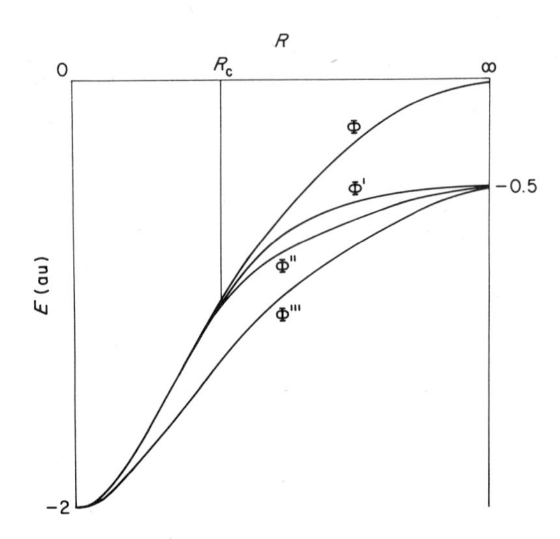

Fig. 4.6. Electronic energies of H_2^+ using simple wave functions. Φ: *restricted* one-centre, scaled floating function; Φ′: *unrestricted* one-centre, scaled floating function; Φ″: *projected unrestricted* one-centre, scaled floating function; Φ‴: *extended* one-centre, scaled floating function.

the total energy, at least for $R > R_c$, and that the energy error (0.5 au ≈ 13.6 eV) and wave function errors for the restricted function Φ are, indeed, catastrophic for large values of R. For values of R ranging from zero to somewhat greater than R_c, the energy lowerings are all quite small, but the changes in the wave function are substantial, for example, at R_c the value of $x \approx (R_c/2)$ for Φ'' is much greater than that for Φ''' ($\approx$0.08 au) (cf. Fig. 3.6).

The steps $\Phi \to \Phi' \to \Phi'' \to \Phi'''$, each accompanied by a lowering of the energy, occur in all instances of the symmetry paradox. The energy lowerings may be very large or very small (Example 3 below), but there is always a substantial change in some property of the wave function.

We note that, even in this very simple case, the final step in the chain ($\Phi'' \to \Phi'''$) involves a substantial increase in the computational effort, in that the determination of the optimum scaled, floating LCAO function requires the evaluation of a moderately complicated 3-centre integral. This, again, is typical; the last step in the chain is always the most difficult to achieve in practice.

EXAMPLE 2. THE LCAO-MO WAVE FUNCTION FOR THE GROUND STATE OF H_2

Since the qualitative features of this example are very similar to those of Example 1, and since the hydrogen molecule is discussed in detail in the next chapter, we merely summarize the principal results. For a detailed discussion, from a somewhat different point of view, the reader is referred to a paper by Coulson and Fischer.[6]

Here, the *restricted* function Φ is the usual simple product LCAO–MO function (unscaled and unfloated)

$$\Phi = (a(1) + b(1))\,(a(2) + b(2)). \tag{4.87}$$

The function Φ is a fair approximation for small values of R, but for $R > 2R_e$ it fails badly, the energy error at $R = \infty$ being 8.5 eV (cf. Section 5.3(a)).

Following Coulson and Fischer we consider the asymmetrical function

$$\Phi' = (a(1) + \lambda b(1))\,(\lambda a(2) + b(2)). \tag{4.88}$$

For $\lambda \neq 1$ the *unrestricted* function (4.88) loses two symmetry properties of the exact, ground-state, spatial, wave function Ψ, namely

$$O_i\Psi = \Psi, \tag{4.89}$$

where i is the simultaneous inversion through the centre of symmetry of the coordinates of both electrons 1 and 2, and

$$P(1, 2)\Psi = \Psi, \tag{4.90}$$

where $P(1, 2)$ is the operator which interchanges the spatial coordinates of electrons 1 and 2. The condition (4.90) characterizes a singlet spatial wave function for a two-electron system (Section 5.1). However, the situation is such that either of the conditions (4.89), (4.90) implies the other. We choose to work in terms of (4.90).

The variation of the optimum value of λ (equation (4.88)) with nuclear separation R is shown by the upper curve of Fig. 4.7. As in Example 1, the unrestricted function (4.88) is symmetrical ($\lambda = 1$) for $R < R_c$ ($\approx$2.3 au) while for $R > R_c$ λ falls rapidly, but smoothly to zero. The energy improvement in the step $\Phi \rightarrow \Phi'$ is very substantial for $R > R_c$; at $R = \infty$, Φ' becomes exact, which is an improvement of 8.5 eV. The projected, unrestricted function Φ'' is given by

$$\begin{aligned}\Phi''(1, 2) &= (1 + P(1, 2))\Phi'(1, 2)\\ &= (a(1) + \lambda b(1))(\lambda a(2) + b(2)) + (a(2) + \lambda b(2))(\lambda a(1) + b(1)).\end{aligned} \tag{4.91}$$

Finally the *extended* function $\Phi'''(1, 2)$ is obtained by revarying the parameter λ in Φ'' to minimize the energy. The energy depressions in the steps $\Phi' \rightarrow \Phi''$ and $\Phi'' \rightarrow \Phi'''$ are quite small ($<$ 0.5 eV), but the change in the wave function is substantial in both steps (Fig. 4.7).

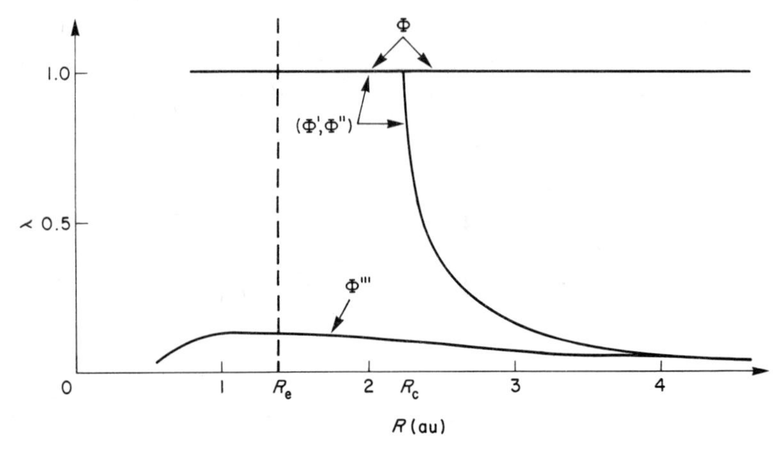

Fig. 4.7. Variation of λ for asymmetric molecular orbitals (ground state of H_2). Φ: restricted function; Φ′: unrestricted function; Φ″: projected unrestricted function; Φ‴: extended function.

Similar, but more complicated, manifestations of the symmetry paradox arise in the LCAO–MO and Hartree–Fock theories of all homonuclear diatomic molecules for large nuclear separations.[7] It should be noted, however, that the most widely used and efficient method for correcting the deficiencies of MO functions for large nuclear separations is *not* to relax the symmetry requirements and use projection operators, but to include configuration-interaction (CI).

EXAMPLE 3. OPEN-SHELL HARTREE-FOCK THEORY OF ATOMS AND MOLECULES

It is in this theory that the manifestations of the symmetry paradox are most troublesome and at the same time most important. Since this theory involves methods for treating the many-electron problem which we have not so far considered (Chapter 6), we confine ourselves here to a few general remarks.

(i) The qualitative situation is very similar to that in Examples 1 and 2. Indeed the terms *restricted, unrestricted, projected unrestricted* and *extended* first arose in discussions of the symmetry properties of Hartree–Fock wave functions.

(ii) All the energy lowerings in the steps $\Phi \rightarrow \Phi' \rightarrow \Phi'' \rightarrow \Phi'''$ are usually quite small (< 1 eV) and frequently very small indeed (< 0.01 eV). However, certain properties of the wave function, for example, the net spin density at the nuclei, change by orders of magnitude.

(iii) The extended Hartree–Fock theory (Φ'''), even for many-electron atoms, involves formidable computational problems and only a few complete calculations have been performed.[8, 9]

(iv) Our use of the term "paradox" is, perhaps, an innovation. Löwdin[10] in a recent discussion refers to the "symmetry dilemma" of Hartree–Fock theory.

References

1. M. Kotani, K. Ohno and K. Kayama (1961), "Quantum Mechanics of Electronic Structure of Simple Molecules", Handbuch der Physik, Vol. XXXVII/2, Springer-Verlag, Berlin.
2. E. P. Wigner (1959), "Group Theory and its Application to the Quantum Mechanics of Atomic Spectra", Academic Press, New York and London.
3. M. Hamermesh (1962), "Group Theory and its Application to Physical Problems", Addison-Wesley, London.
4. R. McWeeny (1960), "Symmetry—An Introduction to Group Theory", Pergamon Press, Oxford.
5. G. G. Hall (1967), "Applied Group Theory", Longmans, London.
5a. A. C. Hurley (1976), "Electron Correlation in Small Molecules", Academic Press, London and New York.
6. C. A. Coulson and I. Fischer (1949), *Phil. Mag.* **40**, 3861.
7. R. K. Nesbet (1961), *Phys. Rev.* **122**, 1497.
8. W. A. Goddard (1967), *Phys. Rev.* **157**, 93.
9. A. P. Jucys (1967), *Int. J. Quantum Chem.* **1**, 311.
10. P. O. Löwdin (1963), *Rev. Mod. Phys.* **35**, 496.

CHAPTER 5

The Hydrogen Molecule

Calculations on the hydrogen molecule have been of central importance in the development of quantum chemistry. On the one hand it has been possible, for this simple system, to calculate very accurate wave functions and hence verify that quantum mechanics provides an adequate basis for a quantitative theory of chemistry. On the other hand simpler, more approximate, calculations have been the starting point for general theories of valency applicable with varying degrees of rigour to much larger molecules.

5.1. Separation of the spin variables

For systems containing two or more electrons the electronic wave function must satisfy the Pauli principle, that is, it must be antisymmetric under the simultaneous exchange of the space and spin coordinates of any two electrons. For two-electron systems this requirement is easily satisfied.

From the four possible spin functions

$$\alpha(1)\alpha(2),\ \alpha(1)\beta(2),\ \beta(1)\alpha(2),\ \beta(1)\beta(2) \tag{5.1}$$

we may form one normalized antisymmetric function

$$\Theta(0,0) = \frac{1}{\sqrt{2}}\{\alpha(1)\beta(2) - \beta(1)\alpha(2)\} \tag{5.2}$$

and three normalized symmetric functions

$$\begin{aligned}\Theta(1,1) &= \alpha(1)\alpha(2)\\ \Theta(1,0) &= \frac{1}{\sqrt{2}}\{\alpha(1)\beta(2) + \beta(1)\alpha(2)\}\\ \Theta(1,-1) &= \beta(1)\beta(2).\end{aligned} \tag{5.3}$$

In equations (5.2) and (5.3), $\Theta(S, M_S)$ denotes a simultaneous eigenfunction of the square of total spin angular momentum $\mathbf{S}^2$ and its z-component S_z. Thus we have

$$\mathbf{S}^2\Theta(S, M_S) = S(S+1)\Theta(S, M_S), \tag{5.4}$$

$$S_z\Theta(S, M_S) = M_S\Theta(S, M_S). \tag{5.5}$$

These properties of the spin functions (5.2) and (5.3) may be verified using the general relations of Section 3.4, and the equations

$$\begin{aligned} S_x &= s_x(1) + s_x(2) \\ S_y &= s_y(1) + s_y(2) \\ S_z &= s_z(1) + s_z(2), \end{aligned} \tag{5.6}$$

which give the components of the total spin in terms of those for the individual electrons.

On the other hand, since the electronic Hamiltonian H is invariant under exchange of the spatial coordinates of the two electrons, the spatial wave functions Φ, obtained as solutions of the Schrödinger equation, must be either symmetric (Φ_S) or antisymmetric (Φ_A) under this exchange.

Total electronic wave functions satisfying the Pauli principle are then obtained by combining a spatially symmetric function with the antisymmetric spin function (5.2)

$$^1\Psi = \Phi_S \cdot \frac{1}{\sqrt{2}}\{\alpha(1)\beta(2) - \beta(1)\alpha(2)\}, \tag{5.7}$$

or by combining a spatially antisymmetric function with any of the three symmetric spin functions (5.3)

$$^3\Psi = \Phi_A \cdot \begin{cases} \alpha(1)\alpha(2) \\ \dfrac{1}{\sqrt{2}}[\alpha(1)\beta(2) + \beta(1)\alpha(2)] \\ \beta(1)\beta(2). \end{cases} \tag{5.8}$$

In equations (5.7) and (5.8) the superscripts on Ψ give the spin degeneracy $2S + 1$. States with $S = 0, \frac{1}{2}, 1, \ldots$ are referred to as singlets, doublets, triplets . . ..

Thus for two electron systems Pauli's principle does not lead to the exclusion of any solutions of the electronic Schrödinger equation; it merely requires that a symmetric space function Φ_S must be combined with a singlet spin function to give a non-degenerate total electronic wave function (if Φ_S is itself non-degenerate) whereas an antisymmetric space function Φ_A is combined with a triplet spin function to give a total electronic wave function which is at least three-fold degenerate.

5.2. The Heitler–London theory of H_2 and its extensions

The first successful treatment of the covalent bond was Heitler and London's[1] theory of the hydrogen molecule. Here, as in the LCAO treatment of H_2^+ in Chapter 3, we start with a wave function which is accurate for large nuclear separations. Let a denote a hydrogen 1*s* orbital on atom a, and b a hydrogen 1*s* orbital on atom b. The coordinates of the two electrons are denoted by 1 and 2. For large nuclear separations the ground state of the system will correspond to one electron on each atom and may be represented by either of the product wave functions $a(1)b(2)$ or $b(1)\,a(2)$. As the atoms approach one another the initial degeneracy between these two functions will be removed. The appropriate linear combinations, found either from symmetry or by solution of the 2 × 2 secular equation are

$$\Phi_{\pm} = \frac{a(1)\,b(2) \pm b(1)\,a(2)}{\sqrt{2(1 \pm S^2)}} \tag{5.9}$$

where the denominator is a normalization constant with

$$S = \int ab \, \mathrm{d}v. \tag{5.10}$$

The electronic Hamiltonian for the system is given by

$$H = -\tfrac{1}{2}\nabla_1^2 - \tfrac{1}{2}\nabla_2^2 - \frac{1}{r_{a1}} - \frac{1}{r_{b2}} - \frac{1}{r_{a2}} - \frac{1}{r_{b1}} + \frac{1}{r_{12}} + \frac{1}{R}. \tag{5.11}$$

The energies of the two functions (5.9) are obtained from the equation

$$E_{\pm} = \int \Phi_{\pm} H \Phi_{\pm} \, \mathrm{d}v_1 \, \mathrm{d}v_2 . \tag{5.12}$$

Making use of the fact that the orbitals a and b are eigenfunctions of the Hamiltonian for a hydrogen atom we may reduce equation (5.12) to the form

$$E_+ = -1 + \frac{Q + A}{1 + S^2}; \qquad E_- = -1 + \frac{Q - A}{1 - S^2} \tag{5.13}$$

where

$$Q = \iint a(1)\,b(2) \left\{ -\frac{1}{r_{b1}} - \frac{1}{r_{a2}} + \frac{1}{r_{12}} + \frac{1}{R} \right\} a(1)\,b(2) \, \mathrm{d}v_1 \, \mathrm{d}v_2, \tag{5.14}$$

and

$$A = \iint a(1)\, b(2) \left\{ -\frac{1}{r_{b1}} - \frac{1}{r_{a2}} + \frac{1}{r_{12}} + \frac{1}{R} \right\} b(1)\, a(2)\, dv_1\, dv_2 . \quad (5.15)$$

The quantities Q and A may be expressed in terms of one- and two-electron integrals over the atomic orbitals a and b. Thus we find

$$Q = -(\mathrm{b}\,|\,aa) - (\mathrm{a}\,|\,bb) + [aa\,|\,bb] + 1/R \quad (5.16)$$

and

$$A = -S(\mathrm{b}\,|\,ab) - S(\mathrm{a}\,|\,ab) + [ab\,|\,ab] + S^2/R \quad (5.17)$$

where

$$(\mathrm{a}\,|\,bb) = (\mathrm{b}\,|\,aa) = \int \frac{b^2}{r_\mathrm{a}}\, dv, \quad (5.18)$$

$$(\mathrm{a}\,|\,ab) = (b\,|\,ab) = \int \frac{ab}{r_\mathrm{a}}\, dv, \quad (5.19)$$

$$[aa\,|\,bb] = \iint a(1)\, a(1) \frac{1}{r_{12}}\, b(2)\, b(2)\, dv_1\, dv_2, \quad (5.20)$$

and

$$[ab\,|\,ab] = \iint a(1)\, b(1) \frac{1}{r_{12}}\, a(2)\, b(2)\, dv_1\, dv_2 . \quad (5.21)$$

We have already evaluated the one-electron integrals S, $(\mathrm{a}\,|\,bb)$ and $(\mathrm{a}\,|\,ab)$ in connection with the LCAO theory of H_2^+ (Section 3.2). The evaluation of the two-electron Coulomb integral (5.20) and especially the two-electron exchange integral (5.21) is more difficult. Indeed, in their original calculation, Heitler and London used an approximate estimate of the exchange integral (5.21). The calculation with accurate values of all the integrals was completed by Sugiura.[(2)]

To this day the difficulty in evaluating molecular integrals remains one of the principal limiting factors in the accurate calculation of wave functions. Progress in this field, largely due to the advent of high speed digital computers, is reviewed in Section 7.4.

All the integrals appearing in equations (5.16)-(5.21) are inherently positive. Calculations show that the first two terms in A outweigh the second two and that A is numerically larger than Q. Consequently the energy curve $E_+(R)$ is a stable one whereas $E_-(R)$ is purely repulsive. These curves are shown in Fig. 5.1 together with the experimental curve for the ground state.

The curves in Fig. 5.1 are labelled by the full symmetry of the two states. Clearly both Φ_+ and Φ_- are of symmetry Σ^+, Φ_+ is even

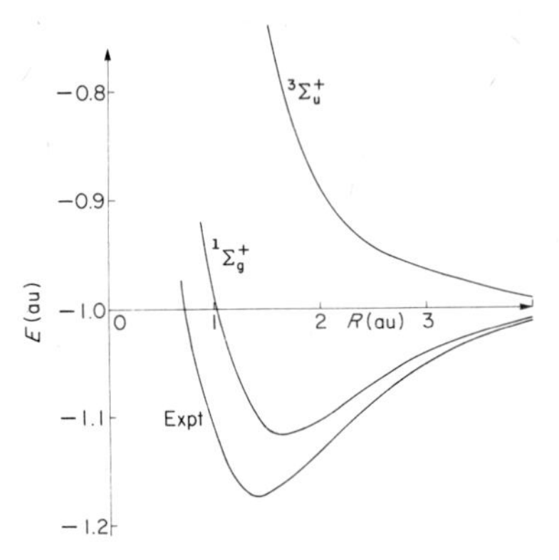

Fig. 5.1. Energy curves for the $^1\Sigma_g^+$ and $^3\Sigma_u^+$ states of H_2 given by the Heitler–London theory (equation (5.9)).

under inversion (g) and Φ_- is odd under inversion (u). Furthermore Φ_+ and Φ_- are, respectively, symmetric and antisymmetric under exchange of the two electrons and hence must be combined with singlet and triplet spin functions (Section 1). The wave functions including spin are

$$\Psi\,(^1\Sigma_g^+) = \Phi_+ \frac{1}{\sqrt{2}}\{\alpha(1)\beta(2) - \beta(1)\alpha(2)\}, \tag{5.22}$$

$$\Psi\,(^3\Sigma_u^+) = \Phi_- \begin{cases} \alpha(1)\alpha(2) \\ \dfrac{1}{\sqrt{2}}\,[\alpha(1)\beta(2) + \beta(1)\alpha(2)] \\ \beta(1)\beta(2). \end{cases} \tag{5.23}$$

From Fig. 5.1 we see that the Heitler–London function Φ_+ provides a qualitatively correct curve for the total electronic energy of H_2. The calculated binding energy and equilibrium nuclear separation are $D_e = 3.20$ eV and $R_e = 1.64$ au, to be compared with experimental values of $D_e = 4.75$ eV and $R_e = 1.40$ au.

If we use the Heitler–London function to calculate the variation of the kinetic and potential energies with nuclear separation, we obtain the results shown in Fig. 5.2. The curves obtained are quite different in form from those deduced in Chapter 2 from the virial theorem (Fig. 2.1). Thus, although the Heitler–London function gives qualitatively correct results for the total electronic energy it gives quite wrong results for the kinetic and potential energies.

The total electronic charge density for the Heitler–London function Φ_+ is given by the formula

$$\rho(1) = 2\int \Phi_+^2(1, 2)\,dv_2 = \frac{a^2(1) + b^2(1) + 2Sa(1)b(1)}{1 + S^2}. \tag{5.24}$$

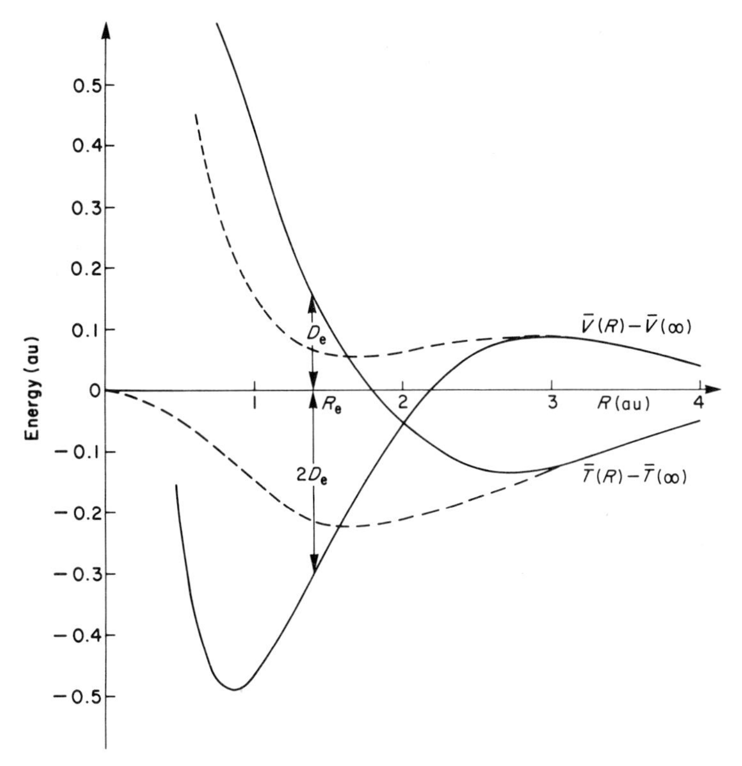

Fig. 5.2. Variation of kinetic and potential energies for ground state of H_2. - - - - -, Heitler-London function. ——, scaled Heitler-London (Wang) function.

If we substitute this charge density into the electrostatic formula (equation (2.30)) we may calculate the forces acting on the nuclei for all nuclear separations.[(3)] This calculation is very similar to that for H_2^+ described in Section 3.2(a). Again the results obtained are rather inaccurate (Fig. 5.3). We do obtain a stable H_2 molecule, but

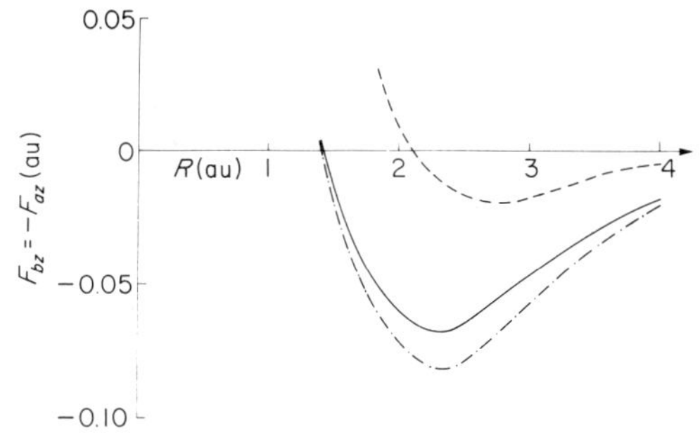

Fig. 5.3. Electrostatic force curves for ground state of H_2. - - - - -, Heitler-London function. ——, floating, scaled Heitler-London function. ·—·—·, exact.

the binding energy $D_e = 0.89$ eV, given by the area under the force curve, is much too small and the equilibrium nuclear separation $R_e = 2.1$ au is too large.

We next consider the two improvements of the Heitler–London wave function that are required to satisfy the virial and electrostatic theorems.

(a) THE SCALED HEITLER–LONDON FUNCTION (WANG[4] FUNCTION)

In order to satisfy the virial theorem, we introduce a scale factor α into the wave function and determine α by minimizing the total electronic energy. The procedure is very similar to that for H_2^+ described in Section 3.2(b). Indeed equations (3.42)–(3.46), which

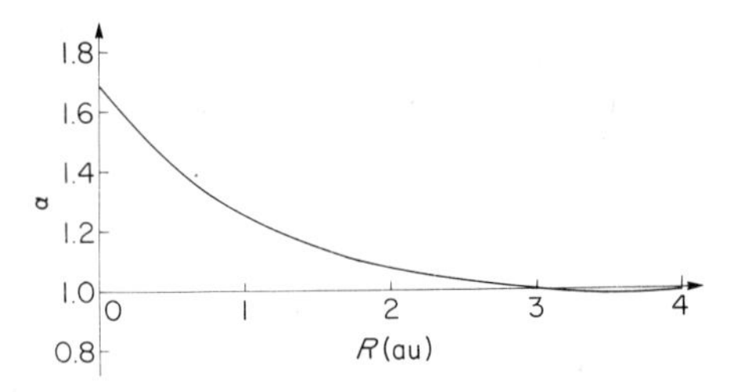

Fig. 5.4. Variation of the scale factor α for Wang's wave function.

determine the optimum value of α, are equally applicable to the present case. This refinement leads to $D_e = 3.78$ eV at $R_e = 1.41$ au, a considerable improvement on the Heitler–London result.

The effect of scaling on the kinetic energy $\overline{T}$ and the potential energy $\overline{V}$ is very much larger than the effect on the total electronic energy. The variation of $\overline{T}$ and $\overline{V}$ with nuclear separation is drastically altered (Fig. 5.2) and is brought into line with the behaviour deduced in Chapter 2 from quite general considerations (Fig. 2.1).

The variation of the parameter α with nuclear separation is shown in Fig. 5.4. As R decreases from infinity, α first falls very slightly below 1, then, at $R \approx 3$ au, α rises rapidly and smoothly towards the value $\alpha = \frac{27}{16} = 1.6875$ appropriate to the united atom (He).

(b) THE FLOATING WANG FUNCTION[5–7]

Again, as for H_2^+, we detach the atomic orbitals from the nuclei and let them take up optimum positions (Fig. 3.5). The wave function now depends on two variable parameters, the scale factor α and the inward displacement of the atomic orbitals x. Minimization of the

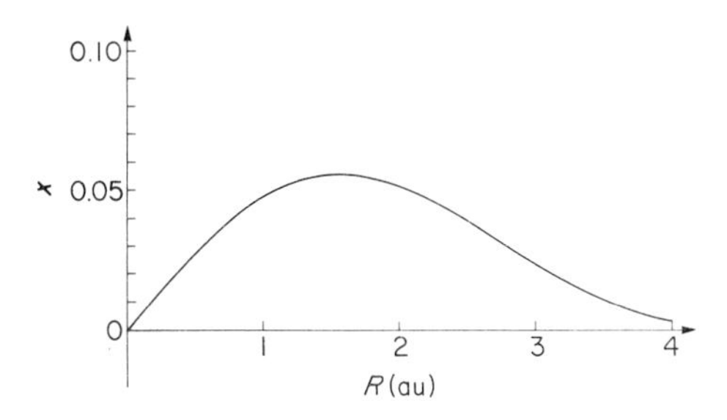

Fig. 5.5. Variation of x for floating Wang function.

total electronic energy leads to a variation of α with R very similar to that for the Wang function (Fig. 5.4) and to a variation x with R shown in Fig. 5.5.

We see from Fig. 5.5 that the displacement x of the atomic orbitals from the nuclei is quite small, never exceeding 0.06 au. The calculated equilibrium constants $D_e = 3.93$ eV, $R_e = 1.42$ au also differ little from the Wang values. However, the use of floating orbitals has a very drastic effect on the force curve calculated from the electrostatic formula (Fig. 5.3). With floating orbitals the electrostatic force curve agrees quite well with the experimental curve. From the general discussion of Chapter 2 we know, in fact, that it must agree exactly with the derivative of the energy curve for the floating Wang function. Conversely, the equilibrium nuclear separation $R_e = 1.42$ au for the floating Wang function is given by the point where the electrostatic force curve crosses the R-axis, and the binding energy $D_e = 3.93$ eV is given by the area under the electrostatic force curve.

The calculations of the last two paragraphs provide striking examples of the way in which application of the variation method can make great differences in some properties of the wave function without making much change in the total energy. This is to be expected from the variational principle; the error in the total energy is of second order in the wave function errors, whereas the errors in other calculated quantities are of the first order. It is evident that the electrostatic formula is particularly sensitive to small changes in the wave function. A recent perturbation calculation by Hirschfelder and Eliason[7a] provides another illustration of this extreme sensitivity. These authors consider the long-range interaction of two ground-state hydrogen atoms. The wave function is expanded in a series in powers of $1/R$,

$$\Psi = \psi_0 + \psi_3 R^{-3} + \psi_4 R^{-4} + \psi_5 R^{-5} + \psi_6 R^{-6} + \psi_7 R^{-7} + \ldots, \tag{5.25}$$

where R is the nuclear separation and the first term ψ_0 is just the

Heitler–London function Φ_+. Corresponding to the expansion (5.25) we have the expansion

$$E_{a,b} = C_6 R^{-6} + C_8 R^{-8} + \ldots \tag{5.26}$$

for the interaction energy and the expansion

$$F_{az} = -6C_6 R^{-7} - 8C_8 R^{-9} \ldots \tag{5.27}$$

for the force on nucleus a in the z-direction (towards nucleus b). Hirschfelder and Eliason show that, if only the first two terms in the expansion (5.25) are used, the expectation value of the electronic Hamiltonian gives the correct value of the leading coefficient C_6 of equation (5.26). However, if the same approximate wave function $\psi_0 + \psi_3 R^{-3}$ is substituted in the electrostatic formula (equation (2.30)) we obtain $C_6 = 0$. Similarly, if we use the approximate wave function $\psi_0 + \psi_3 R^{-3} + \psi_4 R^{-4}$ the electrostatic formula gives only 7% of the correct value of C_6, even though the expectation value of the Hamiltonian gives both C_6 and C_8 exactly. Indeed 93% of the electrostatic calculation of C_6 comes from that part of the electron charge density which arises from the product of ψ_0 and ψ_7.

These calculations verify the conjecture which Feynman[7b] made in his original paper. The electron associated with nucleus a has a charge density which is distorted and shifted towards nucleus b. The positively-charged nucleus together with this negatively-charged electron cloud form a dipole whose moment varies as R^{-7}. The attraction of the nucleus to the centre of its associated electron cloud is the largest contribution to the force on the nucleus.

(c) THE WEINBAUM[8] FUNCTION

The Heitler–London function Φ_+ of equation (5.9) is not the only function of the appropriate symmetry for the ground state of H_2 which can be formed from the two atomic orbitals a and b. We have also the ionic function

$$\Phi_I = \frac{1}{\sqrt{2}}\,[a(1)\,a(2) + b(1)\,b(2)], \tag{5.28}$$

which shows both electrons in the same atomic orbital. An improved wave function may now be obtained by combining this ionic function with the (unnormalized) Heitler–London, or covalent function

$$\Phi_C = \frac{1}{\sqrt{2}}\,[a(1)\,b(2) + b(1)\,a(2)]. \tag{5.29}$$

The wave function Φ for the ground state is then given by

$$\Phi = N[\Phi_C + C\Phi_I], \tag{5.30}$$

and the electronic energy E and the value of the parameter C are determined by the linear variation method (Section 2.2). In this simple case the equations determining E and C reduce to

$$\begin{aligned}(H_{CC} - ES_{CC}) + C(H_{CI} - ES_{CI}) &= 0 \\ (H_{IC} - ES_{IC}) + C(H_{II} - ES_{II}) &= 0\end{aligned} \tag{5.31}$$

and

$$\begin{vmatrix} H_{CC} - ES_{CC}, & H_{CI} - ES_{CI} \\ H_{IC} - ES_{IC}, & H_{II} - ES_{II} \end{vmatrix} = 0. \tag{5.32}$$

The matrix elements appearing in these equations are easily reduced to expressions involving one and two electron integrals over the basic atomic orbitals a and b. In addition to the integrals already encountered in the Heitler–London calculation we require values for the two-electron one centre integral

$$[aa \mid aa] = \iint \frac{a^2(1)\, a^2(2)}{r_{12}}\, dv_1\, dv_2 = \tfrac{5}{8} \tag{5.33}$$

and the so-called hybrid integral

$$[aa \mid ab] = \iint a(1)\, a(1) \frac{1}{r_{12}} a(2)\, b(2)\, dv_1\, dv_2. \tag{5.34}$$

Equation (5.30) provides the simplest example of a phenomenon often referred to as "ionic–covalent resonance". In this description the wave function Φ_C is associated with the "purely covalent" structure H–H and the function Φ_I is associated with the "ionic" structures H^+H^- and H^-H^+. The state represented by equation (5.30) is then said to arise from resonance between these three structures. In using this description it is important to bear in mind that the covalent and ionic structures have no independent existence and that (5.30) is a time-independent wave function.

The improvement in the Heitler–London function represented by equation (5.30) may be combined with those of sub-sections (a) and (b) by scaling and floating the atomic orbitals a and b. The wave function then contains three independent parameters; the scale factor α, the inward displacement of the atomic orbitals x and the mixing parameter C. If all these parameters are chosen to minimize the total electronic energy we find[7] $R_e = 1.428$ au, $D_e = 4.09$ eV, $\alpha = 1.191$, $x = 0.036$, $C = 0.1326$. The variation of the parameters α and x with nuclear separation is similar to that shown in Figs 5.4 and 5.5. As R increases C falls smoothly from a maximum value near R_e and tends to zero for large nuclear separations. Thus as $R \to \infty$, $\alpha \to 1$, $x \to 0$, $C \to 0$ and the three parameter wave function reduces to the Heitler–London function.

From the general discussion of Section 2.4 it is clear that the three parameter wave function will satisfy both the virial and electrostatic theorems for all nuclear separations.

(d) THE COULSON-FISCHER FORM AND ORTHOGONALIZED ATOMIC ORBITALS

There are several alternative ways of expressing the wave function (5.30), which, in the valence bond theory, expresses covalent–ionic resonance. These alternative descriptions lead to no new physical results for H_2, but have proved useful in extensions of the Heitler–London method to larger molecules.

Coulson and Fischer[9] pointed out that, instead of adding an ionic term to the covalent function (5.29), we may achieve the same result by replacing the atomic orbitals a and b by the more flexible polarized atomic orbitals

$$A = \frac{a + \lambda b}{\sqrt{1 + 2\lambda S + S^2}}, \qquad B = \frac{b + \lambda a}{\sqrt{1 + 2\lambda S + S^2}}. \tag{5.35}$$

These expressions are chosen to ensure that the functions A and B, like the original functions a and b, are interchanged under inversion. The denominators ensure that A and B are normalized.

If we substitute the expressions (5.35) into equation (5.29) and collect like terms we obtain the wave function

$$\Phi_C' = N'[(1 + \lambda^2)\Phi_C + 2\lambda\Phi_I]. \tag{5.36}$$

The functions (5.36) and (5.30) are seen to be identical provided that

$$C = \frac{2\lambda}{1 + \lambda^2} \tag{5.37}$$

and the constants N, N' are chosen so that both functions are normalized.

The variation of λ with nuclear separation is of some interest. If λ is chosen to minimize the total energy its behaviour is qualitatively similar to that of C. At infinite separation λ is zero and we have the Heitler–London ground-state function, which is correct in this limit. As R decreases, λ rises smoothly to a maximum value $\lambda \approx 0.12$ near $R_e \approx 1.4$ au.

Two other choices of λ may be used to relate the Coulson–Fischer form (5.36) to other descriptions. If we set $\lambda = 1$ for all nuclear separations, corresponding to $C = 1$ in equation (5.30), we obtain

$$\Phi_{MO} = N''[\Phi_C + \Phi_I] \tag{5.38}$$

$$= N''[a(1) + b(1)]\,[a(2) + b(2)] \tag{5.39}$$

which is just the LCAO molecular orbital function for the ground state of H_2 (cf. Section 5.3 below). Thus the molecular orbital function, like the valence bond function, is a special case of the Coulson–Fischer form.

We see from the above analysis that the Heitler–London function provides a better overall description of the ground state of H_2 than does the LCAO molecular orbital function. Even near equilibrium where λ and C assume their maximum values (≈ 0.12 and 0.24 respectively) the Heitler–London function with $\lambda = C = 0$ is much nearer the optimum function than is the molecular orbital function with $\lambda = C = 1$. We may say that, whereas the Heitler–London function neglects any contribution from the ionic structures H^+H^- and H^-H^+, the molecular orbital function grossly exaggerates their role by including them with a weight equal to that of the covalent structure. The relative advantage of the Heitler–London description increases steadily with increasing nuclear separation and as $R \to \infty$ it gives the correct dissociation products $H(1s) + H(1s)$, whereas the molecular orbital description fails badly at large R values and gives on dissociation an equal mixture of the states $H(1s) + H(1s)$ and $H^+ + H^-(1s)^2$.

These shortcomings of the molecular orbital wave function may also be discussed in terms of electron correlation. Since Φ_{MO} of equation (5.39) is just a product of two orbitals, the relative probability of finding electron 1 at different positions in space is independent of the position of electron 2. In other words there is no correlation between the positions of the two electrons. Physically we expect, however, that if electron 2 is at a certain point, say P, the probability of finding electron 1 near P will be reduced because of the Coulomb repulsion. The covalent function Φ_C describes a strong correlation of this type, for when one electron is near nucleus a (in the a orbital) the other is necessarily in the b orbital near nucleus b. The admixture of the ionic function Φ_I makes the correlation less prominent, and when Φ_C and Φ_I are included with equal weight, as in Φ_{MO}, the correlation disappears.

In making these comparisons between Φ_C and Φ_1 are included with remember that both functions provide rather crude approximations to the true physical situation and that the electron correlation incorporated in Φ_C (left–right correlation along the bond) is only one of several types of correlation manifested by an accurate wave function.

Another interesting way of fixing the Coulson–Fischer parameter λ is to require that the orbitals of equation (5.35) be orthogonal to each other. Denoting these orthogonalized atomic orbitals by A^o and B^o we require

$$\int A^o(1)B^o(1)\, dv_1 = \frac{S + 2\lambda/(1+\lambda^2)}{1 - 2S\lambda/(1+\lambda^2)} = 0. \qquad (5.40)$$

Equation (5.40) leads to two values for λ

$$\lambda = \frac{-1 \pm \sqrt{1 - S^2}}{S}. \qquad (5.41)$$

One of these values of λ is the reciprocal of the other. We therefore obtain the same pair of functions A^o and B^o for each root; if we change from one root to the other we merely interchange A^o and B^o. If we require $|\lambda| < 1$, so that orbital A^o will be located mainly on atom a and B^o on atom b we obtain

$$\lambda = \frac{-1 + \sqrt{1 - S^2}}{S} = -S/2 + \ldots \qquad (5.42)$$

for small values of S.

The use of orthogonalized atomic orbitals like A^o and B^o in valence bond theory was suggested by Löwdin.[10] In many-electron systems the matrix elements required for a calculation of the energy and other physical quantities are very much simpler to evaluate if the basic orbitals are chosen to be orthogonal. For this reason it is of interest to evaluate the energy curve for the ground state of H_2 using the wave function

$$\Phi_C^o = \frac{1}{\sqrt{2}} [A^o(1)B^o(2) + B^o(1)A^o(2)], \qquad (5.43)$$

obtained from the Heitler–London function Φ_C by replacing the atomic orbitals a and b by their orthogonalized counterparts. This calculation has been carried through by Slater.[11] He obtained a purely repulsive curve lying somewhat above the ${}^3\Sigma_u^+$ curve in Fig. 5.1. In order to obtain binding with orthogonal atomic orbitals it is essential to include the ionic term

$$\Phi_I^o = \frac{1}{\sqrt{2}} [A^o(1)A^o(2) + B^o(1)B^o(2)]. \qquad (5.44)$$

The total wave function

$$\Phi = \Phi_C^o + C^o \Phi_I^o$$

with an optimum choice of C^o is then identical with (5.30).

We see then that the use of orthogonalized atomic orbitals has a drastic effect on valence bond wave functions. The ionic terms which were previously a minor refinement become the sole source of binding and cannot be omitted.

The physical reason for this becomes clear if we look at the total electronic charge density (equation (5.24)). For orthogonalized orbitals $S = 0$ and we obtain simply

$$\rho(1) = (A^{\circ}(1))^2 + (B^{\circ}(1))^2. \qquad (5.45)$$

There is no overlap term of the form $A^{\circ}(1)B^{\circ}(1)$ in equation (5.45). Just as in the case of H_2^+ (Chapter 3) it is overlap terms of this type which lead to a concentration of charge in the region of low potential energy between the nuclei and a stable molecule. If orthogonalized atomic orbitals are used, overlap terms of the type $A^{\circ}(1)B^{\circ}(1)$ arise solely as off-diagonal terms linking the covalent and ionic terms in the wave function.

All the qualitative conclusions of this sub-section are unaltered if the basic atomic orbitals are scaled and floated in order to satisfy the virial and electrostatic theorems.

5.3. The molecular orbital method[12–14]

The molecular orbital (MO) theory of the electronic structure of molecules is a natural extension of the orbital theory of atomic structure to molecules. Each electron is pictured as moving independently in an effective one-electron potential due to the nuclei and the averaged effect of the other electrons. Most early applications were of a qualitative nature and were concerned with the possible symmetries and degeneracies of the electronic levels. In such applications the precise specification of the one-electron potential is unnecessary; the major qualitative features follow from the symmetry of the potential, which is assumed to be the same as that of the nuclear framework. The one-electron space wave functions or molecular orbitals must then transform according to the irreducible representations of the molecular symmetry group (Chapter 4). In general this implies that these orbitals will extend around all the nuclei in the molecule. To obtain the wave function of the ground state of the molecule we simply allocate the available electrons to the molecular orbitals of lowest energy, subject to the restriction, imposed by the Pauli exclusion principle, that at most two electrons can occupy each orbital. Excited electronic states are similarly described by promotion of one or more electrons to orbitals of higher energy. As in atomic theory, a specific allocation of the electrons to the available molecular orbitals, which defines an *electron configuration*, leads, in general, to several possible electronic wave functions or *states*.

Let us apply this method to the H_2 molecule. Here the effective one-electron potential will have the same symmetry as that of the nuclear potential in $H_2^+(D_{\infty h})$ and we expect that the two lowest MOs will have the same symmetries σ_g and σ_u as in that case. Abbreviating these orbitals to g and u, we have three possible configurations for the two electrons of H_2, namely g^2, gu and u^2.

The lowest configuration g^2 leads to only one spatial wave function

$$\Phi(g^2) = g(1)g(2), \qquad {}^1\Sigma_g^+ \tag{5.46}$$

which is symmetric under interchange of the two electrons. It must therefore be combined with the antisymmetric spin function (Section 5.1) to yield a total wave function which is of symmetry ${}^1\Sigma_g^+$.

The configuration gu leads to two spatial wave functions, one symmetric (spin-singlet) and the other antisymmetric (spin-triplet) under electron exchange; both are of spatial symmetry Σ_u^+.

$$\Phi^1(gu) = \frac{1}{\sqrt{2}}[g(1)u(2) + u(1)g(2)], \qquad {}^1\Sigma_u^+ \tag{5.47}$$

$$\Phi^3(gu) = \frac{1}{\sqrt{2}}[g(1)u(2) - u(1)g(2)], \qquad {}^3\Sigma_u^+. \tag{5.48}$$

Finally the configuration u^2 leads only to a singlet function, which is of symmetry ${}^1\Sigma_g^+$ since simultaneous inversion of the coordinates of the two electrons leaves the product $u(1)u(2)$ unaltered.

$$\Phi(u^2) = u(1)u(2), \qquad {}^1\Sigma_g^+. \tag{5.49}$$

If we write the electronic Hamiltonian (omitting the nuclear repulsion term) in the form

$$H = h(1) + h(2) + \frac{1}{r_{12}} \tag{5.50}$$

where

$$h(1) = -\tfrac{1}{2}\nabla_1^2 - \frac{1}{r_{a1}} - \frac{1}{r_{b1}} \tag{5.51}$$

$$h(2) = -\tfrac{1}{2}\nabla_2^2 - \frac{1}{r_{a2}} - \frac{1}{r_{b2}}, \tag{5.52}$$

we can express the energies associated with the functions (5.46)–(5.49) in terms of one- and two-electron integrals over the molecular orbitals g and u. These expressions are

$$g^2 \quad {}^1E(g^2) = 2I_g + J_{gg} \qquad {}^1\Sigma_g^+ \tag{5.53}$$

$$gu \begin{cases} {}^3E(gu) = I_g + I_u + J_{gu} - K_{gu} & {}^3\Sigma_u^+ \quad (5.54) \\ {}^1E(gu) = I_g + I_u + J_{gu} + K_{gu} & {}^1\Sigma_u^+ \quad (5.55) \end{cases}$$

$$u^2 \quad {}^1E(u^2) = 2I_u + J_{uu} \qquad {}^1\Sigma_g^+ \tag{5.56}$$

where

$$I_i = (i \mid h \mid i) = \int i^* h i \, \mathrm{d}v \qquad (i = g, u) \tag{5.57}$$

$$J_{ij} = (ij \mid ij) = [ii \mid jj] = \iint \frac{i^*(1)j^*(2)i(1)j(2)}{r_{12}} \, \mathrm{d}v_1 \, \mathrm{d}v_2 \qquad (i, j = g, u) \tag{5.58}$$

$$K_{gu} = (gu \mid ug) = [gu \mid ug] = \iint \frac{g^*(1)u^*(2)u(1)g(2)}{r_{12}} \, \mathrm{d}v_1 \, \mathrm{d}v_2 . \tag{5.59}$$

The two-electron integrals J and K are referred to as Coulomb and exchange integrals, respectively. Note the ordering of the orbitals and electron variables in equations (5.58) and (5.59). The ordering and complex conjugation in the round bracket expressions correspond to those in the Dirac bra-ket expression $\langle pq \mid 1/r_{12} \mid pq \rangle$, whereas in the square bracket notation the vertical bar separates functions of electron 1 from functions of electron 2. For four, distinct, complex orbitals the definitions become

$$(ij \mid kl) = [ik \mid jl] = \iint \frac{i^*(1)j^*(2)k(1)l(2)}{r_{12}} \, \mathrm{d}v_1 \, \mathrm{d}v_2 . \tag{5.60}$$

Both notations are widely used, sometimes with minor variations. For instance, in the definitions of Parr[(15)] the round and square brackets appearing in equation (5.60) are interchanged.

The square bracket notation is most useful for physical interpretation; the integral $[ij \mid kl]$ is just the classical electrostatic interaction between two charge densities $[ij]$ and $[kl]$.

(a) THE LCAO APPROXIMATION

The expressions (5.53)–(5.56) for the energies of the low-lying states of H_2 apply whatever the form of the two molecular orbitals g and u. In order to carry the calculation further we must assume some specific forms for these orbitals. The simplest forms to take are the LCAO approximations used in Section 3.2 for H_2^+. From equations (3.14)–(3.18) we obtain the normalized orbitals

$$g = \frac{a + b}{\sqrt{2(1 + S)}} \tag{5.61}$$

and

$$u = \frac{(a - b)}{\sqrt{2(1 - S)}}. \tag{5.62}$$

Substituting these MOs into equations (5.57)–(5.59) we obtain expressions for I_g, I_u, J_{gg}, J_{gu}, K_{gu} and J_{uu} in terms of integrals over the atomic orbitals a and b. For example, the Coulomb integral J_{gg} appearing in the energy of the ground state ${}^1\Sigma_g^+(g^2)$ (equation (5.53)) reduces to

$$\begin{aligned} J_{gg} = [gg \mid gg] &= \frac{[a + b, a + b \mid a + b, a + b]}{4(1 + S)^2} \\ &= \frac{[aa \mid aa] + 4[aa \mid ab] + [aa \mid bb] + 2[ab \mid ab]}{2(1 + S)^2} \end{aligned}$$

where we have simplified the final expression using obvious equalities among the atomic integrals. The atomic integrals required to complete the calculation are the same as those encountered in Section 5.2. The resulting energy curves are shown in Fig. 5.6. These curves include the nuclear repulsion term $1/R$.

In the short-hand notation that has been developed to describe various approximate wave functions, the curves of Fig. 5.6 result from an LCAO–MO–SCF calculation. The letters SCF stand for self-consistent field and imply that the coefficients of the atomic orbitals in the expansions of the MOs are chosen to minimize the total energy; in this particular case these coefficients are fixed by symmetry. It has recently proved possible to carry through complete calculations at about this level of accuracy for polyatomic molecules of moderate size (ethane, benzene, pyridine, etc.) as well as for many-electron diatomic molecules. It is therefore important to study the results of Fig. 5.6 carefully and compare them with the experimental properties of the various states.

First we consider the ground-state curve ${}^1\Sigma_g^+(g^2)$. Near equilibrium the behaviour is qualitatively correct. The calculations predict a stable molecule with about the right bond length (R_e (calc) = 1.6 au, R_e (exp) = 1.4 au) but the calculated binding energy D_e = 2.65 eV is only about 55% of the observed value and considerably worse than the Heitler–London value of 3.20 eV. At larger nuclear separations the ground-state curve rapidly becomes less accurate and the

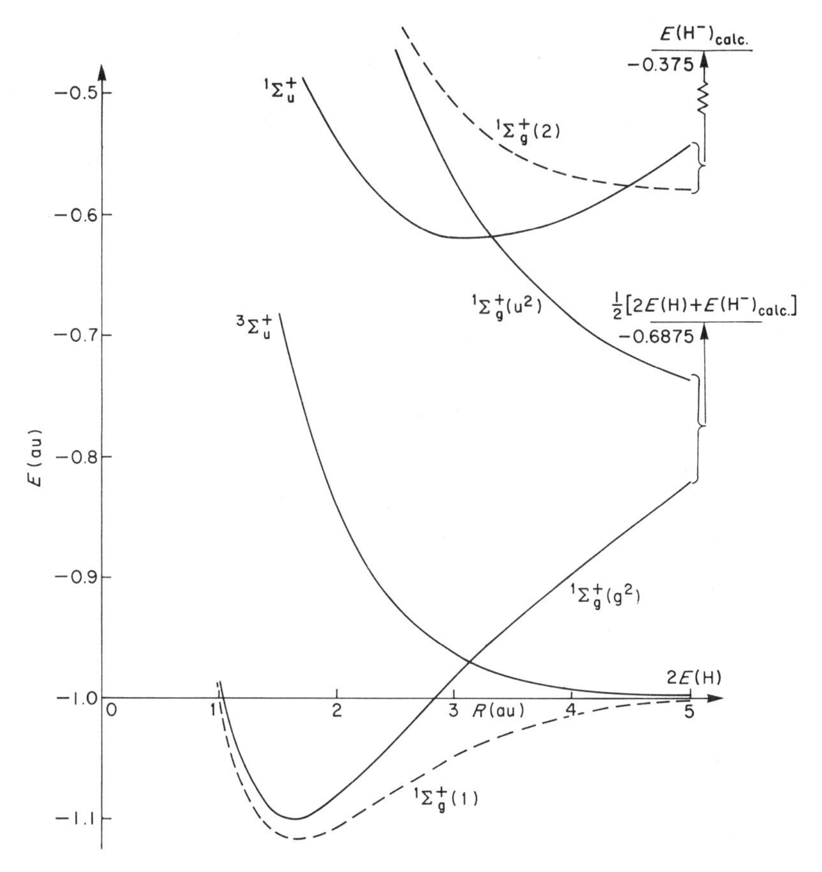

Fig. 5.6. Molecular orbital (MO) energy curves for H_2. ——, without CI. - - - - -, with CI.

description breaks down completely for R larger than about 3 au (twice the bond length). For infinite nuclear separation the curve tends to an incorrect dissociation limit half-way between the correct limit H(1s) + H(1s) and the calculated value of the limit $H^+ + H^-(1s)^2$; the error in the energy at $R = \infty$ being 8.5 eV. As already mentioned this large error is due to the neglect of any correlation between the two electrons. Nearly all the error arises from the molecular orbital form itself (equation (5.46)) and not from the LCAO approximation. If an optimum form is used for the molecular orbital σ_g the error at $R = \infty$ still amounts to 7.74 eV.[16]

The behaviour of the excited state curves ${}^3\Sigma_u^+$ and ${}^1\Sigma_u^+$ is again qualitatively, but not quantitatively, correct. The ${}^1\Sigma_u^+$ state is correctly predicted to be stable with an equilibrium nuclear separation considerably greater than that for the ground state. The calculated excitation energy $T_e = 13.0$ eV and $R_e = 3$ au are in fair agreement with the experimental values $T_e = 11.4$ eV and $R_e = 2.44$ au.[17] The repulsive state $^3\Sigma_u^+$ is also quite well represented; the error near equilibrium is considerably less than for the ground state and the dissociation limit is correct. We notice that the energy curves of the two states $^3\Sigma_u^+$, $^1\Sigma_u^+$ arising from the configuration gu are quite different in shape. This behaviour is exceptional. In general, different states arising from the same electronic configuration have energy curves which are approximately parallel (at least near equilibrium) and have similar R_e and ω_e values.

(b) CONFIGURATION INTERACTION

The worst feature of the MO results obtained above is clearly the incorrect behaviour of $^1\Sigma_g^+$ curves for large nuclear separation. This gross error can be removed by allowing for interaction between the two $^1\Sigma_g^+$ functions $^1\Sigma_g^+(g^2)$, $^1\Sigma_g^+(u^2)$. That is, we should use a wave function of the form

$$\Phi = \Phi(g^2) + \mu\Phi(u^2) \tag{5.63}$$

and determine the coefficient μ by the linear variation method. We then obtain two modified energy curves $^1\Sigma_g^+(1)$ and $^1\Sigma_g^+(2)$ as the roots of the secular equation

$$\begin{vmatrix} {}^1E(g^2) - E, & H_{gu} \\ H_{ug}, & {}^1E(u^2) - E \end{vmatrix} = 0. \tag{5.64}$$

The off-diagonal matrix elements of the Hamiltonian appearing in equation (5.64) are easily obtained in terms of integrals over the molecular and atomic orbitals. Thus we obtain

$$\begin{aligned} H_{gu} = H_{ug} &= \iint g(1)g(2)Hu(1)u(2)\, dv_1\, dv_2 \\ &= [gu\,|\,gu] \\ &= \frac{[aa\,|\,aa] - [aa\,|\,bb]}{2(1-S^2)}. \end{aligned} \tag{5.65}$$

The two roots of the secular equation (5.64) are shown as the dashed curves $^1\Sigma_g^+(1)$ and $^1\Sigma_g^+(2)$ in Fig. 5.6. We see that the configuration interaction (CI) has removed the gross errors for large nuclear separations, and that the dissociation limit of the ground state is now obtained correctly.

The other states ($^1\Sigma_u^+$ and $^3\Sigma_u^+$) shown in Fig. 5.6 have unique symmetries, so that no further configuration interaction effects are possible if we restrict our attention to the two MOs σ_g and σ_u.

The curve $^1\Sigma_g^+(2)$ obtained from the higher root of equation (5.64) bears little relationship to any of the excited $^1\Sigma_g^+$ states of

H_2 which have been observed spectroscopically.[17] This behaviour is typical; a simple LCAO-MO-CI calculation based on the atomic orbitals of the valence shell (here $1s_a$ and $1s_b$) is never adequate for any but the few lowest excited states (here $^3\Sigma_u^+$, $^1\Sigma_u^+$).

(c) RELATIONSHIP WITH VB THEORY

The MO wave functions which we have been using may all be related to the VB functions of Section 5.2. Consider first the $^3\Sigma_u^+$ function (equation (5.48)). Substituting the LCAO forms (5.61) and (5.62), expanding and collecting like terms we obtain

$$\Phi^3(gu) = \frac{\{a(1)+b(1)\}\{a(2)-b(2)\} - \{a(1)-b(1)\}\{a(2)+b(2)\}}{\sqrt{2(1-S^2)}}$$
$$= -\frac{a(1)b(2)-b(1)a(2)}{\sqrt{2(1-S^2)}}. \qquad (5.66)$$

Apart from an unimportant reversal in sign, the result (5.66) is seen to be identical with the VB wave function Φ_- of equation (5.9). The LCAO-MO and VB descriptions of the $^3\Sigma_u^+$ state are thus identical.

Similarly the LCAO-MO function for the $^1\Sigma_u^+$ state (equation (5.47)) yields a unique, purely ionic, VB wave function.

$$\Phi^1(gu) = \frac{a(1)a(2)-b(1)b(2)}{\sqrt{2(1-S^2)}}. \qquad (5.67)$$

For the ground $^1\Sigma_g^+$ state the relationship is a little more complicated. Expansion of the single-configuration LCAO-MO wave function (5.46) leads to an equal mixture of covalent and ionic VB functions

$$\Phi(g^2) = \frac{\{a(1)b(2)+b(1)a(2)\} + \{a(1)a(2)+b(1)b(2)\}}{2(1+S)}. \qquad (5.68)$$

As was pointed out previously, it is this overweighting of the ionic terms which leads to the failure of the MO method for large nuclear separations. However, if we expand the two-configuration wave function (5.63), we find

$$\Phi(g^2) + \mu\Phi(u^2) = N[(1-\mu')\{a(1)b(2)+b(1)a(2)\} + (1+\mu') \times \{a(1)a(2)+b(1)b(2)\}] \qquad (5.69)$$

where

$$\mu' = \mu\frac{1+S}{1-S}. \qquad (5.70)$$

The relative contributions of the ionic and covalent terms are now at our disposal through the parameter μ. As $R \to \infty$ $S \to 0$ and the optimum value of $\mu \to -1$. The ionic term, therefore, drops out and we obtain the correct dissociation limit. Indeed the wave function (5.69) is simply another way of writing the Weinbaum function (5.30) which describes resonance between covalent and ionic VB functions. In terms of electron correlation we may say that admixture of the function $\Phi(u^2)$ to the MO ground configuration $\Phi(g^2)$ introduces a correlation between the two electrons along the bond direction (left–right correlation).

It is often stated that the MO and VB theories of molecular electronic structure are equivalent if each is carried far enough. This statement can be interpreted in two different ways. On the one hand we have the somewhat trivial interpretation that each is a first-order treatment capable, in principle, of being extended to give results of any desired accuracy; hence they must ultimately become equivalent. The equivalence established above for the low-lying states of H_2 is much more precise; it implies a one–one correspondence of the wave functions of the two theories at a relatively low level of accuracy. It applies specifically to the LCAO form of the molecular orbital method (with configuration interaction) and, although the equivalence may be extended to all molecular systems, it is subject to two important provisos:

(i) For each molecular state the same atomic orbitals must be used for all MOs and all VB structures. Thus for the ground $^1\Sigma_g^+$ state of H_2 the simple relationship established by equations (5.30) and (5.69) is destroyed if different scale factors (orbital exponents) are used in the atomic orbitals for the covalent and ionic VB functions. However, the wave functions may be improved by scaling, floating (to satisfy the virial and electrostatic theorems), hybridization, etc. without destroying the simple relationships provided that, for each molecular electronic state, the same AOs are used throughout.

(ii) To establish the equivalence it is, in general, necessary to include *all* molecular orbital configurations and *all* valence bond functions which can be formed from the chosen set of atomic orbitals. In special cases this requirement can be relaxed somewhat by utilizing the symmetry of the molecular electronic state and imposing restrictions of double occupancy on inner shell AOs. Nevertheless, as we shall see in later chapters, the number of functions involved increases very rapidly with the number of valence electrons.

An example of the complications which arise when different atomic orbitals are used in the LCAO expressions for different MOs, is provided by the calculations of Huzinaga[(18)] and Phillipson and Mulliken[(19)] on the $^1\Sigma_u^+$ excited state of H_2. These authors use an

MO wave function of the form given by equations (5.47), (5.61) and (5.62) except that the orbital exponents ζ and ζ' of the σ_g and σ_u orbitals are varied independently to minimize the total electronic energy. Thus their spatial wave function is given by the equations

$$\Phi(gu') = \frac{1}{\sqrt{2}} [g(1)u'(2) + u'(1)g(2)], \tag{5.71}$$

where

$$g = \frac{a + b}{\sqrt{2(1 + S)}}, \qquad S = \int ab \, dv, \tag{5.72}$$

$$u' = \frac{a' - b'}{\sqrt{2(1 - S')}}, \qquad S' = \int a'b' \, dv, \tag{5.73}$$

with

$$a = \left(\frac{\zeta^3}{\pi}\right)^{\frac{1}{2}} e^{-\zeta r_a}, \qquad b = \left(\frac{\zeta^3}{\pi}\right)^{\frac{1}{2}} e^{-\zeta r_b} \tag{5.74}$$

$$a' = \left(\frac{\zeta'^3}{\pi}\right)^{\frac{1}{2}} e^{-\zeta' r_a}, \qquad b' = \left(\frac{\zeta'^3}{\pi}\right)^{\frac{1}{2}} e^{-\zeta' r_b}. \tag{5.75}$$

The energy obtained for the $^1\Sigma_u^+$ state is greatly improved by this refinement. Table 5.1 shows the results obtained by Phillipson and Mulliken at several internuclear separations including $R_e(^1\Sigma_g^+) = 1.40$ au and $R_e(^1\Sigma_u^+) = 2.44$ au.

Table 5.1. Calculated energy curves[a] for $^1\Sigma_u^+$ state of H_2 (19)

R (au)	ζ	ζ'	$E(\zeta \neq \zeta')$	E (exp.)	$E(\zeta = \zeta' = 1)$
1.40	1.450	0.275	8.912	8.0	17.71
1.93	1.244	0.314	7.846	6.7	12.80
2.44	1.147	0.349	7.872	6.5	11.01
3.31	1.010	0.401	8.468	6.8	10.48

[a] Energies are in electron volts relative to the energy of two ground-state H atoms.

From Table 5.1 we see that the error in the energy is reduced from values of from 4–10 eV to a nearly constant value of about 1 eV. The values of the orbital exponents ζ and ζ' are also interesting. The values of ζ indicate a slightly contracted σ_g orbital, while those of ζ' indicate a greatly expanded σ_u' orbital. The "size" of this orbital, estimated from the reciprocals of the ζ' values, is roughly three times that for $\zeta = 1$.

Let us now find the VB equivalent of the wave function (5.71). Substituting the LCAO forms of g and u', expanding and collecting terms we find

$$\Phi(gu') = N[\Psi_I - \Psi_C] \tag{5.76}$$

where

$$\Psi_I = \{a(1)a'(2) + a'(1)a(2)\} - \{b(1)b'(2) + b'(1)b(2)\}, \tag{5.77}$$

and

$$\Psi_C = \{a(1)b'(2) + b'(1)a(2)\} - \{b(1)a'(2) + a'(1)b(2)\}. \tag{5.78}$$

The function Ψ_I is the usual ionic function (equation (5.67)) except that the closed-shell ion $H^-(1s)^2$ has been replaced by the open shell ion $H^-(1s1s')$. This replacement leads to a marked improvement in the calculated energy of the dissociation limit $H^+ + H^-$ shown in Fig. 5.6. We now obtain -0.5133 au for this limit, to be compared with the accurate value of -0.5276 au and the $\zeta = \zeta' = 1$ value of -0.375 au shown in Fig. 5.6. A large fraction of this improvement is due to the in-out radial correlation introduced by the open-shell function.[20]

The second term Ψ_C of equation (5.76) is of a novel type. For $\zeta = \zeta'$ it vanishes, and we have the usual purely ionic wave function. For $\zeta \neq \zeta'$, Ψ_C is unaltered by replacing a' by $a'' = a' - \alpha a$ and b' by $b'' = b' - \alpha b$, where α is chosen so that

$$\int aa''\,dv = \int bb''\,dv = 0. \tag{5.79}$$

It is now clear that Ψ_C describes a covalent bond between two neutral hydrogen atoms; one in the $1s$ state and the other in the $2s$ state. Admittedly a'' and b'' are not exact hydrogenic $2s$ functions, but for optimum values of ζ, ζ' they are accurate approximations. Indeed, for $\zeta = 0.9590$, $\zeta' = 0.2807$, Ψ_C gives a value of -0.6189 au[21] for the dissociation limit $H(1s) + H(2s)$; quite close to the accurate value of -0.625 au.

The MO wave function (5.76) thus leads to a mixture of the ionic state $H^- + H^+$ and the covalent state $H(1s) + H(2s)$ on dissociation. As in the case of the simple MO function for the ground state, this incorrect behaviour can be eliminated by introducing configuration interaction. The second configuration required is clearly

$$\Phi(g'u) = \frac{1}{\sqrt{2}}[g'(1)u(2) + u(1)g'(2)] \tag{5.80}$$

where

$$g' = \frac{a' + b'}{\sqrt{2(1 + S')}}, \qquad u = \frac{a - b}{\sqrt{2(1 - S)}}. \tag{5.81}$$

This molecular orbital configuration interaction calculation is entirely equivalent to using the VB wave function

$$\Psi = \lambda_I \Psi_I + \lambda_C \Psi_C, \tag{5.82}$$

which describes resonance between the ionic structure H^-H^+ and the covalent structure H(1*s*)—H(2*s*). As $R \to \infty$ we find $\lambda_I \to 0$ and the $^1\Sigma_u^+$ state energy curve goes to the correct dissociation limit H(1*s*) + H(2*s*).

The optimum values of the three independent parameters ζ, ζ' and λ_I/λ_C in the wave function (5.82) show interesting behaviour as functions of the nuclear separation R. At $R \approx 4.8$ au and $R \approx 14$ au there are points of discontinuity where the optimum parameters change suddenly.[21a] These sudden changes are most easily interpreted in terms of the energies $E(gu')$ and $E(g'u)$ of the MO wave functions (5.71) and (5.80). For both large (>14 au) and small (<4.8 au) nuclear separations $E(gu') < E(g'u)$, whereas for $4.8 < R < 14$ we have $E(g'u) < E(gu')$. There is a simple physical basis for this seemingly strange behaviour; in the intermediate range $4.8 < R < 14$ it is energetically more favourable to have one "bonding" electron in the expanded σ_g' orbital (which involves appreciable overlap between the atomic orbitals a' and b') and one "non-bonding" electron in σ_u (which involves negligible overlap) than it is to have one "antibonding" electron (in σ_u') and one non-bonding electron (in σ_g). On the other hand it is clear, from a glance at the united atom He($1s2p_\sigma$) and from a simple calculation of the asymptotic behaviour of the optimum parameters,[21b] that $E(gu') < E(g'u)$ both for very small and for very large values of R.

(d) HARTREE-FOCK ORBITALS

As we have already mentioned, equations (5.53)–(5.56) for the energies of the various MO states apply whatever the form of the molecular orbital. For the ground state we have from equation (5.46) and (5.53)

$$\Phi = g(1)g(2) \tag{5.83}$$

$$E = 2I_g + J_{gg}. \tag{5.84}$$

It is natural to seek the best numerical orbital to use in equation (5.83), that is, to vary the function g continuously, subject to the normalization condition

$$\int g^2 \, dv = 1, \tag{5.85}$$

in such a way as to minimize the electronic energy of equation (5.84). This is the self-consistent field method of Hartree and Fock.

In this simple case the original Hartree and more refined Hartree–Fock procedures become identical. Incorporating the normalization condition (5.85) by a Lagrangian multiplier -2ϵ, we require

$$\delta I = \delta\left[2I_g + J_{gg} - 2\epsilon \int g^2 \, dv\right] = 0, \tag{5.86}$$

for unrestricted variations of g.

Assuming, without loss of generality, that g is real, and using the Hermitian property of the one-electron operator h of equations (5.51) and (5.52), we obtain

$$\delta I_g = 2 \int \delta g(1) h(1) g(1) \, dv_1 \tag{5.87}$$

and

$$\delta \int g^2 \, dv = 2 \int \delta g(1) g(1) \, dv_1. \tag{5.88}$$

From the definition (5.58) of J_{gg} there follows

$$\delta J_{gg} = 4 \int \delta g(1) \left[\int \frac{g(2)g(2)}{r_{12}} \, dv_2\right] g(1) \, dv_1, \tag{5.89}$$

$$= 4 \int \delta g(1) J_g(1) g(1) \, dv_1. \tag{5.90}$$

Here we have introduced the *Coulomb operator* J_g, whose effect on any function ϕ is defined by

$$J_g(1)\phi(1) = \left[\int \frac{g(2)g(2)}{r_{12}} \, dv_2\right] \phi(1). \tag{5.91}$$

Substituting the results (5.87), (5.88) and (5.90) into equation (5.86), we obtain

$$\int \delta g(1) [h(1) + J_g(1) - \epsilon] g(1) \, dv_1 = 0. \tag{5.92}$$

Since the variation $\delta g(1)$ is arbitrary, the condition (5.92) leads to the integro-differential equation

$$[h(1) + J_g(1)] g(1) = \epsilon g(1) \tag{5.93}$$

for the determination of the optimum function g. Defining the one-electron Fock operator F by the equation

$$F(1) = h(1) + J_g(1), \tag{5.94}$$

equation (5.93) becomes

$$F(1) g(1) = \epsilon g(1). \tag{5.95}$$

We note that, despite appearances, equation (5.95) is not a simple one-electron Schrödinger equation, since the operator F depends on the solution g via the operator J_g. An iterative method of solution is, therefore, indicated. In atomic problems, where the orbitals depend essentially on a single variable, it is feasible to carry out the iterations by numerical integration of the differential equation; and this is probably the most efficient procedure.(22) However, for molecules the orbitals depend essentially on two or three variables and a direct numerical solution is much more difficult. Nevertheless, approximate solutions have been obtained by expressing the orbitals as analytic functions with a sufficient number of adjustable parameters and determining these parameters from the variation principle. A convenient analytic form is a linear expansion in terms of a fixed set of basis functions χ_p. Thus we take

$$g = \sum_{q=1}^{M} \chi_q(1) c_q. \tag{5.96}$$

If the functions χ_p form a complete set, then we may expect that the expansion (5.96) will converge to the exact solution as $M \to \infty$.

The variation δg is now expressed in terms of changes in the linear expansion coefficients

$$\delta g = \sum_{q=1}^{M} \chi_q \, \delta c_q. \tag{5.97}$$

Substituting equations (5.96) and (5.97) into equation (5.92), and equating to zero the coefficient of δc_q we obtain

$$\sum_q (F_{pq} - \epsilon S_{pq}) c_q = 0, \qquad (p = 1 \ldots M) \tag{5.98}$$

where

$$S_{pq} = \int \chi_p \chi_q \, dv \tag{5.99}$$

and

$$F_{pq} = I_{pq} + \int \chi_p J_g \chi_q \, dv \tag{5.100}$$

$$= I_{pq} + [pq \mid gg] \tag{5.101}$$

$$= I_{pq} + \sum_{r=1}^{M} \sum_{s=1}^{M} [pq \mid rs] \, c_r c_s. \tag{5.102}$$

Equation (5.98) has the appearance of a matrix eigenvector equation. The condition determining the eigenvalues is

$$| F_{pq} - \epsilon S_{pq} | = 0. \tag{5.103}$$

However, equation (5.98), like (5.95), is in reality non-linear since the matrix F_{pq} depends on the solution vector c_q via equation (5.102).

Again an iterative procedure can be used. From a guessed vector c_q^0 $(q = 1, \ldots, M)$ one calculates the matrix F_{pq}^0 from equation (5.102), solves the matrix equations (5.103) and (5.98) and uses the resulting vector c_q^1 $(q = 1, \ldots, M)$ to compute a new matrix F_{pq}^1. This procedure is repeated until the input and output matrices agree to an acceptable degree of accuracy. In practice various special devices, such as extrapolating from the vectors of earlier iterations, may be used to accelerate convergence.

The approximate SCF ground state orbital is obtained by substituting the lowest root of equation (5.103) into equation (5.98). For a basis consisting of M functions there will be $M - 1$ other roots of equation (5.103). These yield the so-called "virtual" orbitals, which, in the case of larger diatomic and polyatomic molecules, have often been used to describe excited states or to improve the description of the ground state by configuration interaction. Since the matrices F and S are Hermitian these virtual orbitals have the convenient property of being orthogonal to each other and to the ground-state orbital. Nevertheless, they are not really suitable either as excited-state orbitals or as "correlation" orbitals for improving the ground-state function. Indeed, there are strong indications from atomic calculations that, for neutral systems, there are *no* discrete solutions of the *numerical* Hartree–Fock equations (5.95) other than the ground-state orbitals; all other eigenvalues lie in the continuum. The virtual orbitals are, therefore, an artifact of using a finite expansion in equation (5.96), and, although, for a severely restricted basis set, they may accidentally mimic the properties of excited state or correlation orbitals they have no chemical or physical significance. These deficiencies in the virtual orbitals are unimportant if *all* configurations formed from virtual orbitals are used in a configuration interaction calculation; in such applications equations (5.103) and (5.98) simply provide a convenient orthogonalization procedure.

Any convenient set of analytic functions χ_q $(q = 1 \ldots M)$ may be used as the basis set in equation (5.96). The set should be complete and it should be capable of representing the numerical Hartree–Fock orbitals accurately with as small a number of terms as possible, that is, it should provide rapid convergence. The first accurate approximation to the Hartree–Fock orbitals of the ground state of H_2 was obtained by Coulson,[23] who used an expansion in elliptical coordinates (μ, ν, φ) of the form

$$g = e^{-\alpha\mu} \sum_{l,m} c_{lm}\, \mu^l \nu^{2m}. \tag{5.104}$$

The orbital he obtained at the equilibrium nuclear separation is

$$g(\mathrm{r}) = 0.87758\, e^{-0.75\mu} \{1 + 0.27787\nu^2 - 0.12863\mu + 0.12503\mu^2 - 0.039589\mu\nu^2\}. \tag{5.105}$$

Coulson's calculation has been extended by Kolos and Roothaan.[24] These authors used more terms in the expansion (5.104) and considered a wide range of nuclear separations. They obtained a binding energy $D_e = 3.64$ eV at the calculated equilibrium separation $R_e = 1.38$ au.

In recent discussions of accurate atomic and molecular wave functions the concept of correlation energy has played an important role. The *correlation energy* is defined[16] as the difference between the energy obtained in a Hartree–Fock calculation and the exact eigenvalue of the non-relativistic Schrödinger equation for the same state. For systems, like H_2, involving light atoms, relativistic effects are very small and the correlation energy may be obtained from a comparison of Hartree–Fock results with experimental energies.

Figure 5.7, taken from the work of Kolos and Roothaan,[24] shows the correlation energy of the ground state of H_2 as a function

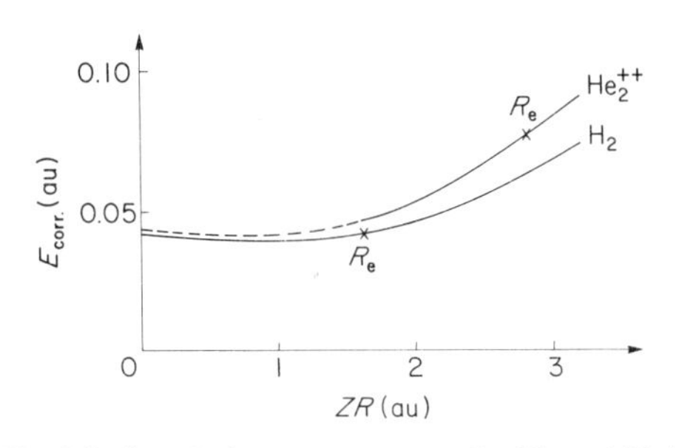

Fig. 5.7. Correlation energy curves for H_2 and He_2^{++}.

of nuclear separation. We see that the correlation energy is almost constant in the range from $R = 0$, corresponding to the united atom $He(1s)^2$, to around $R = R_e = 1.40$ au. For larger nuclear separations the correlation energy starts to rise rapidly and as $R \to \infty$ it approaches the high value 7.74 eV already quoted in Section 5.3(a). Figure 5.7 also shows the corresponding curve for the isoelectronic system He_2^{++}. Notice that the scale of the abscissa has been doubled for the He_2^{++} curve. With this choice of scale the two curves are very similar. Figure 5.7 exhibits a general property of the correlation energy which has been found from the study of a wide variety of atomic and molecular systems (cf. Section 7.4). In the absence of effects arising from the degeneracy or near degeneracy of orbital energies, the correlation energy is rather insensitive to quite large changes in parameters such as nuclear charges and nuclear separations. For a pair of electrons in a well-localized orbital the correlation energy amounts to rather more than 1 eV ($\approx$23 kcal). The steep rise in the correlation energy of H_2 for $R > R_e$ is not inconsistent with this general property of the correlation energy; it is attributed to the approaching degeneracy of the orbitals σ_g and σ_u.

(e) OPTIMUM DOUBLE CONFIGURATIONS; THE (*A*, *B*) FORM

The Hartree–Fock wave function discussed above breaks down for large nuclear separations in just the same way as its LCAO counterpart of Section 5.3(a). The obvious way of removing this deficiency is to extend the Hartree–Fock method by including a contribution from the excited configuration $u(1)u(2)$ of equation (5.49). To accomplish this we assume a wave function of the form

$$\Phi(1,2) = A_1 g(1)g(2) + A_2 u(1)u(2) \tag{5.106}$$

and minimize the expectation value of the Hamiltonian with respect to variations of (i) the linear expansion coefficients A_1 and A_2, (ii) the numerical functions g and u. We assume without loss of generality that A_1, A_2, g and u are real and that the variations of g and u are such as to retain σ_g and σ_u symmetry respectively.

The linear expansion coefficients A_1 and A_2 are subject to the normalization condition

$$A_1^2 + A_2^2 = 1. \tag{5.107}$$

The energy corresponding to the wave function (5.106) is given by the equation

$$E = A_1^2 E(g^2) + A_2^2 E(u^2) + 2A_1 A_2 H_{gu}, \tag{5.108}$$

where from equations (5.53), (5.56) and (5.65)

$$E(g^2) = 2I_g + J_{gg}, \tag{5.109}$$

$$E(u^2) = 2I_u + J_{uu}, \tag{5.110}$$

and

$$H_{gu} = [gu\,|\,gu] = K_{gu}. \tag{5.111}$$

Minimization of E with respect to the linear expansion coefficients A_1 and A_2, subject to the normalization condition (5.107) leads to the usual eigenvector equations

$$\begin{aligned} \{E(g^2) - E\}A_1 + H_{gu}A_2 &= 0 \\ H_{ug}A_1 + \{E(u^2) - E\}A_2 &= 0. \end{aligned} \tag{5.112}$$

We determine the orbitals g and u by minimization of (5.108), subject to the orbital normalization conditions

$$\int g^2\, \mathrm{d}v = \int u^2\, \mathrm{d}v = 1. \tag{5.113}$$

The restrictions (5.113) are incorporated using the method of Lagrange multipliers. That is, we require

$$\delta I = \delta\left(E - 2A_1\epsilon_g \int g^2\, \mathrm{d}v - 2A_2\epsilon_u \int u^2\, \mathrm{d}v\right) = 0 \tag{5.114}$$

for unrestricted variations of the orbitals g and u, which preserve σ_g, σ_u symmetry. In more general applications of this multi-configurational SCF theory off-diagonal Lagrange multipliers must be included to preserve the orthogonality of the orbitals. Here, however, the different symmetries of the orbitals ensure orthogonality and we need only the diagonal Lagrange multipliers ϵ_g and ϵ_u.

For the variations of the first two terms in equation (5.108) we find

$$\delta E(g^2) = 4 \int \delta g(1)\{h(1) + J_g(1)\}g(1)\, dv_1 \tag{5.115}$$

$$\delta E(u^2) = 4 \int \delta u(1)\{h(1) + J_u(1)\}u(1)\, dv_1 \tag{5.116}$$

whereas for the third term we find

$$\delta H_{gu} = \delta K_{gu} = 2 \int \delta g(1)\left\{\int \frac{u(2)g(2)}{r_{12}}\, dv_2\right\}u(1)\, dv_1 + 2 \int \delta u(1)\left\{\int \frac{g(2)u(2)}{r_{12}}\, dv_2\right\} g(1)\, dv_1. \tag{5.117}$$

Equation (5.117) may be re-expressed using *exchange operators* K_g and K_u, whose effect on a general function ϕ is defined by the equations

$$K_g(1)\phi(1) = \left\{\int \frac{g(2)\phi(2)}{r_{12}}\, dv_2\right\} g(1), \tag{5.118}$$

$$K_u(1)\phi(1) = \left\{\int \frac{u(2)\phi(2)}{r_{12}}\, dv_2\right\} u(1). \tag{5.119}$$

We thus obtain

$$\delta K_{gu} = 2 \int \delta g(1)K_u(1)g(1)\, dv_1 + 2 \int \delta u(1)K_g(1)u(1)\, dv_1. \tag{5.120}$$

Substituting the results (5.115), (5.116) and (5.120) into equation (5.114) we find the variational condition

$$0 = \delta I = 4A_1 \int \delta g(1)[F_g(1) - \epsilon_g g(1)]\, dv_1 + 4A_2 \int \delta u(1)[F_u(1) - \epsilon_u u(1)]\, dv_1. \tag{5.121}$$

Here the Fock-type operators F_g and F_u are given by the equations:

$$F_g(1) = A_1\{h(1) + J_g(1)\} + A_2 K_u(1) \tag{5.122}$$

$$F_u(1) = A_2\{h(1) + J_u(1)\} + A_1 K_g(1). \tag{5.123}$$

The variation condition (5.121) leads to two equations which closely resemble the Hartree–Fock equation (5.95), namely

$$F_g(1)g(1) = \epsilon_g g(1) \tag{5.124}$$

$$F_u(1)u(1) = \epsilon_u u(1). \tag{5.125}$$

Equations (5.124) and (5.125) form a pair of integro-differential equations for the orbitals. They are coupled through the exchange operators K_u and K_g of equations (5.122) and (5.123), and the linear expansion coefficients A_1 and A_2 of equations (5.112). In more complicated multi-configuration SCF calculations further couplings arise from off-diagonal Lagrange multipliers. A set of equations equivalent to (5.112), (5.124) and (5.125) was first derived by Lennard-Jones and Pople.[25] Like the Hartree–Fock equations themselves they are difficult to solve directly in this form, since the orbitals g and u depend essentially on two variables.

However, approximate solutions may again be obtained by assuming linear expansions of the orbitals in terms of a fixed set of basis functions. Let us assume that we use M such functions in all; the first M_g functions being of σ_g symmetry and the remaining $M - M_g$ functions being of σ_u symmetry. We then obtain

$$g(1) = \sum_{p=1}^{M_g} \chi_p(1)c_p = \sum_p^g \chi_p(1)c_p, \tag{5.126}$$

$$u(1) = \sum_{p=M_g+1}^{M} \chi_p(1)c_p = \sum_p^u \chi_p(1)c_p. \tag{5.127}$$

In the second form of writing these equations we have abbreviated the summation symbols. The notations Σ_p^g, Σ_p^u indicate that we sum over basis function χ_p of g, u symmetry, respectively.

The variations of the orbitals are now expressed by

$$\delta g(1) = \sum_p^g \chi_p(1)\delta c_p, \tag{5.128}$$

and

$$\delta u(1) = \sum_p^u \chi_p(1)\delta c_p. \tag{5.129}$$

Substituting equations (5.126)–(5.129) into the variational condition (5.121), and equating to zero the coefficients of each δc_p we obtain the matrix equivalents of the Hartree–Fock equations (5.124) and (5.125), namely

$$\sum_q^g (F_{g,pq} - \epsilon_g S_{pq})c_q = 0 \qquad (p = 1, \ldots, M_g), \tag{5.130}$$

and

$$\sum_q^u (F_{u,pq} - \epsilon_u S_{pq})c_q = 0 \qquad (p = M_g + 1, \ldots, M). \quad (5.131)$$

Here S_{pq} is the overlap matrix

$$S_{pq} = \int \chi_p \chi_q \, dv,$$

and the Fock matrices $F_{g,pq}$ and $F_{u,pq}$ may be expanded in terms of integrals over the basis orbitals χ_p thus

$$F_{g,pq} = A_1\left\{I_{pq} + \sum_r^g \sum_s^g [pq \mid rs] c_r c_s\right\} + A_2 \sum_r^u \sum_s^u [pr \mid sq] c_r c_s, \quad (5.132)$$

$$F_{u,pq} = A_2\left\{I_{pq} + \sum_r^u \sum_s^u [pq \mid rs] c_r c_s\right\} + A_1 \sum_r^g \sum_s^g [pr \mid sq] c_r c_s. \quad (5.133)$$

There is a simple expression for the total energy E in terms of the Lagrange multipliers ϵ_g and ϵ_u. Multiplying equations (5.124) and (5.125) by $A_1 g(1)$ and $A_2 u(1)$, respectively, integrating, summing the results and comparing with equation (5.108) we obtain

$$E = A_1(\epsilon_g + A_1 I_g) + A_2(\epsilon_u + A_2 I_u). \quad (5.134)$$

In the special case of the Hartree–Fock theory itself ($A_1 = 1$, $A_2 = 0$, $\epsilon_g = \epsilon$) equation (5.134) reduces to

$$E = \epsilon + I_g. \quad (5.135)$$

To carry through an optimized double-configuration calculation by the expansion method we need to satisfy simultaneously equations (5.112), (5.130) and (5.131). This is achieved by an iterative procedure similar to that employed in Hartree–Fock theory. First one guesses starting vectors c_p^0 ($p = 1, \ldots, M_g$) and c_p^0 ($p = M_g + 1, \ldots, M$) for the coefficients in equations (5.126) and (5.127). The corresponding orbitals g^0 and u^0 are used to calculate $E(g^2)$, $E(u^2)$, H_{ug} and hence approximations A_1^0 and A_2^0, to A_1 and A_2 from equation (5.112). These values and the c_p^0 are then substituted into equations (5.132) and (5.133) to obtain first approximations $F_{g,pq}^0$ and $F_{u,pq}^0$ to the Fock matrices. The final step in the iteration is the solution of the pseudo-eigenvector equations (5.130) and (5.131) to obtain ϵ_g^0, ϵ_u^0 and improved expansion vectors c_p^1.

There is an alternative way of writing the double-configuration function (5.106) which corresponds to the Coulson–Fischer form of the Weinbaum calculation. Calculations show that, if the constant A_1 in equation (5.106) is chosen positive, the optimum value of A_2 is negative. We may therefore write equation (5.106) in the form

$$\Phi(1, 2) = n_1^2 g(1)g(2) - n_2^2 u(1)u(2), \quad (5.136)$$

where n_1, n_2 are real and positive. The normalization condition (5.107) then becomes

$$n_1^4 + n_2^4 = 1. \tag{5.137}$$

We now define two normalized orbitals A and B by the equations

$$A = \frac{n_1 g + n_2 u}{(n_1^2 + n_2^2)^{\frac{1}{2}}} \tag{5.138}$$

$$B = \frac{n_1 g - n_2 u}{(n_1^2 + n_2^2)^{\frac{1}{2}}}. \tag{5.139}$$

The orbital A is mainly localized near nucleus a and B is localized near nucleus b. Under inversion through the molecular mid-point we have

$$g \leftrightarrow g, \qquad u \leftrightarrow -u, \qquad A \leftrightarrow B. \tag{5.140}$$

The orbitals A and B are not orthogonal; their overlap S is given by the equation

$$S = \int AB \, dv = \frac{n_1^2 - n_2^2}{n_1^2 + n_2^2}. \tag{5.141}$$

Inverting the relations (5.138) and (5.139) we have

$$g = \frac{(n_1^2 + n_2^2)^{\frac{1}{2}}}{2n_1}(A + B) = \frac{A + B}{\sqrt{2(1 + S)}}, \tag{5.142}$$

$$u = \frac{(n_1^2 + n_2^2)^{\frac{1}{2}}}{2n_2}(A - B) = \frac{A - B}{\sqrt{2(1 - S)}}, \tag{5.143}$$

and the wave function (5.136) may be written

$$\Phi(1, 2) = \frac{n_1^2 + n_2^2}{4}[\{A(1) + B(1)\}\{A(2) + B(2)\} - \{A(1) - B(1)\}$$

$$\times \{A(2) - B(2)\}] = \frac{A(1)B(2) + B(1)A(2)}{\sqrt{2(1 + S^2)}}, \tag{5.144}$$

which is just the original Heitler–London function (5.9) with optimum orbitals A and B replacing the atomic orbitals a and b.

Thus the optimum double-configuration wave function (5.106), which is determined by equations (5.112), (5.124) and (5.125) (or equations (5.112), (5.130) and (5.131) if the expansion method is used), is entirely equivalent to a Heitler–London wave function with optimum orbitals.

Explicit integro-differential equations determining the optimum orbitals A and B may be derived either directly, by minimizing the expectation value of the Hamiltonian for the function (5.144), or indirectly by substituting the forms (5.142) and (5.143) in the Fock-type equations (5.124) and (5.125). As $R \to \infty$ the optimum orbitals A and B determined by these equations tend to the atomic orbitals a and b.

Numerical calculations by the methods of this section have been carried through by several authors. All the calculations use a linear expansion of the orbitals as in equations (5.126) and (5.127). The ODC (optimum double configuration) wave function of Das and Wahl[26] was determined by solving equations (5.112), (5.130) and (5.131) iteratively as outlined above; the basis orbitals χ_p were an extensive, optimized set of STOs centred on the two nuclei.

Davidson and Jones[27] and Goddard[28] optimize the orbitals A and B of equation (5.144) directly by minimizing the energy; the basis functions employed are elliptical functions of the form (5.104) and STOs respectively.

The total energies obtained by these three sets of authors at the equilibrium nuclear separation of 1.40 au are as follows:

Das and Wahl[26]	−1.1517 au	
Davidson and Jones[27]	−1.1521 au	(5.145)
Goddard[28]	−1.1515 au.	

All three methods of calculation are equivalent. The small differences in the total energies reflect the adequacy of the basis sets employed. The result of Davidson and Jones has probably converged to the number of figures quoted above; it corresponds to a binding energy $D_e = 4.14$ eV. This result is only slightly better than the value $D_e = 4.09$ eV obtained from the scaled, floating, polarized Heitler–London function (Section 5.2(c) and (d)), indicating that, at least for calculations of the energy curve, the differences between scaled, floating, polarized atomic orbitals and optimum orbitals are not very significant. However, the methods of this section, especially the ODC procedure, may be extended without difficulty to include several configurations and are readily applicable to larger molecules. Comparing the ODC binding energy with the Hartree–Fock result, we see that allowance for left–right correlation along the bond direction has lowered the total energy by about 0.50 eV.

Accurate Hartree–Fock and ODC wave functions obtained from the integro-differential equations (5.95) or (5.124), (5.125) will satisfy the Hellmann–Feynman theorem for all parameters and coordinate systems; in particular they will satisfy the virial and electrostatic theorems. However, if an inadequate basis set is used in the linear expansion method, discrepancies may arise. Even accurate Hartree–Fock and ODC wave functions fail to satisfy the integral Hellmann–Feynman theorem, since the constrained forms (5.83) and (5.106) of the wave function are non-linear.

5.4. Extended orbital calculations

Even the best energy obtained for the H_2 ground state so far lies about 0.61 eV (14 kcal) above the experimental energy. This is an unacceptable error for chemical applications, since experimental determinations of dissociation and activation energies are commonly accurate to within about 1 kcal. The principal reason for this discrepancy is the neglect of all correlations between the positions of the two electrons other than left–right correlation along the bond direction. Other types of correlation may be incorporated in the wave function by allowing appropriate higher configurations to interact with the ground configuration.

(a) IN-OUT CORRELATION; THE σ-LIMIT

A simple type of electron correlation neglected in all our previous wave functions is "*in–out*" *correlation* corresponding to radial correlation in atomic systems. As in the case of left–right correlation there are two equivalent ways of writing the simplest wave function for H_2 which introduces in–out correlation. Firstly we have the configuration-interaction form

$$\Phi(1,2) = N\{1\sigma_g(1)1\sigma_g(2) - \lambda 2\sigma_g(1)2\sigma_g(2)\} \qquad (5.146)$$

where $1\sigma_g$, $2\sigma_g$ are two orthonormal orbitals of σ_g symmetry. An entirely equivalent wave function to (5.146) is the open-shell form

$$\Phi(1,2) = N\{\sigma_g(1)\sigma_g'(2) + \sigma_g'(1)\sigma_g(2)\}, \qquad (5.147)$$

where the orbitals σ_g, σ_g' are normalized but not orthogonal. The transformation linking (5.146) and (5.147) is quite analogous to equations (5.142) and (5.143) for the left–right case.

The best orbitals to use in equations (5.146) are again determined by integro-differential equations similar to the ODC equations (5.124) and (5.125). Here, however, we must include an off-diagonal Lagrange multiplier to ensure orthogonality of the orbitals $1\sigma_g$ and $2\sigma_g$. This leads to some complication in the iterative solution of the equations by the expansion method.

An equivalent procedure is to optimize the orbitals σ_g, σ_g' of equation (5.147). Such a calculation has been carried through by Davidson and Jones[27] using a linear expansion of the orbitals σ_g, σ_g' in terms of the elliptical functions (5.104). At $R = 1.40$ au they find $D_e = 3.86$ eV corresponding to an in–out correlation energy of 0.22 eV.

A wave function showing both left–right and in–out correlations may be expressed by the superposition of three σ^2-type configurations

$$\Phi(1, 2) = N\{1\sigma_g(1)1\sigma_g(2) - \lambda_1 2\sigma_g(1)2\sigma_g(2) - \lambda_2 1\sigma_u(1)1\sigma_u(2)\}. \tag{5.148}$$

Again the optimum orbitals are determined by simultaneous integro-differential equations which reduce to matrix pseudo-eigenvector equations if the expansion methods is used. The derivation and solution of these equations are straightforward generalizations of the methods given in Section 5.3(e).

More generally we may include a large number of configurations constructed from σ orbitals.

$$\Phi(1, 2) = \sum_{i=1}^{M} \sum_{j=1}^{M} A_{ij} i\sigma(1) j\sigma(2) \tag{5.149}$$

and determine the expansion coefficients A_{ij} by the linear variation method. In such calculations it is simplest not to optimize the individual orbitals $i\sigma$. Any convenient, complete set of orbitals may be used and the expansion extended ($M \to \infty$) until convergence is reached. Note that, for such an expansion in terms of a fixed set of basis functions, off-diagonal terms $i\sigma(1)j\sigma(2)$ must be included as well as the diagonal terms. For a singlet wave function $\Phi(2, 1) = \Phi(1, 2)$, and the matrix of coefficients must be symmetrical

$$A_{ij} = A_{ji}. \tag{5.150}$$

The limiting energy for the wave function (5.149) (the σ-limit) has been determined[27] as -1.1617 au corresponding to $D_e = 4.40$ eV and a combined in–out and left–right correlation energy of 0.76 eV. This value is approximately the sum of the individual contributions 0.50 eV (left–right) and 0.22 eV (in–out) quoted above. Such approximate additivity of separately calculated correlation energies of different types appears to be a general property of atoms and molecules.

When expressed in elliptical coordinates the σ-limit wave function $\Phi_\sigma(1, 2)$ depends only on four coordinates μ_1, ν_1, μ_2 and ν_2; it does not involve the azimuthal angles φ_1 and φ_2

$$\Phi_\sigma(1, 2) = F(\mu_1, \nu_1, \mu_2, \nu_2). \tag{5.151}$$

This σ-limit wave function retains a number of interesting properties of the exact wave function. Since the restraint on the wave function represented by equation (5.149) is independent of the values of all parameters such as nuclear charges and the internuclear distance, the Hellmann–Feynman theorem is satisfied for all choices of parameters and coordinate systems. Furthermore the restraint is a linear one, that is, a linear combination of two functions of the form (5.149) is itself of the form (5.149). Consequently, the integral Hellmann–Feynman theorem will also be satisfied for all parameters and coordinate systems. The function Φ_σ is, in fact, the only wave functions among those so far considered for H_2 which satisfies the integral theorem.

(b) ANGULAR CORRELATION

The wave function (5.151) is still not the most general function of overall Σ symmetry. In elliptical coordinates the axial components of the angular momenta of the two electrons are represented by the operators $-i(\partial/\partial\varphi_1)$ and $-i(\partial/\partial\varphi_2)$. For the function (5.151) we have

$$-i\frac{\partial\Phi_\sigma}{\partial\varphi_1} = -i\frac{\partial\Phi_\sigma}{\partial\varphi_2} = 0, \tag{5.152}$$

whereas the condition for overall Σ symmetry is the vanishing of the axial component of the total angular momentum, that is

$$\frac{\partial\Phi}{\partial\varphi_1} + \frac{\partial\Phi}{\partial\varphi_2} = 0. \tag{5.153}$$

The condition (5.153) implies, not that Φ is independent of φ_1 and φ_2, but merely that Φ must be a function of the difference $\varphi_1 - \varphi_2$. To ensure continuity Φ must be periodic in $\varphi_1 - \varphi_2$ with period 2π. Such a function may be expanded as a Fourier series

$$\Phi(1, 2) = F_0 + \sum_{m=1}^{\infty} F_m \cos m(\varphi_1 - \varphi_2) + \sum_{m=1}^{\infty} G_m \sin m(\varphi_1 - \varphi_2). \tag{5.154}$$

For Σ^+ symmetry Φ must be unchanged by the substitution $\varphi_1 - \varphi_2 \to \varphi_2 - \varphi_1$, that is, the sine terms must be dropped from equation (5.154) leaving

$$\Phi(1, 2) = F_0 + \sum_{m=1}^{\infty} F_m \cos m(\varphi_1 - \varphi_2). \tag{5.155}$$

In equation (5.155) $F_0, F_1, \ldots$ are all functions of the remaining coordinates μ_1, ν_1, μ_2 and ν_2. Consider the particular term

$$F_m(\mu_1\nu_1\mu_2\nu_2) \cos m(\varphi_1 - \varphi_2). \tag{5.156}$$

Let us expand F_m as a series of products of functions of the coordinates of the two electrons. Thus

$$F_m(\mu_1\nu_1\mu_2\nu_2) = \sum_i \sum_j a_{ij} f_i(\mu_1\nu_1) f_j(\mu_2\nu_2). \tag{5.157}$$

Here $f_i(\mu\nu)$ may be any convenient complete set of functions. Let us choose them such that each function is either even in ν (of g symmetry) or is odd in ν (of u symmetry). Then for overall Σ_g^+ symmetry the expansion (5.157) contains only products of functions of the same symmetry type; each term is either a product of two g functions or a product of two u functions.

The final symmetry requirement is that $\Phi(1, 2)$ should represent a singlet state, that is

$$\Phi(1, 2) = \Phi(2, 1). \tag{5.158}$$

This requirement will be satisfied if the matrix of coefficients in equation (5.157) is symmetric

$$a_{ij} = a_{ji}. \tag{5.159}$$

With these restrictions equations (5.155) and (5.157) represent the most general spatial wave function of symmetry type $^1\Sigma_g^+$. The functions f_i and the matrix a_{ij} may, of course, differ for different values of m.

Substituting the expansions (5.157) into equation (5.155) and expanding the cosine functions according to the equation

$$\cos m\,(\varphi_1 - \varphi_2) = \cos m\varphi_1 \cos m\varphi_2 + \sin m\varphi_1 \sin m\varphi_2 \tag{5.160}$$

the total wave function is obtained as a series of orbital products

$$\Phi(1, 2) = \sum_i \sum_j A_{ij}\phi_i(1)\phi_j(2). \tag{5.161}$$

Each of the orbitals ϕ_i of equation (5.161) belongs to one of the symmetry types $\sigma_g, \sigma_u, \pi_g, \pi_u, \delta_g, \delta_u, \ldots$ encountered in the study of H_2^+ (Chapter 3). The degenerate orbitals $\pi, \delta \ldots$ appear in the real form with angular factors $\cos m\varphi$, $\sin m\varphi$. In each term of the expansion, ϕ_i and ϕ_j are of the same symmetry type and the terms involving degenerate orbitals appear in pairs corresponding to the two terms in equation (5.160). The matrix of coefficients in equation (5.161) must be symmetric

$$A_{ij} = A_{ji}, \tag{5.162}$$

to ensure that $\Phi(1, 2)$ represents a singlet state.

The above symmetry analysis is confirmed by a glance at Table 4.8 which gives the reduction of direct products for symmetry $D_{\infty h}$. We see that the direct products $\sigma_g \times \sigma_g$, $\sigma_u \times \sigma_u$, $\pi_g \times \pi_g$, $\pi_u \times \pi_u, \ldots$ (and no others) all contain the representation Σ_g^+.

The simplest angular dependent wave function Φ_a of the form (5.161) involves the two configurations $(\sigma_g)^2$ and $(\pi_u)^2$

$$\Phi_a(1, 2) = A_{11}\sigma_g(1)\sigma_g(2) + A_{22}\,[\pi_{uc}(1)\pi_{uc}(2) + \pi_{us}(1)\pi_{us}(2)]. \tag{5.163}$$

Here the subscripts c and s refer to the two components of the degenerate π_u orbital with angular dependence $\cos\varphi$ and $\sin\varphi$.

For fixed values of the coordinates $\mu_1\nu_1\mu_2\nu_2$ the angular dependence of $\Phi_a(1, 2)$ is given by the equation

$$\Phi_a(1, 2) = A - B(\cos\varphi_1 \cos\varphi_2 + \sin\varphi_1 \sin\varphi_2) \tag{5.164}$$

that is

$$\Phi_a(1, 2) = A - B\cos(\varphi_1 - \varphi_2). \tag{5.165}$$

Equation (5.165) shows that, if A and B are positive, $\Phi_a(1, 2)$ is largest for $\varphi_1 - \varphi_2 = \pi$ and smallest for $\varphi_1 - \varphi_2 = 0$. That is, the wave function (5.163) introduces angular correlation; if electron 1 is on one side of the internuclear axis, electron 2 is most likely to be found on the other side.

There have been a number of calculations on the ground state of H_2 using wave functions of the form (5.161). In some of these the different orbitals ϕ_i are all chosen orthonormal, in others open-shell terms of the form (5.147) are included.

McLean, Weiss and Yoshimine[(29)] write their wave function in the form

$$\Phi(1, 2) = c_1(\sigma_g 1s\sigma_g 1s') + c_2(\sigma_g 2s\sigma_g 2p) + c_3(\sigma_u 1s\sigma_u 1s') + c_4(\pi_u 2p)^2 + c_5(\pi_g 2p)^2. \tag{5.166}$$

Here $\sigma_g 1s$, $\pi_g 2p$, . . . are symmetry orbitals each constructed as the sum or difference of two identical STOs located on the two nuclei. Thus the function (5.166) contains five non-linear parameters, the orbital exponents ζ_{1s}, ζ'_{1s}, ζ_{2s}, $\zeta_{2p\sigma}$, $\zeta_{2p\pi}$ of the STOs, as well as the linear parameters $c_1, \ldots, c_5$. All of these parameters were determined over a wide range of nuclear separations by minimizing the total energy. At the equilibrium nuclear separation 1.40 au the function (5.166) gives a binding energy $D_e = 4.54$ eV.

The first two terms in equation (5.166) describe in–out correlation, the third left–right correlation, and the last two angular correlation. Thus all important correlations are allowed for, at least crudely, by the function (5.166).

Das and Wahl[(26,30)] use a wave function of the form

$$\Phi(1, 2) = A_1(1\sigma_g)^2 + A_2(2\sigma_g)^2 + A_3(3\sigma_g)^2 + A_4(1\sigma_u)^2 + A_5(1\pi_u)^2 + A_6(1\pi_g)^2. \tag{5.167}$$

The orbitals of equation (5.167) are expressed as linear combinations of the symmetry orbitals of McLean *et al.* (for $R = 1.40$ au) together with $\pi_u 3d$, $\pi_g 3d$. This basis comprises four σ_g functions, four σ_u functions, two π_u functions and two π_g functions. We thus obtain

$$\begin{aligned}
1\sigma_g(1) &= \sigma_g 1s(1)c_{11} + \sigma_g 1s'(1)c_{21} + \sigma_g 2s(1)c_{31} + \sigma_g 2p(1)c_{41} \\
2\sigma_g(1) &= \sigma_g 1s(1)c_{12} + \sigma_g 1s'(1)c_{22} + \sigma_g 2s(1)c_{32} + \sigma_g 2p(1)c_{42} \\
3\sigma_g(1) &= \sigma_g 1s(1)c_{13} + \sigma_g 1s'(1)c_{23} + \sigma_g 2s(1)c_{33} + \sigma_g 2p(1)c_{43} \quad (5.168) \\
1\sigma_u(1) &= \sigma_u 1s(1)c_{14} + \sigma_u 1s'(1)c_{24} + \sigma_u 2s(1)c_{34} + \sigma_u 2p(1)c_{44} \\
1\pi_u(1) &= \pi_u 2p(1)c_{55} + \pi_u 3d(1)c_{65} \\
1\pi_g(1) &= \pi_g 2p(1)c_{76} + \pi_g 3d(1)c_{86}.
\end{aligned}$$

The coefficients $A_1 \ldots A_6$ in equation (5.167) and $c_{11} \ldots c_{86}$ in equations (5.168) are varied simultaneously to minimize the total energy, by an extension of the procedure described in Section 5.3(e). The resulting wave function, which Das and Wahl refer to as an OVC (optimized valence configuration) function, gives a binding energy of 4.62 eV, only 0.08 eV better than the result of McLean *et al.* This small improvement arises from the limited flexibility of the expansions (5.168); in particular the π orbitals describing the angular correlation are almost fixed by symmetry.

A much more extensive configuration interaction calculation has been carried out by Hagstrom and Shull.[31] Their wave function may also be put in the form (5.161), where the (unnormalized) orbitals ϕ_i are given by

$$\phi(\zeta, n, j, m) = \mu^n \nu^j \, e^{-\zeta\mu} [(\mu^2 - 1)(1 - \nu^2)]^{m/2} \begin{cases} \cos m\varphi \\ \sin m\varphi. \end{cases} \quad (5.169)$$

For $m = 0, 1, 2, \ldots$ the orbital (5.169) is of symmetry type $\sigma, \pi, \delta \ldots$. For $m + j$ even (odd) we have an orbital of g(u) symmetry under inversion. The configurations considered by Hagstrom and Shull are of the forms

$$\sigma_g\text{-type;} \qquad (\sigma_g, \sigma_g) = \sigma_g(1)\sigma_g(2) \quad (5.170)$$

$$(\sigma_g, \sigma_g') = \sigma_g(1)\sigma_g'(2) \quad (5.171)$$

σ_u-type; as for σ_g type, u replacing g

$$\pi_u\text{-type;} \qquad (\pi_u, \pi_u) = \pi_{uc}(1)\pi_{uc}(2) + \pi_{us}(1)\pi_{us}(2) \quad (5.172)$$

$$(\pi_u, \pi_u') = \pi_{uc}(1)\pi_{uc}'(2) + \pi_{us}(1)\pi_{us}'(2) \quad (5.173)$$

π_g-type; as for π_u type, g replacing u.

δ_g-type; as for π_u type, δ_g replacing π_u.

Because of the symmetry of the matrix of expansion coefficients (equation (5.162)) the open-shell configurations, such as (5.171) and

(5.173), appear in pairs with equal coefficients. Counting such a pair as a single configuration, the final wave function contains altogether 15 configurations of σ_g-type, 6 of σ_u-type, 6 of π_u-type, 3 of π_g-type and 3 of δ_g-type. The calculated binding energy is $D_e = 4.71$ eV. This differs by only 0.04 eV (0.9 kcal) from the experimental value.

It is clear from the results of this section that a configuration interaction function such as (5.161) may be used to calculate an energy curve as accurately as we please, provided that sufficient orbitals ϕ_i, of all possible symmetry types, are included in the expansion. The resulting wave function is extremely complicated and awkward to use for the calculation of other properties of interest. In Section 5.6 we will show how such a wave function may be greatly simplified without significant loss of accuracy.

5.5. The James-Coolidge method

Although the configuration-interaction method can, in principle, be used to obtain results of any desired accuracy, the convergence of the expansion is rather slow. The principal reason for this poor convergence is related to the difficulty found in Chapter 3 for single-centre expansions of the H_2^+ wave functions. In that case the poor convergence was caused by the singularity (cusp) in the orbital at a nucleus remote from the centre of expansion. Many terms are needed to represent adequately such a cusp by an expansion in terms of smooth functions.

There is a similar singularity in the H_2 wave function at points in configuration space where the interelectronic separation r_{12} tends to zero. The nature of this singularity can be inferred by retaining only those terms in the Hamiltonian which are dominant for very small values of r_{12}. This leads to the approximate equation

$$\left(-\tfrac{1}{2}\nabla_1^2 - \tfrac{1}{2}\nabla_2^2 + \frac{1}{r_{12}}\right)\Phi = E\Phi. \tag{5.174}$$

Equation (5.174) is very similar to the Schrödinger equation for the hydrogen atom before separation of the motion of the centre of mass. Proceeding as in that case we define "centre of mass" coordinates $\mathbf{R}$ and $\mathbf{r}$, by the equations

$$\mathbf{R} = \tfrac{1}{2}(\mathbf{r}_1 + \mathbf{r}_2), \qquad \mathbf{r} = \mathbf{r}_1 - \mathbf{r}_2, \qquad r = r_{12}. \tag{5.175}$$

Transforming equation (5.174) to those coordinates we find that the dependence of Φ on $\mathbf{r}$, the vector distance between the two electrons, is given by

$$\left(-\tfrac{1}{2}\nabla_r^2 + \frac{1}{2r}\right)\Phi = E\Phi. \tag{5.176}$$

Any bounded solution of equation (5.176) may be expanded in terms of spherical harmonics as follows

$$\Phi(\mathbf{r}) = \sum_{l=0}^{\infty} \sum_{m=-l}^{l} r^l f_{lm}(r) Y_{lm}(\theta, \varphi). \tag{5.177}$$

Furthermore, for a singlet state, the wave function is unaltered by interchanging the coordinates $\mathbf{r}_1$ and $\mathbf{r}_2$, so that

$$\Phi(-\mathbf{r}) = \Phi(\mathbf{r}). \tag{5.178}$$

This implies that the odd harmonics $l = 1, 3, \ldots$ are absent from the expansion (5.177). Hence we obtain, putting $f(r) = f_{00}(r)$,

$$\Phi(\mathbf{r}) = f(r) + 0(r^2). \tag{5.179}$$

Expanding $f(r)$ as a power series

$$f(r) = a + br + cr^2 \ldots, \tag{5.180}$$

substituting in equation (5.176), and equating to zero the coefficient of the leading term (which involves $1/r$) we find

$$b = \tfrac{1}{2}a. \tag{5.181}$$

Hence, in the neighbourhood of the singularity at $r_{12} = 0$, the wave function is given by

$$\Phi = \Phi_{(r_{12}=0)}\{1 + \tfrac{1}{2}r_{12} + 0(r_{12})^2\}. \tag{5.182}$$

This derivation of the relation (5.182) is somewhat heuristic. Kato[32] has proved that as $r_{12} \to 0$,

$$\left\langle \frac{\partial \Phi}{\partial r_{12}} \right\rangle_{\text{AV}} = \tfrac{1}{2}\Phi \tag{5.183}$$

where the average is taken over all directions of the vector $\mathbf{r}_1 - \mathbf{r}_2$. Pack and Byers Brown[32a] have investigated the special forms which Kato's result assumes in various cases. They find, in fact, that the simple result (5.182) is correct for the ground state of H_2.

Equation (5.182) shows that the accurate wave function for H_2 has a cusp at the singularity $r_{12} = 0$ similar, but opposite in sign, to the cusps of the orbitals at the nuclei. This suggests that inclusion of terms linear in r_{12} would improve the convergence of an expansion of the wave function. Such terms are absent from the configuration interaction expansion (5.155), since r_{12} and $\cos(\varphi_1 - \varphi_2)$ are related by the equation

$$\cos(\varphi_1 - \varphi_2) = \frac{\mu_1^2 + \mu_2^2 + \nu_1^2 + \nu_2^2 - 2 - 2\mu_1\mu_2\nu_1\nu_2 - (2/R)^2 r_{12}^2}{2\sqrt{(\mu_1^2 - 1)(\mu_2^2 - 1)(1 - \nu_1^2)(1 - \nu_2^2)}}. \tag{5.184}$$

An expansion of the H_2 ground-state wave function including terms linear in r_{12} was first used by James and Coolidge in 1933.[33] These authors used r_{12} directly, in place of $\varphi_1 - \varphi_2$, as the fifth coordinate in the wave function and assumed an expansion of the form

$$\Phi(1, 2) = e^{-\delta(\mu_1+\mu_2)} \sum_{k,l,m,n,p} c_{klmnp}\mu_1^k \mu_2^l \nu_1^m \nu_2^n r_{12}^p. \qquad (5.185)$$

The coefficients c_{klmnp} were determined by the linear variation method.

The expansion (5.185) is much more rapidly convergent than the CI expansion (5.161). Again, for a state of symmetry $^1\Sigma_g^+$, there are restrictions on the parameter values ($m + n$ even) and on the expansion coefficients

$$c_{klmnp} = c_{mnklp}. \qquad (5.186)$$

Using only 5 independent terms in the expansion (5.185) James and Coolidge obtained the simple wave function

$$\Phi(1,2) = \frac{1}{2\pi} e^{-0.75(\mu_1+\mu_2)}\{2.23779 + 0.80483\ (\nu_1^2 + \nu_2^2)$$

$$- 0.5599\nu_1\nu_2 - 0.60985\ (\mu_1 + \mu_2) + 0.56906\ r_{12}\}, \qquad (5.187)$$

which predicts a dissociation energy of 4.53 eV; comparable with the much more elaborate CI wave functions of Section 5.4(b). Their best function, containing 13 independent terms, gave $D_e = 4.72$ eV. By extrapolating their results they estimated a theoretical binding energy $D_e = 4.73 \pm 0.02$, which was consistent with the current experimental value of 4.74 ± 0.04 eV.

Since 1933 experimental determinations of D_e and R_e by spectroscopic methods have been greatly refined, as have the theoretical calculations. In 1960 Kolos and Roothaan[24] using 50 terms in the James–Coolidge expansion obtained values $D_e = 4.7467$ eV, $R_e = 1.40081$ au $= 0.74127$ Å; to be compared with spectroscopic values $D_e = 4.7466 \pm 0.0007$ eV and $R_e = 0.74116$. As these authors point out, the excellent agreement with the experimental binding energy is somewhat surprising in view of the neglect of corrections for finite nuclear masses and relativistic effects. A really thorough test of the theory requires a careful calculation of these small terms. It is also preferable to carry out the comparison in terms of the dissociation energy D_0, the vibrational quanta $\Delta G_{v+\frac{1}{2}} = F(v + 1, 0) - F(v, 0)$, and the rotational quanta

$$S_0(J) = F_0(J + 2) - F_0(J), \qquad (5.188)$$

and

$$Q_1(J) = \nu_{1-0} + (B_1 - B_0)J(J+1) - (D_1 - D_0)J^2(J+1)^2, \tag{5.189}$$

since these quantities are obtained more directly from the spectroscopic observations than are the extrapolated constants D_e and R_e. The quantities appearing in these equations are defined in Section 1.2; ν_{1-0} in equation (5.189) is the vibrational quantum $F(1, 0) - F(0, 0)$.

Two careful comparisons of this type have been carried out by Kolos and Wolniewicz[34-37] and are described below.

(a) THE ADIABATIC CALCULATION

This calculation falls into three parts; (i) the determination of an accurate potential energy curve for infinite nuclear mass over a wide range of nuclear separations, (ii) calculation of the diagonal corrections for nuclear motion and (iii) numerical determination of the vibrational and rotational energy levels for the resulting potential energy curve.

For part (i) of the calculation Kolos and Wolniewicz use the following modified form of the James and Coolidge expansion

$$\Phi = e^{-\alpha(\mu_1+\mu_2)} \cosh \beta(\nu_1 - \nu_2) \sum_{k,l,m,n,p} c_{klmnp}\mu_1^k \mu_2^l \nu_1^m \nu_2^n r_{12}^p. \tag{5.190}$$

The conditions on the parameters for a $^1\Sigma_g^+$ state are the same as for the expansion (5.185). The addition of the factor $\cosh \beta(\nu_1 - \nu_2)$ improves the convergence of the expansion for large nuclear separations. Indeed, for appropriate values of α and β, the first term in (5.190) with $k = l = m = n = p = 0$ becomes equal to the asymptotic wave function in the limit $R \to \infty$

$$\Phi_0 = \exp[-\tfrac{1}{2}R(\mu_1 + \mu_2)] \cosh [-\tfrac{1}{2}R(\nu_1 - \nu_2)]. \tag{5.191}$$

This asymptotic wave function is simply the Heitler–London function Φ_+ of equation (5.9) expressed in elliptical coordinates.

Using up to 80 terms in the expansion (5.190) Kolos and Wolniewicz computed an accurate potential energy curve in the range $0.4 < R < 10.0$ au.

The diagonal corrections for nuclear motion were obtained using a modified form of the Born method described in Section 1.1. If we apply that method in the centre of mass coordinate system and use atomic units, the total Hamiltonian appears in the form

$$H = H^0 + H_1 + H_2 \tag{5.192}$$

where H^0 is the Hamiltonian (5.11) of the clamped nuclei approximation,

$$H_1 = -\frac{1}{M}\nabla_{\mathbf{R}}^2$$

and

$$H_2 = -\frac{1}{4M}(\nabla_{\mathbf{r}_1}^2 + \nabla_{\mathbf{r}_2}^2 + 2\nabla_{\mathbf{r}_1}\cdot\nabla_{\mathbf{r}_2}).$$

Here $\mathbf{R}$ is the relative position vector of the nuclei, each of mass M, and $\mathbf{r}_1, \mathbf{r}_2$ are the position vectors of the two electrons.

The diagonal correction for nuclear motion (the term $\Delta_n(R)$ of equation (1.10) now appears in the form

$$\Delta(R) = \int \Phi^*(\mathrm{r}, \mathbf{R})(H_1 + H_2)\Phi(\mathrm{r}, \mathbf{R})\, \mathrm{d}^3\mathbf{r} \tag{5.193}$$

where $\Phi(\mathrm{r}, \mathbf{R})$ is the wave function in the clamped nuclei approximation.

The results obtained in this way are shown in Fig. 5.8.

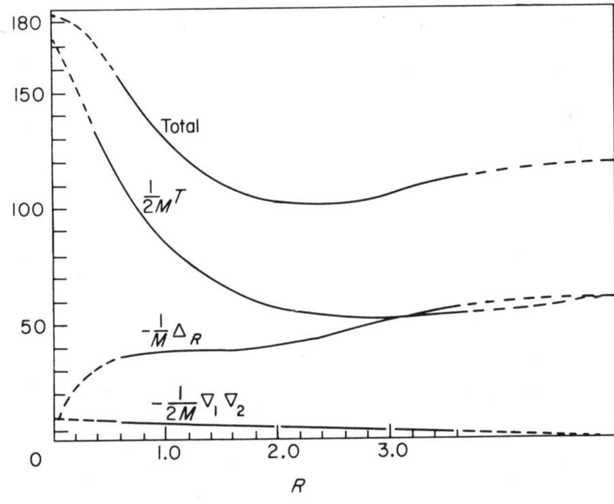

Fig. 5.8. Diagonal corrections for nuclear motion (ground state of H_2) (cm^{-1}).

We see that the contribution (5.193) to the potential energy curve is quite appreciable, being of the order 100–150 cm^{-1}. Kolos and Wolniewicz also considered relativistic corrections to the potential energy curve. These were found to be much smaller ($\approx 0.5\ \mathrm{cm}^{-1}$ near $R = R_e$) and we shall neglect them here.

At this stage we may compare the theoretical curve with that obtained from the spectroscopic data by the RKR method (Section 1.5). Such a comparison is shown in Table 5.2, which shows the deviations of the theoretical curve, with and without the correction term $\Delta(R)$ of equation (5.193), from the RKR results of Weissman, Vanderslice and Battino.[38]

Table 5.2. Deviations of computed potential-energy curves for the ground state of H_2 from the classical turning points obtained from the RKR method[a]

v	$R_{min.}$	ΔR_1	ΔR_2 [b]	$R_{max.}$	ΔR_1	ΔR_2
0	0.633	0	0	0.883	0	0
1	0.571	0	0	1.013	0	1
2	0.535	−1	0	1.120	0	0
3	0.509	0	0	1.219	0	0
4	0.489	0	0	1.316	0	0
5	0.473	0	0	1.413	0	0
6	0.460	0	0	1.513	0	0
7	0.449	0	0	1.618	−1	0
8	0.439	0	0	1.730	−1	0
9	0.432	0	0	1.853	−1	0
10	0.425	0	0	1.992	−1	0
11	0.420	0	0	2.158	−1	0
12	0.416	0	0	2.370	−1	—
13	0.413	0	0	2.675	−2	—
14	0.411	0	0	3.260	0	—

[a] R in Angstroms, ΔR in 10^{-3} Å.
[b] ΔR_2 includes diagonal corrections for nuclear motion.

We see from Table 5.2 that the agreement between the theoretical and RKR results is excellent and is significantly improved by including the correction term $\Delta(R)$ of equation (5.193). However, as Kolos and Wolniewicz point out, the significance of the agreement shown in Table 5.2 should not be over-estimated. It would not be affected by deviations of the order 20–100 cm^{-1} between the theoretical and RKR curves at fixed values of R.

A much more critical test of the theory is obtained by substituting the theoretical potential energy in the radial equation determining the vibrational and rotational energy levels (equation (1.17)) and solving this radial equation numerically. Wolniewicz[36] has carried through this calculation, using numerical techniques introduced by Cooley[39] for solving the radial wave equation. A comparison of his calculated dissociation energy D_0, zero-point vibrational energy $D_e - D_0$, vibrational quanta $\Delta G_{v+\frac{1}{2}}$, and rotational quanta $S_0(J)$ and $Q_1(J)$ (equations (5.188) and (5.189)) with spectroscopic values is given in Tables 5.3, 5.4 and 5.5. Values are given for the ground states of the three isotopic molecules H_2, HD, D_2. The potential curves for these three molecules differ only in the small corrections for nuclear motion (equation (5.193)).

Table 5.3. Computed and experimental dissociation and zero-point energies (cm^{-1})

		H_2	HD	D_2
D_0	Exp.[a]	36 113.6 ± 0.3	36 400.5 ± 1.0	36 744.2 ± 0.5
	B[b]	36 112.2	36 401.5	36 745.6
	C	36 118.0	36 405.7	36 748.3
	N	36 114.7	36 402.9	36 746.2
$D_e - D_0$	Exp.[a]	2179.2 ± 0.2	1890.4 ± 0.6	1547.0 ± 0.4
	B	2180.5	1891.2	1547.1
	C	2179.6	1890.3	1546.9

[a] Herzberg and Monfils (Ref. 40).
[b] Cases B (C) neglect (include) corrections for nuclear motion; case N is the non-adiabatic calculation (Section 5.5(b) following).

Table 5.4. Deviations from experimental vibrational quanta $\Delta G_{v+\frac{1}{2}}$ (cm^{-1})

	v	Exp.[a]	B[b]	C[b]
H_2	0	4161.13	2.33	0.91
	1	3925.97	1.96	0.80
	2	3695.24	2.17	1.26
	3	3468.01	1.50	0.85
	4	3241.56	1.13	0.75
	5	3013.73	0.95	0.82
	6	2782.18	0.48	0.66
	7	2543.14	0.10	0.58
	8	2292.96	−0.27	0.46
	9	2026.26	−0.60	0.28
	10	1736.66	−1.13	−0.27
	11	1414.98	−1.33	−0.94
HD	0	3632.14	1.78	0.85
	1	3454.73	1.57	0.77
	2	3280.68	1.72	1.06
	3	3109.31	1.40	0.89
D_2	0	2993.55	1.35	0.82

[a] Herzberg and Howe (Ref. 41).
[b] Cases B(C) neglect (include) corrections for nuclear motion; case N is the non-adiabatic calculation (Section 5.5(b) following).

Table 5.5. Comparison of rotational quanta with experiment (cm^{-1})

		$Q_1(J)$			$S_0(J)$		
	J	Exp.[a]	Theoret.	dif.	Exp.[a]	Theoret.	dif.
H_2	0	4161.13	4162.04	0.91	354.38	354.39	0.01
	1	4155.20	4156.13	0.93	587.06	587.07	0.01
	2	4143.39	4144.34	0.95	814.41	815.48	0.07
	3	4125.83	4126.75	0.92	1034.65	1034.75	0.10
HD	0	3632.05	3632.99	0.94	267.09	267.12	0.03
	1	3628.20	3629.13	0.93	443.08	443.17	0.05
	2	3620.51	3621.43	0.92	616.09	616.21	0.12
	3	3608.99	3609.92	0.93	784.99	785.15	0.16
	4				948.82	948.98	0.16
D_2	0	2993.55	2994.37	0.82	179.06	179.12	0.06
	1	2991.45	2992.26	0.81	297.52	279.63	0.11
	2	2987.23	2988.04	0.81	414.66	414.78	0.12
	3	2980.88	2981.73	0.85	529.91	530.07	0.16
	4	2972.56	2973.34	0.78	642.81	643.01	0.20

[a] Stoicheff (Ref. 42).

The overall agreement shown in Tables 5.3, 5.4 and 5.5 is extremely good; however some discrepancies occur. Consider first the rotational quanta shown in Table 5.5. The $S_0(J)$ values are essentially exact, and the small deviations in $Q_1(J)$ are independent of J, which means that the disagreement is due to inaccuracies in the vibrational levels; the rotational energies being very accurate (cf. equation (5.189)).

Table 5.4 shows small errors in the vibrational quanta which are reduced, but not entirely eliminated, by the corrections for nuclear motion. These small errors accumulate to produce the largest discrepancy between theory and experiment; the difference between the calculated and experimental dissociation energies D_0 of Table 5.3. If we correct the theoretical results by -0.5 cm^{-1}, which is the relativistic correction for the ground-state dissociation energy,[37] and if we combine an estimated 1 cm^{-1} numerical uncertainty in the calculated values with the experimental errors, we arrive at the conclusion that the computed ground-state energies lie below the experimental levels by 3.9 ± 1.3, 4.7 ± 2.0 and 3.6 ± 1.5 cm^{-1} for H_2, HD and D_2, respectively. This discrepancy is definitely outside the combined error limits. Furthermore the disagreement is not caused by the adiabatic approximation. In fact, if the diagonal correction for nuclear motion is included in the computations, the ground-state energy is equal to the expectation value of the total non-adiabatic Hamiltonian, and thus gives an upper limit to the exact eigenvalue. This means that any improvement of the electronic wave function would increase the disagreement. (cf. footnote p. 153).

(b) THE NON-ADIABATIC CALCULATION[37]

In this calculation no use is made of the adiabatic approximation. The complete non-relativistic Hamiltonian for two identical nuclei and two electrons is used. The trial form of the vibronic wave function is

$$\Psi(R, \mathbf{r}) = \sum_{mn} c_{mn} \Psi_m (R, \mathbf{r}) \chi_m (R). \tag{5.194}$$

Here the $\Phi_m (R, \mathbf{r})$ are electronic basis functions of the James–Coolidge form (5.185) and the $\chi_n (R)$ are harmonic oscillator functions of the form

$$\chi_n (R) = (4\pi)^{-\frac{1}{2}} R^{-3} \exp(-x^2/2) \mathscr{H}_n (x), \tag{5.195}$$

with

$$x = \gamma(R - R_e) \tag{5.196}$$

and $\mathscr{H}_n (x)$ the nth Hermite polynomial.

The variable parameters are the exponent δ in equation (5.185) (assumed the same for all the functions Φ_m), γ and R_e of equation (5.196), and the linear coefficients c_{mn}. The latter are determined by the linear variation method. The total wave function (5.194) was obtained by combining 54 electronic basis functions of the James–Coolidge form with vibrational functions (5.195) with $n \leqslant 5$. Many of the resulting terms proved unimportant. The final wave function includes 147 terms and the computed energy is probably very close to the best value obtainable from a function of the form (5.194) with the following restrictions on the indices; $k + m$ and $l + n$ of (5.185) both $\leqslant 4$, $p \leqslant 2$ and $n \leqslant 5$.

The resulting values of D_0 are given in the row labelled N in Table 5.3. We see that these non-adiabatic calculations also give dissociation energies significantly larger than the experimental values. The discrepancies are smaller than for case C but, according to Kolos and Wolniewicz, this simply reflects the lesser flexibility of the non-adiabatic wave function; if more terms were included in equation (5.194) it is to be expected that the non-adiabatic results would approach those of case C. The results of case B in Table 5.3 confirm and extend the result of Kolos and Roothaan;[24] if the effects due to nuclear motion are neglected entirely the calculated D_0 (or D_e) values are in excellent agreement with experimental values. When corrections to this adiabatic calculation are included a significant discrepancy is introduced.

The cause of this discrepancy is, at present, unknown. Wolniewicz[36] considers three possibilities; (i) the radiative correction (Lamb shift), (ii) relativistic corrections connected with the nuclear motion, and (iii) corrections for finite dimensions of the nuclei. It seems, however, that these possibilities are not very promising. The Lamb shift correction cannot decrease the theoretical dissociation energy by more than 0.39 cm^{-1}, and the other two effects are estimated to be still smaller. A careful study of the long-range interaction between $1s$ and $2s$ or $2p$ hydrogen atoms,[43] which is relevant to a correct interpretation of the spectroscopic results on which the experimental D_0 values are based, served only to confirm the discrepancy.‡

For the lowest excited states of $^1\Pi_u$ and $^1\Sigma_u^+$ symmetry similar calculations[35, 44] give dissociation energies in satisfactory agreement with experiment. Furthermore there is no discrepancy between theory and experiment for the ground state of the united atom (He).

5.6. The natural orbital expansion

In the previous two sections we have seen that wave functions of any desired accuracy may be constructed either by an extensive configuration-interaction (CI) or by an expansion of the James–Coolidge type. The resulting wave functions are extremely complicated and it is natural to try and set up a general scheme for interpreting the wave functions and, if possible, simplifying them without significant loss of accuracy. Such a scheme is provided by the natural orbital expansion. This expansion is a special case of the natural spin-orbital analysis of determinantal wave functions for many electron systems introduced by Löwdin[45] (cf. Ref. 45a). For a singlet state of a two-electron system there are a number of special properties of the natural expansion which make it especially useful.

Suppose that the set $\phi_i(1)$, $i = 1, 2, \ldots \infty$ constitutes a complete set of, real, orthonormal functions in the space of electron 1. The exact spatial wave function for the ground state of H_2 may then be expressed by a CI expansion containing an infinite number of terms

$$\Phi(1, 2) = \sum_{i=1}^{\infty} \sum_{j=1}^{\infty} A_{ij}\phi_i(1)\phi_j(2). \tag{5.197}$$

‡ This discrepancy has now been removed. New measurements of the absorption spectra of H_2, HD and D_2 by Herzberg[43a] lead to the dissociation energies $D_0^0(H_2) = 36\,117 \pm 1.5$ cm^{-1}, $D_0^0(HD) = 36\,406.2 \pm 0.4$ cm^{-1} and $D_0^0(D_2) = 36\,748.9 \pm 0.4$ cm^{-1} in agreement with the best theoretical estimates (Table 5.3, calculation C).

The matrix A_{ij} of expansion coefficients is real and symmetric

$$A_{ij} = A_{ji}. \tag{5.198}$$

Since the orbitals ϕ_i are orthonormal, the normalization condition

$$\iint \Phi^2(1, 2)\, dv_1\, dv_2 = 1 \tag{5.199}$$

reduces to

$$\sum_i \sum_j A_{ij}^2 = 1. \tag{5.200}$$

The real, symmetric matrix **A** may be reduced to diagonal form by a real orthogonal transformation **U**

$$\mathbf{U}^{\mathrm{T}}\mathbf{U} = 1 \tag{5.201}$$

$$\mathbf{U}^{\mathrm{T}}\mathbf{A}\mathbf{U} = \mathrm{a} = \mathrm{diag}\{c_1, c_2, \ldots\}. \tag{5.202}$$

We choose the matrix **U** in such a way that the real constants c_i (the eigenvalues of **A**) appear in decreasing order of magnitude

$$c_1^2 \geqslant c_2^2 \geqslant c_3^2 \ldots. \tag{5.203}$$

We now define the *natural orbitals* $\chi_1(1), \chi_2(1), \ldots$ by the equation

$$\boldsymbol{\chi} = \boldsymbol{\phi}\mathbf{U}, \tag{5.204}$$

that is

$$\chi_i(1) = \sum_j \phi_j(1) U_{ji}. \tag{5.205}$$

In terms of these natural orbitals the expansion (5.197) of the wave function becomes

$$\Phi(1, 2) = \boldsymbol{\phi}(1)\mathbf{A}\boldsymbol{\phi}^{\mathrm{T}}(2) \tag{5.206}$$

$$= \boldsymbol{\phi}(1)\mathbf{U}\mathrm{a}\mathbf{U}^{\mathrm{T}}\boldsymbol{\phi}^{\mathrm{T}}(2) \tag{5.207}$$

$$= \boldsymbol{\chi}(1)\mathrm{a}\boldsymbol{\chi}^{\mathrm{T}}(2). \tag{5.208}$$

That is, the expansion (5.197) reduces to

$$\Phi(1, 2) = \sum_k c_k \chi_k(1)\chi_k(2). \tag{5.209}$$

The normalization condition (5.200) becomes simply

$$\sum_k c_k^2 = 1. \tag{5.210}$$

Equation (5.209) is the natural orbital expansion of the exact wave function. We see that it is much simpler than the original expansion in that all off-diagonal terms have been eliminated. Furthermore the first, second, . . . natural orbitals $\chi_1, \chi_2, \ldots$ and their occupation numbers defined by

$$n_k = c_k^2 \tag{5.211}$$

are inherent properties of the exact wave function, and are quite independent of the choice of the original basis functions ϕ_i.

We note that the diagonal form of the wave function (5.209) leads also to a diagonal expansion for the (spinless) first-order density matrix $\rho(1', 1)$ defined by the equation

$$\rho(1', 1) = 2 \int \Phi(1', 2)\Phi(1, 2)\, dv_2. \tag{5.212}$$

Thus, substituting (5.209) into the definition (5.212) we obtain

$$\begin{aligned}\rho(1', 1) &= 2 \int \sum_k \sum_j c_k c_j \chi_k(1')\chi_k(2)\chi_j(1)\chi_j(2)\, dv_2 \\ &= 2 \sum_k \sum_j c_k c_j \chi_k(1')\chi_j(1)\delta_{kj} \\ &= 2 \sum_k c_k^2 \chi_k(1')\chi_k(1)\end{aligned}$$

that is

$$\rho(1', 1) = 2 \sum_k n_k \chi_k(1')\chi_k(1). \tag{5.213}$$

The normalization condition on $\rho(1, 1')$ is

$$\int \rho(1, 1)\, dv_1 = 2 \sum_k n_k = 2, \tag{5.214}$$

corresponding to the fact that 2 electrons are present.

In the many-electron case it is the diagonalization of the first-order density matrix (including spin) rather than the wave function which defines the natural (spin) orbitals (cf. Ref. 45a). Thus equation (5.213) serves to link the natural orbitals used in this section with the natural spin-orbitals appropriate to the general case of N electrons.

The natural orbitals have several interesting properties related to the rapidity of convergence of the series (5.197) and (5.209).

(i) The series $\Sigma_k c_k^2$ is term-by-term more rapidly convergent than the series $\Sigma_k (\Sigma_l A_{kl}^2)$ for any other superposition of configurations such as (5.197). This property may be generalized to the many-electron case (cf. Ref. 45a).

(ii) The renormalized, truncated natural orbital expansion

$$\Phi_r(1, 2) = \frac{\sum_{k=1}^{r} c_k \chi_k(1)\chi_k(2)}{\left(\sum_{k=1}^{r} n_k\right)^{\frac{1}{2}}} \tag{5.215}$$

has the largest possible overlap

$$S = \iint \Phi(1, 2)\Phi_r(1, 2)\, dv_1\, dv_2 \tag{5.216}$$

with the exact wave function $\Phi(1, 2)$, of all CI wave functions constructed from r linearly independent orbitals. It is here assumed that the phase of Φ is chosen so that S is real and positive. This property, which follows simply from the diagonalization (5.202) and the ordering of the eigenvalues (5.203),[46] is peculiar to the two-electron case.

As a special case of (ii) we see that the first natural orbital χ_1 yields that simple MO wave function

$$\Phi_1(1, 2) = \chi_1(1)\chi_1(2) \tag{5.217}$$

which is the best approximation to $\Phi(1, 2)$ in the sense of maximum overlap. This maximum overlap criterion for the "best" molecular orbital differs somewhat from the minimum energy criterion used in Hartree–Fock theory. Nevertheless we expect a very close resemblance between the first natural orbital and the Hartree–Fock orbital. Numerical calculations have indeed shown a close resemblance in most circumstances (see below). Exceptions may occur in cases of degeneracy or near degeneracy. For example, it is easy to see that, as $R \to \infty$, the first natural orbital for the ground state of H_2 tends to the LCAO form

$$\chi_1 \to \frac{1}{\sqrt{2}}(a + b), \tag{5.218}$$

whereas the asymptotic form of the Hartree–Fock orbital is less simple (and less physically meaningful).[16]

A similar close relationship is to be expected between multi-configuration SCF wave functions such as (5.167) and the corresponding truncated natural orbital expansion. Like the MOs in a multi-configuration SCF function, the natural orbitals (NOs) are automatically symmetry adapted provided there is no accidental degeneracy between the eigenvalues of equation (5.202), and even in this case the NOs may always be chosen in symmetry adapted form. These symmetry properties are evident from the discussion leading up to the most general CI expansion (5.161). The matrix of coefficients A_{ij} of equation (5.161) is already partially diagonal in that there are no terms in the expansion containing products of orbitals of different symmetry species. The full diagonalization of the matrix A, which yields the NOs, does not involve the mixing of different symmetry species, but merely the separate diagonalization of the sub-matrices for each distinct symmetry species.

The NOs of the degenerate symmetry species $\pi_u, \pi_g, \delta_g, \ldots$ will occur in pairs in the natural expansion, the two members of a degenerate pair having equal eigenvalues c_k and occupation numbers n_k. We choose the real forms of these degenerate NOs; the dependence on azimuthal angle being

$$\left.\begin{matrix}\cos\varphi \\ \sin\varphi\end{matrix}\right\} \text{for a degenerate pair of symmetry species } \pi$$

$$\left.\begin{matrix}\cos 2\varphi \\ \sin 2\varphi\end{matrix}\right\} \text{for a degenerate pair of symmetry species } \delta$$

etc.

We now obtain the NO expansion of two of the more accurate wave functions discussed in Section 5.4 and 5.5; the extensive CI wave function of Hagstrom and Shull[31] and the 50-term James–Coolidge type wave function of Kolos and Roothaan.[24] As in all numerical applications of the method, the infinite complete basis ϕ_i of equation (5.197) is curtailed to a large, finite basis ϕ_i $(i = 1, 2, \ldots, M)$ chosen from some convenient complete orthonormal set. The required diagonalization of the matrix A of equation (5.197) is then a straight-forward task. Since only a finite basis is used, and since the original wave functions are only approximate, the NOs which result from the analysis are approximations to the true NOs defined above. This involves some obvious modifications of the properties established above; for example, the truncated NO expansion (5.215) is now the best (maximum overlap) CI wave function which can be constructed using r linearly independent combinations of the M basis functions.

In both cases a preliminary transformation is required to express the original wave function in the form (5.197) (with M replacing ∞), where the ϕ_i form an orthonormal set. For the CI wave function of Hagstrom and Shull we merely have to orthonormalize their basis set (5.169) and express the total wave function in terms of the orthonormal orbitals. A convenient way of carrying out the orthonormalization is the Schmidt procedure described in Appendix 1.

In the NO analysis of the Kolos–Roothaan wave function the preliminary transformation is somewhat more complicated.[47, 48] The basis functions are chosen to be the elliptical functions

$$\phi(n, j, m) = e^{-\alpha\mu}\mu^n \nu^j (\mu^2 - 1)^{m/2}(1 - \nu^2)^{m/2} \begin{matrix}\cos m\varphi \\ \sin m\varphi,\end{matrix} \quad (5.219)$$

where n is limited to the range 0–4, j to the range 0–5 and m to the range 0–5.

The orthonormal basis ϕ_i $(i = 1, \ldots, M)$ is now formed by a

Schmidt orthonormalization of the functions (5.219). The M^2 product functions

$$\phi_i(1)\phi_j(2) \qquad (i, j = 1, 2, \ldots, M), \tag{5.220}$$

now form (part of) a complete orthonormal set of functions in the space of the two electrons. The Kolos–Roothaan wave function $\Phi(1, 2)$ may now be expressed (approximately) in terms of this basis via the equations

$$\Phi(1, 2) = \sum_{i=1}^{M} \sum_{j=1}^{M} A_{ij}\phi_i(1)\phi_j(2) \tag{5.221}$$

where

$$A_{ij} = \iint \phi_i(1)\Phi(1, 2)\phi_j(2)\, dv_1\, dv_2. \tag{5.222}$$

In practice the orthonormalization step and the expansion (5.221) may be conveniently combined with the transformations (5.201) and (5.202) defining the natural orbitals.(47, 48)

Before looking at the results of these natural orbital analyses, we draw attention to the great simplification of a CI wave function such as (5.221), which results from the transformation to natural orbitals. The expansion (5.221) contains M^2 terms, and if we take account of the symmetry condition (5.198), we find that $M(M + 1)/2$ of these terms are independent. By contrast the natural expansion of the same wave function contains only M terms. If the original CI wave function is constructed from symmetry orbitals, the reduction in length of the CI series is less spectacular, since the original wave function is already partially diagonalized. For example, the natural orbital expansion of the 33-term CI wave function of Hagstrom and Shull contains 15 independent terms. Unfortunately, this shortening of the CI series by the natural orbital transformation occurs only for 2-electron systems. In the N electron case it is the first-order density matrix rather than the wave function which is diagonalized, and the length of the CI series may, in fact, increase as a result of the transformation to natural orbitals.

Table 5.6 shows the coefficients c_k of the terms in the natural expansion (5.209) of the 50-term James-Coolidge function of Kolos and Roothaan and the 33-term CI wave function of Hagstrom and Shull. The NOs are listed in decreasing order of $n_k = c_k^2$ for the former, more accurate wave function; the degenerate $\pi, \delta, \ldots$ configurations each yield a pair of NOs with equal n_k, their order in the sequence being determined by $2n_k$. The last row of Table 5.6 gives the sum of the occupation numbers n_k for the listed NOs. The difference between unity and this sum is a measure of that part of the wave function omitted from the truncated NO expansions of

Table 5.6. Coefficients c_k of natural expansions and the multi-configurational self-consistent-field expansion for the ground state of H_2

NO[d]	Symmetry type	NO-expansions: J-C function[a]	NO-expansions: CI function[b]	MC-SCF expansion[c]
1	$1\sigma_g$	0.9911	0.9909	0.9910
2	$1\sigma_u$	−0.0995	−0.1008	−0.0996
3	$1\pi_{uc}$	−0.0460	−0.0466[e]	−0.0462[e]
4	$1\pi_{us}$	−0.0460	−0.0466[e]	−0.0462[e]
5	$2\sigma_g$	−0.0548	−0.0551	−0.0593
6	$1\pi_{gc}$	−0.0084	−0.0089[e]	−0.0091[e]
7	$1\pi_{gs}$	−0.0084	−0.0089[e]	−0.0091[e]
8	$3\sigma_g$	−0.0100	−0.0103	−0.0100
9	$2\sigma_u$	−0.0097	−0.0094	—
10	$1\delta_{gc}$	−0.0069	−0.0066[e]	—
11	$1\delta_{gs}$	−0.0069	−0.0066[e]	—
12	$2\pi_{uc}$	−0.0066	−0.0067[e]	—
13	$2\pi_{us}$	−0.0066	−0.0067[e]	—
14	$4\sigma_g$	−0.0065	−0.0074	—
$\Sigma c_k^2 = \Sigma n_k$		0.999894	0.999962	1.0000

[a] The 50-term James–Coolidge function of Ref. 24 as analysed in Ref. 47. $R = 1.4009$.

[b] The 33-term CI wave function of Ref. 31. $R = 1.40$.

[c] The 6-term OVC wave function of Ref. 26. $R = 1.40$.

[d] The NOs are listed in decreasing order of $c_k^2 = n_k$ ($2n_k$ for the degenerate π, δ ... configurations) for the calculation of Ref. 47.

[e] These numbers are $2^{-\frac{1}{2}}$ times the numbers in Refs 31 and 26. This converts the definitions of Refs 31 and 26 to those of the text and Ref. 47.

the table; Davidson and Jones[47] derive a total of 180 natural orbitals nearly all of which are quite negligible. We see that the 14 natural orbitals of Table 5.6 (10 of which are independent) account for nearly all of the 33-term CI wave function, whose NO expansion contains only 15 independent terms. However, for the Kolos–Roothaan function there is an appreciable contribution (≈ 0.0001) to the occupation number sum from higher natural orbitals.

The agreement between the two NO analyses shown in Table 5.6 is strikingly good. Starting from two accurate wave functions expressed in completely different form we end up with almost identical sets of coefficients. The ordering of the NOs is the same for the two calculations except for the trio $1\delta_g$, $2\pi_u$ and $4\sigma_g$, which appears in the order $4\sigma_g$, $2\pi_u$, $1\delta_g$ for the CI wave function; the values of the coefficients show that these NOs are almost degenerate in both analyses. The dominance of the first natural orbital configuration σ_g^2 is clearly shown by both analyses; the coefficients for the other configurations are smaller by at least an order of magnitude, and they all appear with opposite sign to that of the main configuration. This is typical of all singlet two-electron systems in the absence of degeneracy or near degeneracy of the leading NOs, the same effect being apparent in the natural analysis of the ground-state wave function of He. By contrast, for the ground-state wave function of H_2 at large nuclear separations, there is near degeneracy between $1\sigma_g^2$, $1\sigma_u^2$, and we find $c_1 \approx 2^{-\frac{1}{2}}$, $c_2 \approx -2^{-\frac{1}{2}}$, whereas the other coefficients tend to zero.

Equally striking, and perhaps less expected, is the close agreement between the coefficients c_k of the NO expansions and the coefficients A_i in the multi-configuration SCF calculation of Das and Wahl[26] (cf. equation (5.167)). The criteria of best fit used in these two types of calculation are different; the NO expansion maximizes the overlap with the exact wave function, whereas the MC–SCF expansion is determined by minimizing the expectation value of the energy. Table 5.6 shows that these two criteria are almost equivalent, at least for the determination of the linear expansion coefficients.

All three calculations show that, after the main configuration $1\sigma_g^2$, the dominant configurations are $1\sigma_u^2$, $1\pi_u^2$ and $2\sigma_g^2$ which describe left-right, angular and in-out correlation, respectively. The coefficients of the higher NOs are smaller by almost an order of magnitude; the dominance of these early NOs is, of course, even more clearly demonstrated by the occupation numbers $n_k = c_k^2$. Davidson and Jones[47] find that there is another sharp break in the natural expansion which occurs at the foot of Table 5.6. The NOs omitted from the table all have much smaller occupation numbers than those listed.

These conclusions are confirmed by Table 5.7 which lists the energies of several truncated natural orbital expansions and the corresponding multi-configuration self-consistent field (MC–SCF) calculations. The energies listed for the James–Coolidge type function are taken directly from the work of Davidson and Jones.[47] Hagstrom and Shull[31] and Das and Wahl[30] list energies for somewhat different selections of configurations from those in Table 5.7. We have, therefore, estimated some of the entries in columns 5 and 6 by assuming additivity of the energy lowerings caused by the less important configurations. These estimated values, enclosed in parentheses, are probably correct to within 1 or 2 in the last figure quoted. The values in the final column of the table are also from Davidson and Jones;[47] apart from the first two entries, these values are obtained by revarying the coefficients c_k in the truncated natural expansion (5.215) to minimize the total energy. They are, therefore, upper bounds to accurate MC–SCF results as indicated in the table.

We see from Table 5.7 that the two NO analyses are again in excellent agreement. They also agree very well with the accurate MC-SCF results, indicating that minimum energy and maximum overlap with the exact wave function are almost equivalent criteria for determining the wave function. The somewhat poorer MC-SCF results of Das and Wahl (column 6) clearly reflect the inadequacy of their basis set, especially for the π orbitals (cf. equations (5.168)).

Table 5.7. Energies of truncated expansions for the ground state of H_2 (au)

		Natural Orbital Results			MC-SCF results	
Number of orbitals	Added configurations	Expectation of $\delta(\mathbf{r}_{12})$[a]	J-C function[a]	CI function[b]	Approximate[c]	Accurate[d]
1	$1\sigma_g^2$	0.0440	−1.1335	−1.1334	−1.1332	−1.1336
2	$1\sigma_u^2$	0.0357	−1.1519	−1.1519	−1.1517	−1.1521
4	$1\pi_u^2$	0.0314	−1.1628	−1.1627	(−1.1604)[e]	$\leqslant -1.1629$
5	$2\sigma_g^2$	0.0273	−1.1699	−1.1698	(−1.1682)	$\leqslant -1.1700$
7	$1\pi_g^2$	0.0265	−1.1707	(−1.1706)[e]	(−1.1689)	$\leqslant -1.1707$
8	$3\sigma_g^2$	0.0260	−1.1712	(−1.1711)	(−1.1689)	$\leqslant -1.1712$
9	$2\sigma_u^2$	0.0252	−1.1717	(−1.1716)	–	$\leqslant -1.1717$
11	$1\delta_g^2$	0.0247	−1.1722	(−1.1721)	–	$\leqslant -1.1722$
13	$2\pi_u^2$	0.0239	−1.1727	(−1.1726)	–	$\leqslant -1.1727$
14	$4\sigma_g^2$	0.0234	−1.1730	(−1.1729)	–	$\leqslant -1.1730$
∞		0.0170	−1.1744			−1.1744

[a] See footnote a Table 5.6.
[b] See footnote b Table 5.6.
[c] See footnote c Table 5.6.
[d] From Ref. 47.
[e] Values in parentheses estimated assuming additivity of effects of less important orbitals (see text).

The dominance of the first four configurations $1\sigma_g^2$, $1\sigma_u^2$, $1\pi_u^2$ and $2\sigma_g^2$ is again striking. The first three "correlation orbitals" left-right ($1\sigma_u^2$), angular ($1\pi_u^2$) and in-out ($2\sigma_g^2$) together account for 89% of the correlation error; the remaining six configurations in the table give a further 8% of the correlation energy. After this point Davidson and Jones find that the configuration interaction expansion, even in natural orbital form, converges very slowly. This slow convergence is associated with the difficulty in fitting the cusp at $r_{12} = 0$ (equation (5.182)) by an expansion in terms of smooth functions. The poor fitting of the cusp is dramatically shown by the very poor convergence of the expectation value of the operator $\delta(\mathrm{r}_{12})$. Contributions to this operator arise only from the region of the cusp. The behaviour

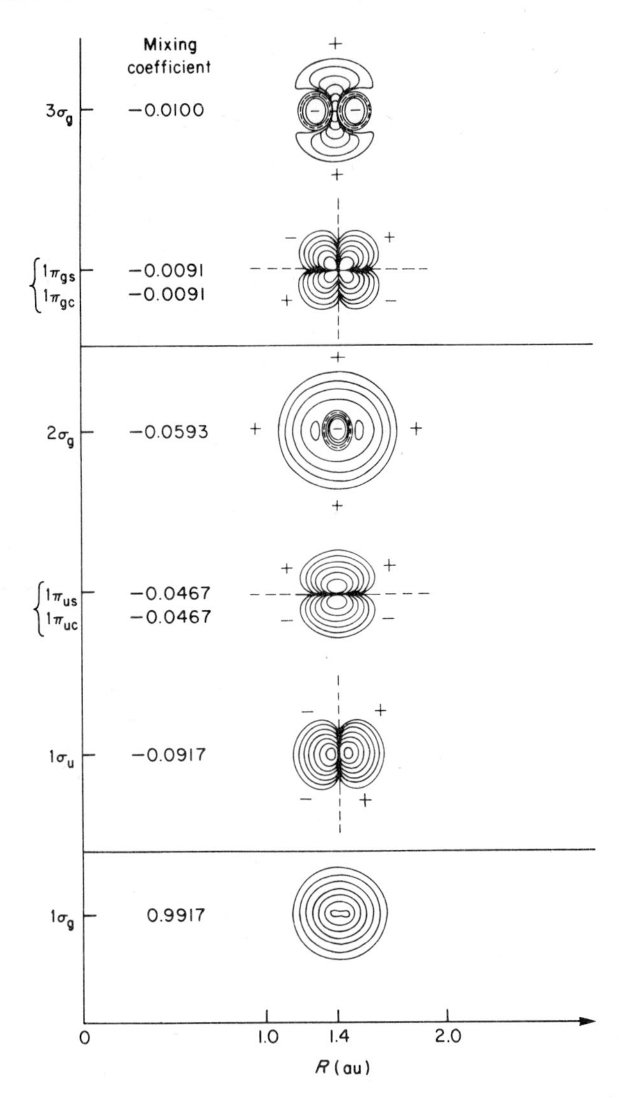

Fig. 5.9. Contour diagrams for the ground orbital $1\sigma_g$ and some correlation orbitals ($1\sigma_u$, $1\pi_u$, $2\sigma_g$; $1\pi_g$, $3\sigma_g$) for the ground state of H_2. The contours are for the charge densities ρ (everywhere positive). The outside contour represents 6.1×10^{-5} e^-/bohr^3 successive inner contours represent increases of ρ by factors of four. The signs (±) are for the orbitals themselves. – – – – –, represents a nodal surface.

of this operator serves as a warning that other quantities of physical interest may converge very much more slowly than the total energy.

The nature of the ground configuration $1\sigma_g^2$ and of the three principal correlating configurations $1\sigma_u^2$, $1\pi_u^2$ and $2\sigma_g^2$ is shown in Fig. 5.9, where contours of the electronic charge density are given for each orbital. The charge density is given by the square of the orbital (for the real orbitals used here) and is everywhere positive; the orbitals themselves change sign whenever we cross one of the nodal surfaces where the orbital and the charge density both vanish.

We see from Fig. 5.9 that the correlating orbitals have about the same spatial extension as the ground orbital $1\sigma_g$; that is, the regions of space where they differ appreciably from zero are about the same. The main difference between the ground orbital and the correlating orbitals is the presence of one or more nodes in the latter. The correlating effect of the orbitals $1\sigma_u$, $1\pi_u$ and $2\sigma_g$ can be understood in terms of these nodal surfaces. It is easy to see that the admixture of a small negative contribution from a correlating configuration to the ground configuration tends to segregate the two electrons on opposite sides of the nodal surface of the correlating orbital. Left–right, angular and in–out correlation are all special cases of this general segregation across nodal surfaces.

The orbitals shown in Fig. 5.9 are those of the MC–SCF (or OVC) calculation of Das and Wahl.[30] Almost identical contours would be obtained from the corresponding natural orbitals (Tables 5.6, 5.7). The higher natural orbitals, some of which appear in these tables, also have either about the same or even smaller extent. They contain an increasing number of nodal surfaces (two each for $1\pi_g$, $3\sigma_g$, $2\sigma_u$, $1\delta_g$ and $2\pi_u$, three for $4\sigma_g$) which result in a more refined description of correlation effects.

The constant or even decreasing spatial extent of the series of natural orbitals is in marked contrast to the behaviour of the molecular orbitals needed to describe excited electronic states. The simplest example of these excited state orbitals is provided by the wave functions of the hydrogen atom. In this case the spatial extent of the orbitals $1s$, $2s$, $2p$, $3s$, $3p$, $3d$, . . . is proportional to the square of the principal quantum number. These orbitals, therefore, become rapidly more diffuse and they are quite inappropriate for describing correlation effects in, say, the ground state of $He((1s)^2\ {}^1S)$. Indeed, even if *all* the (infinite number of) ${}^1S$ functions formed from the discrete H orbitals $1s$, $2s$, $2p$ . . . are included in a configuration interaction calculation, the energy will converge, not to the ground-state energy of He, but to a higher value. This is because the H orbitals do not form a complete set, unless the unquantized continuum orbitals are included.

A molecular example of the contrast between "correlating" orbitals (either NOs or orbitals from MC–SCF calculations) and the orbitals for excited electronic states, is provided by the correlating orbital $1\sigma_u$ and the orbital u' appearing in the description of the excited state ${}^1\Sigma_u^+$ (equation (5.71)). Although $1\sigma_u$ and u' have the same symmetry and nodal properties, the spatial extent of u', determined by the reciprocal of the orbital exponent ζ' (Table 5.1), is about three times larger than that of $1\sigma_u$ and $1\sigma_g$.

In the molecular orbital theory of H_2, and larger molecules, it is useful to distinguish between the *three* types of "excited" orbitals we have encountered. Firstly we have the higher natural orbitals or correlating orbitals needed for a refined CI description of the ground state; these all have about the same spatial extent as the corresponding ground-state MOs but possess additional nodal surfaces. Secondly we have the excited state orbitals needed for a single configuration description of excited electronic states; the optimum form for these excited-state orbitals is determined by separate self-consistent field calculations, one for each electronic state. These excited-state orbitals have similar symmetry and nodal properties to the correlating orbitals but are more diffuse. Thirdly we have the "virtual" orbitals which are obtained from the higher eigenvalues of the LCAO–SCF equations (5.98) and (5.103). As already pointed out in Section 5.3(d) these latter orbitals have no physical or chemical significance and are very dependent on the particular basis set used in the SCF expansion. Nevertheless, for larger molecules one is often forced to use these virtual orbitals, both for the description of excited states, and for improving the ground state by configuration interaction. Such calculations can yield useful qualitative and semi-quantitative results, but the distinction between the "virtual" orbitals and the true excited state or correlating orbitals should always be borne in mind.

It is also useful to distinguish between two levels of configuration interaction in molecular orbital theory. If two (or more) electronic states of the same symmetry type arising from different configurations are degenerate or near-degenerate, it is essential to include them all in the wave function even for a rough qualitative discussion. An example is provided by the H_2 ground state at large nuclear separations; the near-degenerate $1\sigma_g^2$ and $1\sigma_u^2$ configurations must both be included with nearly equal weights to obtain qualitatively correct results. We call this a case of *first-order configuration interaction.* For first-order configuration-interaction calculations the distinction between excited-state orbitals and correlation orbitals is often unimportant. In the example quoted the excited-state orbital $1\sigma_u$ obtained by minimizing the energy of the state $(1\sigma_u)^2$, ${}^1\Sigma_g^+$ is, in fact, similar in extent to the correlating orbital $1\sigma_u$ (the second natural orbital) for large nuclear separations. For more complex molecules it is often necessary to include some first-order configuration interaction even at, or near, the equilibrium nuclear configuration, especially in the treatment of excited states.

The *second-order CI* needed to provide an accurate description of correlation effects is, in general, of quite a different type and is much more extensive. Here the one-electron energies of the correlating orbitals are usually much greater than that of the ground orbital and their relative magnitudes are not of crucial importance; what is important is the strength of their interaction with the ground configuration. This interaction is maximized by localizing the correlation orbitals in the same regions of space as the ground orbitals. To obtain a really accurate description of the correlation a large number of correlating configurations must be used. These all appear in the wave function with quite small coefficients. A typical example of second-order CI is provided by the natural orbital analysis of the ground state of H_2 near equilibrium (Table 5.6). Some authors prefer the term "superposition of configurations" to "second-order CI" to describe this situation.

Of course, intermediate situations arise where it is difficult to assign certain configurations to first-order or second-order configuration interaction in a clear-cut way.

The above results reveal a very close relationship between MC–SCF calculations and the natural orbital analysis at least for the two-electron singlet case. There is an important distinction which must be made, however. The MC–SCF method provides a complete scheme both for the calculation of the wave function and for its analysis. On the other hand the natural orbital method, as given above, is only a method for analysing and interpreting wave functions which have already been obtained by some other method.

This weakness in the NO approach has been partially overcome in several ways. Kutzelnigg[(49)] has derived a set of integro-differential equations for the direct determination of the natural orbitals of two-electron systems. His method has been applied to the ground states of He[(50)] and H_2[(51)]. Alternatively, natural orbital analyses may be incorporated into the process of building up a CI wave function; one carries out a limited CI calculation, simplifies the resulting wave-function by a transformation to natural orbitals, adds more configurations, redetermines the NOs etc. For many-electron systems a modification of the NO scheme involving pseudo-natural orbitals has proved useful.[(45a)]

5.7. Summary

Some calculations on the ground state of H_2 described in this chapter are summarized in Table 5.8. This list is far from complete. McLean *et al.*[(29)] give a comprehensive annotated bibliography of some 60 calculations carried out before 1960 and there have been, perhaps, equally many reported since then.

For each wave function we list the dissociation energy D_e in electronvolts, and the discrepancy from the exact energy in kcal mole^{-1}. The level of accuracy obtainable by chemical methods for such quantities as bond-dissociation energies, activation energies of reactions, etc. is about 1 kcal mole^{-1}. We see that only the best few

Table 5.8. Summary of calculations on the ground state of hydrogen

		Wave function[b]	Eqn.
VAO[a] calculations	(1)	Heitler–London	9
	(2)	S(1)	—
	(3)	FS(1)	—
	(4)	FS(1) + Ionic	30
	(5)	LCAO–MO	46, 61
	(6)	FS(5)	—
	(7)	(6) + CI ≡ (4)	63
Optimum orbital calculations	(8)	Hartree–Fock	46, 95, 105
	(9)	MC–SCF (2 configs)	106, 112, 124. 125
	(10)	(A, B) ≡ (9)	144
	(11)	NOE (2 configs)	215
	(12)	σ-limit	149
	(13)	NOE (4 configs)	215
	(14)	MC–SCF (4 configs)	—
	(15)	NOE (10 configs)	215
	(16)	MC–SCF (10 configs)	—
James–Coolidge type calculations	(17)	5-terms	187
	(18)	13-terms	185
	(19)	50-terms	185
	(20)	80-terms (mod.)	190
			Exact

[a] VAO = valence atomic orbitals, $1s_a$ and $1s_b$.

[b] S = scaled, F = floating, LCAO–MO = linear combination of atomic orbitals–molecular orbital. CI = configuration interaction, MC–SCF = multi-configuration self-consistent-field. (A, B) = optimum open shell form, NOE = natural orbital expansion.

[c] Calculated either at $R_e(\text{exp}) = 1.40$ au or at $R_e(\text{calc.})$

[d] V.T. = virial theorem, E.T. = electrostatic theorem, I.H.F.T. = integral Hellmann–Feynman theorem.

D_e[c] (eV)	Δ (kcal)	$R = \infty$	V.T.[d]	E.T.[d]	I.H.F.T.[d]
3.20	36	✓	x	x	x
3.78	22	✓	✓	x	x
3.93	19	✓	✓	✓	x
4.09	15	✓	✓	✓	x
2.65	48	x	x	x	x
3.58	27	x	✓	✓	x
4.09	15	✓	✓	✓	x
3.64	26	x	✓	✓	x
4.14	14	✓	✓	✓	x
4.14	14	✓	✓	✓	x
4.13	14	✓	(✓)[e]	(✓)[e]	x
4.40	8	✓	✓	✓	✓
4.62	3	✓	(✓)[e]	(✓)[e]	x
4.63	3	✓	✓	✓	x
4.71	1	✓	(✓)[e]	(✓)[e]	x
4.71	1	✓	✓	✓	x
4.53	5	x	✓	x	x
4.72	0.69	x	✓	(✓)[f]	(✓)[f]
4.75	−0.01	x	✓	(✓)[f]	(✓)[f]
4.75	−0.01	✓	✓	(✓)[f]	(✓)[f]
4.75	0	✓	✓	✓	✓

[e] The truncated natural orbital expansions satisfy the virial and electrostatic theorems to the extent that minimum energy and maximum overlap with the exact wave function are equivalent criteria.

[f] These functions satisfy the electrostatic, and integral Hellmann–Feynman theorems approximately, since they differ little from the exact wave function.

wave functions listed in Table 5.8 reach or surpass this level of accuracy. We note that the last two wave functions listed ((19) and (20)) actually overshoot the mark and give calculated energies below the experimental (spectroscopic) value. Although this discrepancy (−0.01 kcal) is of considerable interest in itself (Sections 5.5(a), (b)), since it provides a critical test of quantum mechanics, it is quite negligible for chemical purposes. The final four columns give various qualitative properties of the wave functions; their behaviour on dissociation ($R = \infty$), and the status of the virial, electrostatic and integral Hellmann–Feynman theorems (cf. Chapter 2). We see that, whereas most of the wave functions, including all the better ones ($D_e > 4.5$ eV), satisfy the first two theorems (at least approximately), very few of them satisfy the integral theorem. In this connection it must be pointed out that the selection of wave functions in Table 5.8 is biased in favour of the electrostatic theorem, especially for the VAO calculations. Very few of the calculations listed by McLean[(29)] *et al.* and omitted from Table 5.8 satisfy this theorem.

The calculations fall naturally into three categories separated by horizontal gaps in Table 5.8. Firstly we have valence atomic orbital (VAO) calculations; here the wave function is built up from the atomic orbitals in the valence shell of the atoms, by either valence bond (VB) or LCAO–MO methods. The VAOs (in this case $1s_a$ and $1s_b$) may also be varied by scaling and/or floating. In the second category we have wave functions constructed from optimum orbitals whose form is (in principle) continuously variable. In practice such orbitals are approximated by linear expansions (Section 5.3(d), (e)). Finally, in the third category, we have James–Coolidge type wave functions including r_{12} explicitly.

The three categories give a rough indication of the computational difficulty of the various calculations and of the current state of complete *ab initio* calculations on larger molecules. Complete (all-electron) calculations at about the VAO level of accuracy have recently been carried through for quite a large number of polyatomic molecules of moderate size (e.g. ethane (C_2H_6), formaldehyde (CH_2O), benzene (C_6H_6), pyridine (C_5NH_5), ammonium chloride (NH_4Cl) etc.). Calculations in the optimum orbital category (mostly Hartree–Fock calculations) have been completed for most diatomic molecules (or radicals) formed from first or second row atoms (Li_2, . . ., Cl_2) and, at a slightly lower level of accuracy, for some linear polyatomic molecules (e.g. acetylene C_2H_2, CO_2, etc.). By contrast, progress in extending the results of category three wave functions to larger systems has been almost non-existent; Conroy and Bruner's[(32b)] monumental study of the H_3 energy surface is the only calculation of this type on a molecule with more than two electrons whose accuracy is comparable with that of the last three wave functions listed in Table 5.8 (cf. Ref. 45a).

References

1. W. Heitler and F. London (1927), *Z. Physik* **44**, 455.
2. Y. Sugiura (1927), *Z. Physik* **45**, 484.
3. A. C. Hurley (1954), *Proc. Roy. Soc.* **A226**, 170.
4. S. C. Wang (1928), *Phys. Rev.* **31**, 579.
5. E. F. Gurney and J. L. Magee (1950), *J. Chem. Phys.* **18**, 142.
6. A. C. Hurley (1954), *Proc. Roy. Soc.* **A226**, 179.
7. H. Shull and D. D. Ebbing (1958), *J. Chem. Phys.* **28**, 866.
7a. J. O. Hirschfelder and M. A. Eliason (1967), *J. Chem. Phys.* **47**, 1164.
7b. R. P. Feynman (1939), *Phys. Rev.* **56**, 340.
8. S. Weinbaum (1933), *J. Chem. Phys.* **1**, 593.
9. C. A. Coulson and I. Fischer (1949), *Phil. Mag.* **40**, 386.
10. P. O. Löwdin (1950), *J. Chem. Phys.* **18**, 365.
11. J. C. Slater (1951), *J. Chem. Phys.* **19**, 220.
12. E. Hund (1927), *Z. Physik* **40**, 742.
13. J. E. Lennard-Jones (1929), *Trans. Faraday Soc.* **25**, 668.
14. R. S. Mulliken (1928), *Phys. Rev.* **32**, 186.
15. R. G. Parr (1963), "The Quantum Theory of Molecular Electronic Structure", Benjamin, New York, p. 26.
16. P. O. Löwdin (1959), "Advances in Chemical Physics", Vol. II, p. 207. Interscience, London.
17. G. Herzberg (1951), "Spectra of Diatomic Molecules", D. Van Nostrand Company, Inc., London.
18. S. Huzinaga (1958), *Progr. Theoret. Phys.* **20**, 15.
19. P. E. Phillipson and R. S. Mulliken (1958), *J. Chem. Phys.* **28**, 1248.
20. H. Shull and P. O. Löwdin (1956), *J. Chem. Phys.* **25**, 1035.
21. H. Preuss (1955), *Z. Naturforsch.* **10**, 267.
21a. E. Brändas (1968), *Int. J. Quantum Chem.* **2**, 37.
21b. A. C. Hurley (1960), *J. Chem. Phys.* **33**, 301; **33**, 1872.
22. D. R. Hartree (1957), "The Calculation of Atomic Structures", Chapman and Hall, London.
23. C. A. Coulson (1938), *Proc. Camb. Phil. Soc.* **34**, 204.
24. W. Kolos and C. C. J. Roothaan (1960), *Rev. Mod. Phys.* **32**, 219.
25. J. E. Lennard-Jones and J. A. Pople (1951), *Proc. Roy. Soc.* **A210**, 190.
26. G. Das and A. C. Wahl (1966), *J. Chem. Phys.* **44**, 87.
27. E. R. Davidson and L. L. Jones (1962), *J. Chem. Phys.* **37**, 1918.
28. W. A. Goddard (1967), *Phys. Rev.* **157**, 81.
29. A. D. McLean, A. Weiss and M. Yoshimine (1960), *Rev. Mod. Phys.* **32**, 211.
30. G. Das and A. C. Wahl (1967), *J. Chem. Phys.* **47**, 2934.
31. S. Hagstrom and H. Shull (1963), *Rev. Mod. Phys.* **35**, 624.
32. T. Kato (1957), *Comm. Pure Appl. Math.* **10**, 151.
32a. R. T. Pack and W. Byers-Brown (1966), *J. Chem. Phys.* **45**, 556.
32b. H. Conroy and B. L. Bruner (1967), *J. Chem. Phys.* **47**, 921.
33. H. M. James and A. S. Coolidge (1933), *J. Chem. Phys.* **1**, 825.
34. W. Kolos and L. Wolniewicz (1964), *J. Chem. Phys.* **41**, 3663.
35. W. Kolos and L. Wolniewicz (1965), *J. Chem. Phys.* **43**, 2429.
36. L. Wolniewicz (1966), *J. Chem. Phys.* **45**, 515.
37. W. Kolos and L. Wolniewicz (1964), *J. Chem. Phys.* **41**, 3674.
38. S. Weissman, J. T. Vanderslice and R. Battino (1963), *J. Chem. Phys.* **39**, 222.

39. J. W. Cooley (1961), *Math. Computation* **15**, 363.
40. G. Herzberg and A. Monfils (1960), *J. Mol. Spectroscopy* **5**, 482.
41. G. Herzberg and L. L. Howe (1959), *Can. J. Phys.* **37**, 636.
42. B. P. Stoicheff (1957), *Can. J. Phys.* **35**, 730.
43. W. Kolos (1967), *Int. J. Quantum Chem.* **1**, 169.
43a. G. Herzberg (1970), *J. Mol. Spectroscopy* **33**, 147.
44. W. Kolos and L. Wolniewicz (1966), *J. Chem. Phys.* **45**, 509.
45. P. O. Löwdin (1955), *Phys. Rev.* **97**, 1474.
45a. A. C. Hurley (1976), "Electron Correlation in Small Molecules", Academic Press, London and New York.
46. P. O. Löwdin and H. Shull (1956), *Phys. Rev.* **101**, 1730.
47. E. R. Davidson and L. L. Jones (1962), *J. Chem. Phys.* **37**, 2966.
48. E. R. Davidson (1962), *J. Chem. Phys.* **37**, 577.
49. W. Kutzelnigg (1963), *Theoret. Chim. Acta* **1**, 327.
50. R. Ahlrichs, W. Kutzelnigg and W. Bingel (1966), *Theoret. Chim. Acta* **5**, 289.
51. R. Ahlrichs, W. Kutzelnigg and W. Bingel (1966), *Theoret. Chim. Acta* **5**, 305.

CHAPTER 6

The Determinantal Method

Many of the methods used in the previous chapter to obtain approximate wave functions for the hydrogen molecule may be extended to atoms and molecules with more than two electrons. The principal tool is again the variational method, which is used to optimize the values of variable parameters in some assumed trial wave function.

6.1. The antisymmetry principle

Some complication is caused by Pauli's principle, which requires the total wave function to be antisymmetric under the simultaneous interchange of the space and spin coordinates of any two electrons. For $N > 2$, this principle leads to the exclusion from consideration of some solutions of the N-electron Schrödinger equation

$$H\Phi(\mathrm{r}_1, \mathrm{r}_2, \ldots, \mathrm{r}_N) = E\Phi(\mathrm{r}_1, \mathrm{r}_2, \ldots, \mathrm{r}_N). \tag{6.1}$$

The simplest example of this exclusion is provided by a completely symmetric solution $\Phi(\mathrm{r}_1, \mathrm{r}_2, \mathrm{r}_3)$ of equation (6.1) in the case $N = 3$.

$$\Phi(\mathrm{r}_1, \mathrm{r}_2, \mathrm{r}_3) = \Phi(\mathrm{r}_2, \mathrm{r}_1, \mathrm{r}_3) = \Phi(\mathrm{r}_3, \mathrm{r}_2, \mathrm{r}_1) = \ldots. \tag{6.2}$$

On general grounds we expect that the lowest energy solution of equation (6.1) for $N = 3$ will have this symmetry. To obtain a total wave function

$$\Psi(x_1, x_2, x_3) = \Phi(\mathrm{r}_1, \mathrm{r}_2, \mathrm{r}_3)\Theta(\omega_1, \omega_2, \omega_3) \tag{6.3}$$

which satisfies Pauli's principle, we then require a totally antisymmetric spin function

$$\Theta(\omega_1, \omega_2, \omega_3) = -\Theta(\omega_2, \omega_1, \omega_3) = -\Theta(\omega_3, \omega_2, \omega_1) = \ldots. \tag{6.4}$$

However, it is clearly impossible to construct a three-electron spin function of this symmetry using only the two elementary spin functions $\alpha(\omega)$ and $\beta(\omega)$ introduced in Section 3.4. The totally symmetric solution (6.2) of the Schrödinger equation cannot, therefore, satisfy Pauli's principle and has no physical significance. As the number of electrons increases beyond three, more and more non-physical solutions of the Schrödinger equation arise in this way.

There are two main methods of overcoming this difficulty. The first is to construct all possible spin functions for an N-electron system, investigate their transformations under permutations and consider only those solutions of equation (6.1) whose transformations are the duals of those of the spin functions; that is, those which can combine with the known spin functions to produce a total wave function which is antisymmetric under permutations. A good account of this method is given by Kotani *et al.*[(1)]

A simpler procedure, originally due to Slater,[(2)] is to abandon the strategy of separating the spatial and spin dependence of the wave functions and to combine these two features in the most intimate way from the beginning of the calculations, despite the fact that the Schrödinger equation has no reference to electron spin. We follow this method here.

6.2. Determinantal wave functions

Suppose that $\phi_i(\mathbf{r})(i = 1, 2, \ldots, t)$ is some set of orbitals, that is functions of the spatial coordinates of a single electron. By combining these functions with the spin functions α and β we obtain a set of $2t$ spin-orbitals

$$\begin{aligned}
\psi_1(x) &= \phi_1(\mathbf{r})\alpha(\omega) \\
\psi_2(x) &= \phi_1(\mathbf{r})\beta(\omega) \\
&\vdots \\
\psi_{2t-1}(x) &= \phi_t(\mathbf{r})\alpha(\omega) \\
\psi_{2t}(x) &= \phi_t(\mathbf{r})\beta(\omega).
\end{aligned} \tag{6.5}$$

Wave functions for an N-electron system ($N \leqslant 2t$) are now constructed as determinants containing these spin orbitals. The rows of a determinant are labelled by the electronic space-spin variables x_i, abbreviated by i, and the columns by the subscripts of N spin orbitals. Wave functions of this type are usually referred to as Slater determinants. A typical Slater determinant constructed from the set (6.5) is given by

$$D_K = \frac{1}{\sqrt{N!}} \begin{vmatrix} \psi_{k_1}(1), \psi_{k_2}(1), \ldots, \psi_{k_N}(1) \\ \psi_{k_1}(2), \psi_{k_2}(2), \ldots, \psi_{k_N}(2) \\ \cdots\cdots\cdots\cdots\cdots\cdots \\ \psi_{k_1}(N), \psi_{k_2}(N), \ldots, \psi_{k_N}(N) \end{vmatrix}. \tag{6.6}$$

Here the subscript K specifies a selection $(k_1, k_2, \ldots, k_N)$ of N spin-orbital subscripts from the set $(1, 2, \ldots, 2t)$ of equation (6.5), and, as we shall see below, the factor $1/\sqrt{N!}$ ensures that D_K is normalized if the spin-orbitals ψ_i form an orthonormal set, that is, if

$$\int \psi_i^*(1)\psi_j(1)\,d\tau_1 = \delta_{ij} \qquad (i, j = 1, 2, \ldots, 2t). \tag{6.7}$$

Since the determinant (6.6) changes sign when any two rows are interchanged, a function of this form automatically satisfies Pauli's principle, as does any function formed as a linear combination of Slater determinants. The determinant (6.6) vanishes if any two columns are the same, so that any spin orbital can appear at most once and any spatial orbital at most twice, once associated with α spin and once with β spin. The antisymmetry requirement thus includes earlier versions of the exclusion principle as special cases.

Slater determinants are usually expressed in abbreviated forms, which show only the leading diagonal, and the electronic variables $1, 2, \ldots, N$ are often suppressed. Thus, writing k_i for ψ_{k_i} to eliminate double subscripts, the determinant (6.6) might appear in either of the forms

$$D_K = \frac{1}{\sqrt{N!}} \det\{k_1(1), k_2(2), \ldots, k_N(N)\} \tag{6.8}$$

or

$$D_K = |k_1 k_2 \ldots k_N|. \tag{6.9}$$

Notice that the normalization constant $1/\sqrt{N!}$ is absorbed in the notation of equation (6.9).

It is sometimes convenient to introduce explicitly an antisymmetrizing operator $\mathscr{A}$, which transforms a simple product of spin orbitals into the corresponding Slater determinant. We then have

$$D_K = \mathscr{A} k_1(1)k_2(2) \ldots k_N(N), \tag{6.10}$$

that is

$$D_K = \frac{1}{\sqrt{N!}} \sum_P \epsilon_P P k_1(1)k_2(2) \ldots k_N(N), \tag{6.11}$$

where the operator P permutes the space-spin variables of the N electrons, the summation is over the $N!$ permutations of the symmetric group of degree N and $\epsilon_P = +1$ for even permutations and $\epsilon_P = -1$ for odd permutations.

All the wave functions for H_2 discussed in Chapter 5, except those containing r_{12} explicitly, may be expressed in terms of 2×2 Slater determinants by combining the spin functions of Section 5.1 with the spatial parts of the wave functions. For example, the

Heitler–London function (equation (5.22) and the CI expansion (equation (5.197)) may be written

$$\Psi(^1\Sigma_g^+) = \frac{|a\alpha, b\beta| - |a\beta, b\alpha|}{\sqrt{2(1+S^2)}} = \frac{|a\bar{b}| - |\bar{a}b|}{\sqrt{2(1+S^2)}} \tag{6.12}$$

and

$$\Psi(1,2) = \sum_{i=1} A_{ii}|\phi_i\bar{\phi}_i| + \sum_{i<j} A_{ij}\{|\phi_i\bar{\phi}_j| - |\bar{\phi}_i\phi_j|\}. \tag{6.13}$$

In equation (6.13) and the second form of equation (6.12) we have used another common abbreviation; the spin functions are omitted and the orbitals associated with β spin are distinguished by a bar.

6.3. Matrix elements between Slater determinants

In order to use determinantal functions in molecular theory, we need to evaluate matrix elements of the form

$$Q_{KL} = \int D_K^* \, Q D_L \, d\tau, \tag{6.14}$$

where the integration is over the space and spin coordinates of the N electrons, D_K and D_L are two Slater determinants and Q is some operator depending on the electronic coordinates. This problem was first solved by Slater[2,3] in the case where the spin orbitals appearing in D_K and D_L were chosen from an orthonormal set. We consider the general case of arbitrary spin orbitals and use a simplified form of the original derivation by Löwdin.[4]

All physically significant operators are symmetrical with respect to permutations of the N electrons. This permits an important simplification of the matrix element (6.14). From equation (6.8) we have

$$Q_{KL} = \frac{1}{N!} \int \det\{k_1^*(1), \ldots, k_N^*(N)\} \, Q \det\{l_1(1), \ldots, l_N(N)\} \, d\tau. \tag{6.15}$$

Expanding the two determinants by equation (6.11) we obtain

$$Q_{KL} = \frac{1}{N!} \sum_P \sum_{P'} \int \epsilon_P\{Pk_1^*(1) \ldots k_N^*(N)\} \, Q \epsilon_{P'}\{P'l_1(1) \ldots l_N(N)\} \, d\tau. \tag{6.16}$$

Now the value of the integral in equation (6.16) is unaffected if all the variables of integration are subjected to the permutation P^{-1}. Therefore

$$Q_{KL} = \frac{1}{N!} \sum_P \sum_{P'} \int P^{-1}[\epsilon_P\{Pk_1^*(1) \ldots k_N^*(N)\} Q \epsilon_{P'}\{P'l_1(1) \ldots l_N(N)\} \, d\tau]. \tag{6.17}$$

Using the fact that the symmetric operator Q is unaffected by the permutation P^{-1}, we may write equation (6.17) in the form

$$Q_{KL} = \frac{1}{N!}\sum_P \int k_1^*(1)\ldots k_N^*(N)Q \sum_{P'} \epsilon_P \epsilon_{P'}\{P^{-1}P'l_1(1)\ldots l_N(N)\}\,\mathrm{d}\tau. \tag{6.18}$$

As P' runs over all permutations so does $P^{-1}P'$. Furthermore the parity, even or odd, of the permutation $P^{-1}P'$ is correctly given by $\epsilon_P\epsilon_{P'}$. The sum over P' in equation (6.18) is, therefore, independent of P, giving simply the original determinant $\det\{l_1(1),\ldots, l_N(N)\}$. All the $N!$ terms in the P summation are equal and equation (6.18) reduces to

$$Q_{KL} = \int k_1^*(1)\ldots k_N^*(N)Q\,\det\{l_1(1),\ldots, l_N(N)\}\,\mathrm{d}\tau. \tag{6.19}$$

Consider first the case of a constant operator Q, which we may as well take as the unit operator. Equation (6.19) then gives for the overlap integral S_{KL} between the two Slater determinants D_K and D_L the expression

$$S_{KL} = \int k_1^*(1)\ldots k_N^*(N)\,\det\{l_1(1),\ldots, l_N(N)\}\,\mathrm{d}\tau. \tag{6.20}$$

Since multiplication of all the elements in a row of a determinant by a certain factor multiplies the value of the determinant by the same factor, the factors $k_1^*(1),\ldots, k_N^*(N)$ in equation (6.20) may be allocated one to each row of the determinant to give

$$S_{KL} = \int D_{KL}\,\mathrm{d}\tau, \tag{6.21}$$

where D_{KL} is the determinant

$$D_{KL} = \begin{vmatrix} k_1^*(1)l_1(1), & k_1^*(1)l_2(1),\ldots, & k_1^*(1)l_N(1) \\ k_2^*(2)l_1(2), & k_2^*(2)l_2(2),\ldots, & k_2^*(2)l_N(2) \\ \vdots & & \\ k_N^*(N)l_1(N), & k_N^*(N)l_2(N),\ldots, & k_N^*(N)l_N(N) \end{vmatrix}. \tag{6.22}$$

Substituting equation (6.22) into equation (6.21) and carrying out the integrations over the space and spin coordinates of all the electrons we obtain

$$S_{KL} = \begin{vmatrix} (k_1,l_1), (k_1,l_2),\ldots, (k_1,l_N) \\ (k_2,l_1), (k_2,l_2),\ldots, (k_2,l_N) \\ \vdots \\ (k_N,l_1), (k_N,l_2),\ldots, (k_N,l_N) \end{vmatrix}, \tag{6.23}$$

where

$$(k_i, l_j) - \int k_i^*(1)l_j(1)\,\mathrm{d}\tau_1. \tag{6.24}$$

Thus the overlap integral between two Slater determinants equals the determinant formed from the overlaps of the spin orbitals of one Slater determinant with those of the other.

In order to evaluate the matrix elements of one- and two-electron operators we consider expansions of the determinants D_{KL} and S_{KL} in terms of their signed minors (or cofactors) and their double minors.

Expansion of D_{KL} in terms of the ith row gives

$$D_{KL} = \sum_{m=1}^{N} k_i^*(i) l_m(i) D_{KL}(i \mid m), \tag{6.25}$$

where $D_{KL}(i \mid m)$, the (i, m)th signed minor of D_{KL}, is $(-1)^{i+m}$ times the determinant obtained by removing the ith row and mth column of D_{KL}. We note that $D_{KL}(i \mid m)$ is a function of all the electronic variables except i.

Expanding $D_{KL}(i \mid m)$ in turn and collecting terms we obtain an expansion of D_{KL} in terms of the ith and jth rows;

$$D_{KL} = \sum_{m<n}^{N} \begin{vmatrix} k_i^*(i) l_m(i), & k_i^*(i) l_n(i) \\ k_j^*(j) l_m(j), & k_j^*(j) l_n(j) \end{vmatrix} D_{KL}(ij \mid mn) \tag{6.26}$$

that is

$$D_{KL} = \sum_{m<n}^{N} k_i^*(i) k_j^*(j) \{l_m(i) l_n(j) - l_n(i) l_m(j)\} D_{KL}(ij \mid mn). \tag{6.27}$$

Here $D_{KL}(ij \mid mn)$, a signed double minor of D_{KL}, is $(-1)^{i+j+m+n}$ times the determinant obtained by removing the ith and jth rows and the mth and nth columns of D_{KL}. Here again $D_{KL}(ij \mid mn)$ depends on all the electronic variables except i and j.

From the definitions (6.22) and (6.23) of D_{KL} and S_{KL} we see that the integration of the minors and double minors of D_{KL} over all their variables gives the corresponding minors and double minors of the overlap determinant S_{KL};

$$\int D_{KL}(i \mid m)\, d\tau/d\tau_i = S_{KL}(i \mid m). \tag{6.28}$$

$$\int D_{KL}(ij \mid mn)\, d\tau/d\tau_i\, d\tau_j = S_{KL}(ij \mid mn). \tag{6.29}$$

In equation (6.25) we have separated out the dependence of D_{KL} on the electronic coordinate i. This is just what we need to evaluate the matrix element Q_{KL} of equation (6.19) when Q is a symmetric one-electron operator of the form

$$F = \sum_{i=1}^{N} f(i). \tag{6.30}$$

Thus

$$F_{KL} = \sum_{i=1}^{N} \sum_{m=1}^{N} \int k_i^*(i)f(i)l_m(i)D_{KL}(i \mid m)\, d\tau$$

$$= \sum_{i=1}^{N} \sum_{m=1}^{N} (k_i \mid f \mid l_m) S_{KL}(i \mid m) \tag{6.31}$$

where

$$(k_i \mid f \mid l_m) = \int k_i^*(i)f(i)l_m(i)\, d\tau_i \tag{6.32}$$

and we have used equation (6.28).

Similarly for a symmetric two-electron operator

$$G = \sum_{i<j}^{N} g(i, j) \tag{6.33}$$

we obtain from equation (6.27)

$$G_{KL} = \sum_{i<j}^{N} \sum_{m<n}^{N} \int k_i^*(i)k_j^*(j)g(i, j)\{l_m(i)l_n(j) - l_n(i)l_m(j)\}D_{KL}(ij \mid mn)\, d\tau$$

$$= \sum_{i<j}^{N} \sum_{m<n}^{N} [(k_ik_j \mid g \mid l_m l_n) - (k_ik_j \mid g \mid l_n l_m)] S_{KL}(ij \mid mn) \tag{6.34}$$

where we have used equation (6.29) and the notation

$$(k_ik_j \mid g \mid l_m l_n) = [k_il_m \mid g \mid k_jl_n] = \iint k_i^*(1)k_j^*(2)g(1, 2)l_m(1)l_n(2) \times d\tau_1\, d\tau_2. \tag{6.35}$$

Notice that equation (6.35), unlike the corresponding definition in equation (5.60), involves integration over the spin coordinates of the two electrons as well as the spatial coordinates. For a spin free operator g, the integrals in equation (6.35) vanish unless the spins of k_i, l_m and k_j, l_n match, in which case the integrals reduce to those over spatial coordinates only.

The equations (6.23), (6.31) and (6.34) for matrix elements between Slater determinants are quite general and we may obtain the diagonal matrix elements S_{KK}, F_{KK} and G_{KK} simply by replacing l by k in all the formulae: They are extremely complex and tedious to work with, however, since in general none of the minors and double minors of the overlap determinant vanish. If the overlap determinant $S_{KL} \neq 0$ some saving in labour results from the use of Jacobi's theorem,[4] which expresses the double minors in terms of the minors

$$S_{KL}(ij \mid mn) = \frac{S_{KL}(i \mid m)S_{KL}(j \mid n) - S_{KL}(i \mid n)S_{KL}(j \mid m)}{S_{KL}}. \tag{6.36}$$

Very great simplifications are brought about by selecting the spin orbitals from an orthonormal set. We consider this case now, distinguishing the determinants and matrix elements by a superscript zero.

Suppose that D_K^0 and D_L^0 are Slater determinants constructed from orthonormal spin orbitals. The first step in evaluating the matrix elements is to rearrange the spin orbitals in one determinant, say D_L^0, so that as many as possible of the spin orbitals coincide, that is we have the same spin orbital in the same column of both determinants. If the required permutation is odd this will lead to a change in sign of the matrix element.

After the rearrangement we always have $k_i \neq l_j$ for $i \neq j$ so that, since the spin orbitals are orthonormal

$$(k_i, l_j) = 0 \qquad \text{for all } i \neq j. \tag{6.37}$$

The overlap determinant S_{KL}^0 of equation (6.23) is, therefore, diagonal, the diagonal elements being 1 for coinciding spin orbitals and 0 for non-coincidences of spin-orbitals. It is immediately clear that distinct Slater determinants constructed from orthonormal orbitals are normalized and orthogonal to each other. Furthermore all off-diagonal minors and double minors of S_{KL}^0 vanish and, if there are non-coinciding spin orbitals some of the diagonal ones as well. In fact we have for the minors

$$S_{KL}^0\,(i \mid m) = \Delta_i \delta_{im} \tag{6.38}$$

where

$$\begin{aligned} \Delta_i &= 0 \text{ if there are any non-coincidences other than at } i \\ &= 1 \text{ otherwise} \end{aligned}$$

and for the double minors

$$S_{KL}^0\,(ij \mid mn) = \Delta_{ij} \delta_{im}\, \delta_{jn} \tag{6.39}$$

where

$$\begin{aligned} \Delta_{ij} &= 0 \text{ if there are any non-coincidences other than at } i \text{ and } j \\ &= 1 \text{ otherwise.} \end{aligned}$$

Substituting these expressions (6.38) and (6.39) for the minors and double minors into the general formulae (6.31) and (6.34) we obtain greatly simplified formulae for the matrix elements, namely,

$$F_{KL}^0 = \sum_{i=1}^{N} \Delta_i (k_i \mid f \mid l_i) \tag{6.40}$$

and

$$G_{KL}^0 = \sum_{i<j}^{N} \Delta_{ij} [(k_i k_j \mid g \mid l_i l_j) - (k_i k_j \mid g \mid l_j l_i)]. \tag{6.41}$$

These expressions also include the diagonal matrix elements F^0_{KK}, G^0_{KK} as special cases. For these diagonal elements all the quantities Δ_i and Δ_{ij} are unity and we simply replace l by k in both formulae. For off-diagonal matrix elements the factors Δ_i and Δ_{ij} lead to further reductions.

For one spin orbital non-coincidence $(k_i \to l_i)$ we obtain

$$\begin{aligned} F^0_{KL} &= (k_i \mid f \mid l_i) \\ G^0_{KL} &= \sum_{j=1,(j\neq i)}^{N} [(k_i k_j \mid g \mid l_i k_j) - (k_i k_j \mid g \mid k_j l_i)] \end{aligned} \tag{6.42}$$

whereas for two non-coincidences $(k_i \to l_i,\, k_j \to l_j)$

$$\begin{aligned} F^0_{KL} &= 0 \\ G^0_{KL} &= (k_i k_j \mid g \mid l_i l_j) - (k_i k_j \mid g \mid l_j l_i). \end{aligned} \tag{6.43}$$

If there are more than two non-coincidences all matrix elements of one and two-electron operators vanish.

The most important application of the above formulae is to the calculation of the matrix elements of the Hamiltonian

$$H = \sum_i \left(-\tfrac{1}{2}\nabla_i^2 - \sum_\alpha \frac{Z_\alpha}{r_{i\alpha}} \right) + \sum_{i<j} \frac{1}{r_{ij}} + \sum_{\alpha<\beta} \frac{Z_\alpha Z_\beta}{R_{\alpha\beta}} \tag{6.44}$$

which appears in the electronic Schrödinger equation. Here the final term is a constant which comes outside the integrations. The matrix elements of the other terms are given by the above formulae with

$$f(i) = -\tfrac{1}{2}\nabla_i^2 - \sum_\alpha \frac{Z_\alpha}{r_{i\alpha}} \tag{6.45}$$

and

$$g(i,j) = \frac{1}{r_{ij}}. \tag{6.46}$$

Since the formulae for the matrix elements are so much simpler for orthonormal orbitals, there is a great advantage in working with such orbitals whenever possible. Even if one is primarily interested in matrix elements between determinants of non-orthogonal orbitals, as in valence bond theory, it is often easier to obtain these indirectly via a transformation to an orthonormal basis.

Suppose, for example, we wish to find matrix elements of the Hamiltonian between *all* the independent determinants constructed from the $2t$ spin orbitals of equation (6.5) in the non-orthogonal case. There will be a total of

$$\binom{2t}{N} = \frac{(2t)!}{N!\,(2t-N)!} \tag{6.47}$$

determinants which we denote by

$$D_I \qquad \left(I = 1, 2, \ldots, n = \binom{2t}{N}\right). \tag{6.48}$$

First we orthogonalize the spin-orbitals, say by the Schmidt process (Appendix 1), to obtain the orthonormal set

$$\psi_1^0(x), \psi_2^0(x), \ldots, \psi_{2t}^0(x). \tag{6.49}$$

We now construct all possible determinants

$$D_I^0 \qquad (I = 1, 2, \ldots, n) \tag{6.50}$$

from these orthonormal spin orbitals.

Since the orthonormal spin orbitals (6.49) are linear combinations of the original spin orbitals and vice versa, and since (6.50) includes all determinants constructed from the $\psi_i^0(x)$, it must be possible to express the original determinants D_I as linear combinations of the orthonormal determinants D_I^0;

$$D_I = \sum_{J=1}^{n} D_J^0 T_{JI} \qquad (I = 1, \ldots, n). \tag{6.51}$$

The coefficients T_{JI} may be found by evaluating the overlap integrals

$$\int D_L^{0*} D_I \, d\tau \qquad (L, I = 1, \ldots, n) \tag{6.52}$$

by means of equation (6.23). Since the determinants D_I^0 are orthonormal we find

$$\int D_L^{0*} D_I \, d\tau = \sum_{J=1}^{n} \delta_{LJ} T_{JI} = T_{LI} \qquad (L, I = 1, \ldots, n). \tag{6.53}$$

The matrix elements of the Hamiltonian between the original determinants are then given by

$$H_{IJ} = \int D_I^* H D_J \, d\tau \tag{6.54}$$

$$= \int \sum_{L=1}^{n} (D_L^0 T_{LI})^* H \left(\sum_{M=1}^{n} D_M^0 T_{MJ} \right) d\tau$$

that is

$$H_{IJ} = \sum_{L=1}^{n} \sum_{M=1}^{n} T_{LI}^* H_{LM}^0 T_{MJ}, \qquad (I, J = 1, \ldots, n) \tag{6.55}$$

where

$$H_{LM}^0 = \int D_L^{0*} H D_M^0 \, d\tau. \tag{6.56}$$

Equation (6.55) expresses the required matrix elements in the non-orthogonal basis (6.48) in terms of those in the orthogonal basis (6.50); the latter are easily evaluated using the formulae (6.40) and (6.41).

The transformation equations (6.51) and (6.55) are conveniently expressed in a condensed matrix notation;

$$D = D^0 T, \tag{6.57}$$

$$H = T^{\dagger} H^0 T, \tag{6.58}$$

where $T^{\dagger}$ is the Hermitian conjugate of the matrix T.

The process outlined above is often very much more efficient than a direct application of the general formulae (6.31) and (6.34). In practice it is not necessary to include all possible determinants in the sets (6.48) and (6.50). We can always confine attention to determinants, or sums of determinants, with some particular spatial symmetry (Chapter 4) or spin multiplicity (Section 6.5 following). Furthermore if Schmidt orthogonalization is used we can impose the condition of double occupancy on, say, the first r orbitals of the set $\phi_1, \ldots, \phi_t$ without destroying the linear dependence of the sets (6.48) and (6.50) (cf. Appendix 1).

6.4. The determinantal solution of the electronic Schrödinger equation

The results of the preceding section are already sufficient to provide a simple formal solution of the electronic Schrödinger equation

$$H\Psi = E\Psi \tag{6.59}$$

with the Hamiltonian H given by equation (6.44).

To derive this solution we choose the orbitals $\phi_i(\mathrm{r})$ $(i = 1, 2, \ldots, \infty)$ to be a complete, orthonormal set of functions in the spatial co-ordinates of a single electron. The spin-orbitals $\psi_k(x)$ of equation (6.5) (with $t = \infty$) are then also a complete orthonormal set, so that any normalizable function $\psi(x)$ of the space-spin coordinates of a single electron may be expanded in the form

$$\psi(x) = \sum_k \psi_k(x) c_k \tag{6.60}$$

with

$$c_k = \int \psi_k^*(x)\psi(x)\,\mathrm{d}x. \tag{6.61}$$

The determinants

$$D_K^0 = \frac{1}{\sqrt{N!}} \det\{k_1(1), k_2(2), \ldots, k_N(N)\} \tag{6.62}$$

are then also orthonormal, from equations (6.7) and (6.23). Furthermore if the index K in equation (6.62) runs over all ordered configurations, that is those with $k_1 < k_2 < \ldots k_N$, then the determinants D_K^0 constitute a complete set for all antisymmetric functions of the electronic coordinates $x_1, x_2 \ldots x_n$.[4] Hence any normalized, antisymmetric wave function may be expressed in the form

$$\Psi(x_1, x_2 \ldots x_N) = \sum_K D_K^0 C_K, \tag{6.63}$$

where

$$C_K = \int D_K^{0*} \Psi \, d\tau \tag{6.64}$$

and the normalization condition is

$$\sum_K C_K^* C_K = 1. \tag{6.65}$$

The formal solution to equation (6.59) is now obtained by applying the linear variation method to the wave function (6.63). Since the functions D_K^0 are orthonormal this leads to the secular equation

$$\det\{H_{KL}^0 - E\delta_{KL}\} = 0 \tag{6.66}$$

for the eigenvalues E_i, and the linear equations

$$\sum_L (H_{KL}^0 - E_i \delta_{KL}) C_L = 0 \tag{6.67}$$

for the expansion coefficients C_K (cf. Section 2.2). All the matrix elements H_{KL}^0 appearing in equations (6.66) and (6.67) may be expressed in terms of one- and two-electron integrals using the results of the previous section.

Although it is not feasible to include an infinite number of terms in the expansion (6.63) and so obtain an exact solution of the electronic Schrödinger equation, approximate solutions of progressively increasing accuracy can be obtained by including more and more terms in a large finite expansion. Indeed the majority of explicit wave functions available for molecules larger than H_2 are of the form considered here and differ only in the choice of the orbitals $\phi_i(\mathbf{r})$ and of the configurations K for inclusion in the expansion (6.63).

6.5. Spin eigenfunctions

The secular equation (6.66) may be considerably simplified by exploiting the spatial symmetry and spin properties of the determinantal functions. The simplifications arising from spatial symmetry vary from molecule to molecule and are treated by the methods of Chapter 4. To exploit the spin properties we transform the determinantal functions into spin eigenfunctions, that is, simultaneous eigenfunctions of $\mathbf{S}^2$, the square of the total spin angular momentum, and one of its components, say S_z. Since the electronic Hamiltonian H does not involve the spin variables, both $\mathbf{S}^2$ and S_z commute with H and we know that the matrix elements of H linking spin eigenfunctions with different eigenvalues must vanish. The secular equation therefore factorizes into independent, simpler equations, one for each pair of eigenvalues of $\mathbf{S}^2$ and S_z.

For an N-electron system the components S_x, S_y and S_z of the total spin are simply the sums of the corresponding operators for the individual electrons. Thus

$$S_z = s_z(1) + s_z(2) + \ldots + s_z(N) \tag{6.68}$$

with analogous equations for S_x, S_y, $S_+ = S_x + iS_y$ and $S_- = S_x - iS_y$.

It is easy to verify that the operators S_x, S_y and S_z satisfy the basic commutation relations for angular momentum operators, considered in Section 3.4, namely

$$\begin{aligned} [S_x, S_y] &= iS_z \\ [S_y, S_z] &= iS_x \\ [S_z, S_x] &= iS_y. \end{aligned} \tag{6.69}$$

We may therefore apply all the results given in that section to our present problem, with S replacing J in all the formulae.

The set

$$\Psi(S, M_S) \qquad (M_S = S, S-1, \ldots, -S+1, -S) \tag{6.70}$$

of simultaneous eigenfunctions of $\mathbf{S}^2$ and S_z satisfying

$$\mathbf{S}^2\Psi(S, M_S) = S(S+1)\Psi(S, M_S) \tag{6.71}$$

and

$$S_z\Psi(S, M_S) = M_S\Psi(S, M_S), \tag{6.72}$$

is called a (spin) *multiplet*. The quantum number S is restricted to integral (half-integral) values for systems containing an even (odd) number of electrons, respectively. Since the number of functions in the multiplet (6.70) is $2S + 1$, we have, for an even number of electrons, the possibilities $S = 0$ (singlets), $S = 1$ (triplets), $S = 2$ (quintets), . . .; and, for an odd number of electrons, $S = \frac{1}{2}$ (doublets), $S = 3/2$ (quartets), ...

As in the general case, the $2S + 1$ functions of a spin multiplet are linked by the shift operators S_+ and S_-;

$$S_+\Psi(S, M_S) = \sqrt{(S - M_S)(S + M_S + 1)}\,\Psi(S, M_S + 1), \tag{6.73}$$

$$S_-\Psi(S, M_S) = \sqrt{(S - M_S + 1)(S + M_S)}\,\Psi(S, M_S - 1), \tag{6.74}$$

and we have the convenient expressions

$$\mathbf{S}^2 = S_+S_- + S_z^2 - S_z \tag{6.75}$$

$$\mathbf{S}^2 = S_-S_+ + S_z^2 + S_z, \tag{6.76}$$

for $\mathbf{S}^2$ in terms of S_z and the shift operators.

As we have already mentioned, there are no matrix elements of the electronic Hamiltonian H linking spin eigenfunctions with different values of S or M_S. This is not the only simplification of the secular equation which is brought about by transforming the basis functions into a set of multiplets. We also have the result that, for two sets of functions belonging to the same multiplet system, the matrix elements of H are independent of M_S, that is

$$\int \Phi^*(S, M_S)H\Psi(S, M_S)\,\mathrm{d}\tau = \int \Phi^*(S, M_S')H\Psi(S, M_S')\,\mathrm{d}\tau \tag{6.77}$$

for all values of M_S' lying in the range $S, S-1, \ldots, -S$.

To establish equation (6.77) we use the relation (6.74) to relate the matrix elements for neighbouring members of the multiplet system;

$$\int \Phi^*(S, M_S - 1)H\Psi(S, M_S - 1)\,\mathrm{d}\tau = \frac{\int \{S_-\Phi(S, M_S)\}^* HS_-\Psi(S, M_S)\,\mathrm{d}\tau}{(S + M_S)(S - M_S + 1)}. \tag{6.78}$$

The numerator of the right-hand side of this equation is now rewritten and transformed using the Hermitian character of S_x and S_y (note that S_+ and S_- are *not* Hermitian);

$$\int \{HS_-\Psi(S, M_S)\}(S_x^* + iS_y^*)\Phi^*(S, M_S)\,\mathrm{d}\tau$$
$$= \int \Phi^*(S, M_S)(S_x + iS_y)HS_-\Psi(S, M_S)\,\mathrm{d}\tau. \tag{6.79}$$

But, since S_x and S_y commute with H,

$$\begin{aligned}(S_x + iS_y)HS_- &= H(S_x + iS_y)(S_x - iS_y)\\ &= H(S_x^2 + S_y^2 - i[S_x, S_y])\\ &= H(\mathbf{S}^2 - S_z^2 + S_z)\end{aligned} \tag{6.80}$$

where we have used the first of the commutation relations (6.69). Furthermore

$$\begin{aligned}(\mathbf{S}^2 - S_z^2 + S_z)\Psi(S, M_S) &= \{S(S+1) - M_S^2 + M_S\}\Psi(S, M_S)\\ &= (S + M_S)(S - M_S + 1)\Psi(S, M_S).\end{aligned} \tag{6.81}$$

Substituting these results back into equation (6.78) we obtain

$$\int \Phi^*(S, M_S - 1)H\Psi(S, M_S - 1)\,\mathrm{d}\tau = \int \Phi^*(S, M_S)H\Psi(S, M_S)\,\mathrm{d}\tau. \tag{6.82}$$

Successive applications of equation (6.82) for $M_S = S, S - 1, \ldots$ lead to the desired result (6.77). Because of this result we usually need only one representative function for each multiplet; the other members of the multiplet simply lead to a repetition of identical blocks down the main diagonal of the secular equation. The most convenient representative to choose is, in most cases, the function with the highest value of M_S, that is $\Psi(S, S)$.

We consider first the simple case of a *closed shell*, which, for a molecule without spatial symmetry, is defined as a determinant containing only doubly-occupied orbitals;

$$\Delta_0 = |\phi_1(1)\bar{\phi}_1(2)\phi_2(3)\bar{\phi}_2(4)\ldots\phi_r(2r-1)\bar{\phi}_r(2r)|. \tag{6.83}$$

For a molecule with spatial degeneracy we have the additional requirement that each degenerate set of orbitals is either completely full or completely empty. However, this additional requirement is relevant only when the effect of the spatial symmetry operators is considered.

Now the symmetric one-electron operators, S_z, S_+ and S_-, all commute with the antisymmetrizing operator $\mathscr{A}$ of equation (6.10). Consequently, the effect of one of these operators on a Slater determinant, such as (6.83), is just the same as if the determinant were a simple product of the spin orbitals appearing on the principal diagonal. Therefore we clearly have

$$S_z\Delta_0 = 0, \tag{6.84}$$

and, since

$$s_-(i)\phi(i) = \bar{\phi}(i), \tag{6.85}$$

$$s_-(i)\bar{\phi}(i) = 0, \tag{6.86}$$

each term in S_-, operating on Δ_0, either produces zero directly from equation (6.86) or leads to a determinant with two identical columns. The expression (6.75) for $\mathbf{S}^2$ then shows immediately that a closed shell is a singlet function ($S = 0$). Furthermore, in considering the effect of the spin operators on Slater determinants, we may ignore the presence of closed shells and consider explicitly only those orbitals which are singly occupied.

6.6. The branching diagram

Suppose that there are N singly-occupied orbitals. We may then construct a total of 2^N Slater determinants

$$\Delta_K, \qquad (K = 1, 2, \ldots, 2^N) \tag{6.87}$$

since, for each singly-occupied orbital, we have the choice of α or β spin.

Two simple properties of determinantal functions enable us to determine what multiplets occur in the set (6.87) for any value of N;

(i) Each Slater determinant is already an eigenfunction of S_z, the eigenvalue being half the excess of orbitals with α spin over those with β spin.

(ii) The application of a shift operator to any function of the set (6.87) gives either zero or a linear combination of functions in the set. Consequently, this set must span a number of complete multiplets.

Consider, for example, the case $N = 3$. We then have $2^3 = 8$ determinants with the following values of M_S;

$$\begin{array}{llll}
M_S = \frac{3}{2} & \Delta_1 = |\phi_1\phi_2\phi_3| & & \\
M_S = \frac{1}{2} & \Delta_2 = |\phi_1\phi_2\bar{\phi}_3|, & \Delta_3 = |\phi_1\bar{\phi}_2\phi_3|, & \Delta_4 = |\bar{\phi}_1\phi_2\phi_3| \\
M_S = -\frac{1}{2} & \Delta_5 = |\phi_1\bar{\phi}_2\bar{\phi}_3|, & \Delta_6 = |\bar{\phi}_1\phi_2\bar{\phi}_3|, & \Delta_7 = |\bar{\phi}_1\bar{\phi}_2\phi_3| \\
M_S = -\frac{3}{2} & \Delta_8 = |\bar{\phi}_1\bar{\phi}_2\bar{\phi}_3|. & &
\end{array} \tag{6.88}$$

The function Δ_1 with the highest value of $M_S\,(3/2)$ must belong to a quartet with $S = 3/2$, as may be verified directly using equation (6.76). Successive applications of the shift operator S_- to Δ_1 generate the other three functions of this quartet (equation (6.74)) with $M_S = \frac{1}{2}, -\frac{1}{2}, -3/2$ respectively. The remaining functions spanned by the set (6.88) are then two linearly independent combinations of $\Delta_2, \Delta_3, \Delta_4$ with $M_S = \frac{1}{2}$ and two linearly independent combinations of $\Delta_5, \Delta_6, \Delta_7$ with $M_S = -\frac{1}{2}$. The complete multiplets spanned by these four functions are clearly two doublets.

The results obtained in this way for up to eight singly occupied orbitals are shown in Fig. 6.1.

Figure 6.1 is referred to as the branching diagram, since the results it gives are consistent with a simple vector model for coupling spins. According to this model, the coupling of an extra electron with spin $\frac{1}{2}$ to a multiplet for N electrons with spin S yields two multiplets for $N + 1$ electrons, one with spin $S + \frac{1}{2}$, the other with spin $S - \frac{1}{2}$ (except when $S = 0$, in which case only one multiplet with $S = \frac{1}{2}$ is obtained). Using this rule the numbers f_S^N in the branching

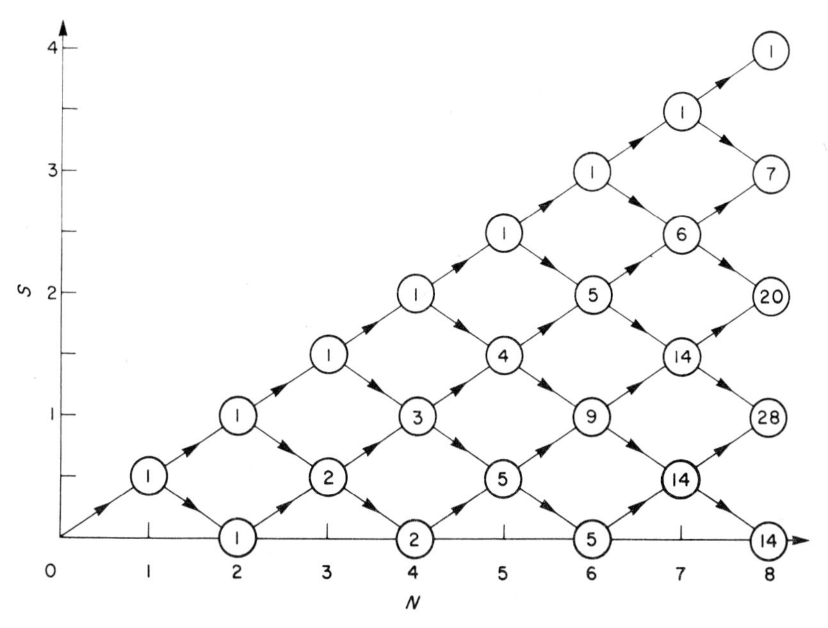

Fig. 6.1. Branching Diagram. The circled numbers are f_S^N, the number of multiplets with total spin S (and multiplicity $2S + 1$) for N singly occupied orbitals.

diagram may be built up successively by forming the sums indicated by the arrows. Another way of expressing this result is to say that the number in a given circle is the number of distinct paths to that circle, starting at the origin and following the arrows.

We may obtain a general expression for f_S^N using the same arguments as for the three electron case. Consider first the total number $n(N, M_S = S + 1)$ of functions with $M_S = S + 1$ in the case of N singly-occupied orbitals. To obtain this value of M_S we must have

$$\tfrac{1}{2}(n_\alpha - n_\beta) = M_S = S + 1 \tag{6.89}$$

where n_α, n_β are the numbers of orbitals with α, β spin respectively.

Since also

$$n_\alpha + n_\beta = N \tag{6.90}$$

we must have

$$n_\alpha = \tfrac{1}{2}N + S + 1, \tag{6.91}$$

and $n(N, M_S = S + 1)$ is given by the number of ways of selecting these n_α orbitals from the total of N, singly-occupied orbitals, that is

$$n(N, M_S = S + 1) = \binom{N}{\tfrac{1}{2}N + S + 1} = \frac{N!}{(\tfrac{1}{2}N + S + 1)!(\tfrac{1}{2}N - S - 1)!}. \tag{6.92}$$

Similarly

$$n(N, M_S = S) = \binom{N}{\tfrac{1}{2}N + S}. \tag{6.93}$$

Now, of the $n(N, M_S = S)$ functions with $M_S = S$, $n(N, M_S = S + 1)$ may be generated from functions with $M_S = S + 1$ by application of S_- (equation (6.74)). These functions, therefore, belong to multiplets with total spin greater than S. The remaining functions with $M_S = S$ must belong to multiplets with total spin S, that is

$$f_S^N = \binom{N}{\tfrac{1}{2}N + S} - \binom{N}{\tfrac{1}{2}N + S + 1} = \frac{(2S + 1)N!}{(\tfrac{1}{2}N + S + 1)!(\tfrac{1}{2}N - S)!}. \tag{6.94}$$

A check on this result is provided by enumerating the total number of functions for N unpaired orbitals. Since there are $2S + 1$ functions in a multiplet with spin S we must have

$$\sum_{S=0 \text{ or } \frac{1}{2}}^{N/2} (2S + 1)f_S^N = 2^N, \tag{6.95}$$

which is easily verified using equation (6.94).

The branching diagram tells us immediately what multiplets to expect in any given situation. The explicit construction of the corresponding spin eigenfunctions is more difficult and several systematic schemes have been developed for this purpose.

If there are only a few singly occupied orbitals and we wish to construct *all* the spin eigenfunctions, the simplest procedure is to start with the function of maximum M_S and use the shift operator S_-, just as we did in deriving the numbers in the branching diagram.

Thus for three singly occupied orbitals, the function Δ_1 of equation (6.88) is already a spin eigenfunction with $S = 3/2$, $M_S = 3/2$.

$$\Psi(\tfrac{3}{2}, \tfrac{3}{2}) = \Delta_1. \tag{6.96}$$

Using equation (6.74) repeatedly we generate the other functions in the quartet;

$$\Psi(\tfrac{3}{2}, \tfrac{1}{2}) = \frac{1}{\sqrt{3}} (\Delta_2 + \Delta_3 + \Delta_4), \tag{6.97}$$

$$\Psi(\tfrac{3}{2}, -\tfrac{1}{2}) = \frac{1}{\sqrt{3}} (\Delta_5 + \Delta_6 + \Delta_7), \tag{6.98}$$

and

$$\Psi(\tfrac{3}{2}, \tfrac{3}{2}) = \Delta_8.$$

Functions belonging to the two doublets ($S = \frac{1}{2}$) with $M_S = \frac{1}{2}$ may be chosen as any independent combinations of Δ_2, Δ_3 and Δ_4 orthogonal to $\Psi(\frac{3}{2}, \frac{1}{2})$, for example the functions

$$\Psi_1(\tfrac{1}{2}, \tfrac{1}{2}) = \frac{1}{\sqrt{2}}(\Delta_2 - \Delta_3) = \frac{1}{\sqrt{2}}(|\phi_1\phi_2\bar{\phi}_3| - |\phi_1\bar{\phi}_2\phi_3|)$$

and (6.99)

$$\Psi_2(\tfrac{1}{2}, \tfrac{1}{2}) = \frac{1}{\sqrt{2}}(\Delta_3 - \Delta_4) = \frac{1}{\sqrt{2}}(|\phi_1\bar{\phi}_2\phi_3| - |\bar{\phi}_1\phi_2\phi_3|).$$

Finally the functions in the two doublets with $M_S = -\frac{1}{2}$ are obtained by applying S_- to the functions (6.99);

$$\Psi_1(\tfrac{1}{2}, -\tfrac{1}{2}) = \frac{1}{\sqrt{2}}(\Delta_5 + \Delta_6 - \Delta_5 - \Delta_7) = \frac{1}{\sqrt{2}}(\Delta_6 - \Delta_7),$$
$$\Psi_2(\tfrac{1}{2}, -\tfrac{1}{2}) = \frac{1}{\sqrt{2}}(\Delta_5 + \Delta_7 - \Delta_6 - \Delta_7) = \frac{1}{\sqrt{2}}(\Delta_5 - \Delta_6). \tag{6.100}$$

We note that the pair of functions (6.99) and likewise the pair (6.100) are not mutually orthogonal. A possible choice of orthonormal pairs is, in each case, Ψ_1 and

$$\Psi_2' = \frac{2}{\sqrt{3}}(\Psi_2 + \tfrac{1}{2}\Psi_1). \tag{6.101}$$

6.7. Extended Rumer diagrams

The procedure used in the previous section for constructing spin eigenfunctions becomes tedious for a large number of singly occupied orbitals, especially since we are usually interested in the states of low multiplicity and require only one representative function for each multiplet. Two systematic methods for constructing single spin eigenfunctions of low multiplicity are the valence-bond method and the projection-operator method. Most applications of valence-bond theory involve the neglect of many integrals and the empirical estimation of others but the technique for forming spin eigenfunctions remains useful in non-empirical calculations on small molecules. The method we follow is based on an extension[5] of the method of Rumer diagrams.

To introduce the method we first re-express the two doublet functions (6.99) in a form which separates the spatial orbitals from the spin functions. Thus

$$\Psi_1(\tfrac{1}{2}, \tfrac{1}{2}) = \mathscr{A}\phi_1(\mathbf{r}_1)\phi_2(\mathbf{r}_2)\phi_3(\mathbf{r}_3)\cdot\alpha(\omega_1)\frac{\alpha(\omega_2)\beta(\omega_3) - \beta(\omega_2)\alpha(\omega_3)}{\sqrt{2}} \tag{6.102}$$

and

$$\Psi_2(\tfrac{1}{2}, \tfrac{1}{2}) = \mathscr{A}\, \phi_1(\mathbf{r}_1)\phi_2(\mathbf{r}_2)\phi_3(\mathbf{r}_3) \cdot \alpha(\omega_3) \frac{\alpha(\omega_1)\beta(\omega_2) - \beta(\omega_1)\alpha(\omega_2)}{\sqrt{2}}, \tag{6.103}$$

where $\mathscr{A}$ is the antisymmetrizer.

The form of the spin functions appearing in equations (6.102) and (6.103) is reminiscent of that for the ground state of H_2 in the Heitler–London theory (equation (5.22)). We might say that the function $\Psi_1(\tfrac{1}{2}, \tfrac{1}{2})$ represents a "bond" between the functions ϕ_2 and ϕ_3, whereas $\Psi_2(\tfrac{1}{2}, \tfrac{1}{2})$ represents a "bond" between the functions ϕ_1 and ϕ_2. These functions may be represented graphically by drawing a circle on which we locate one special point, the pole, and the subscripts of the orbitals in clockwise order starting from the pole.

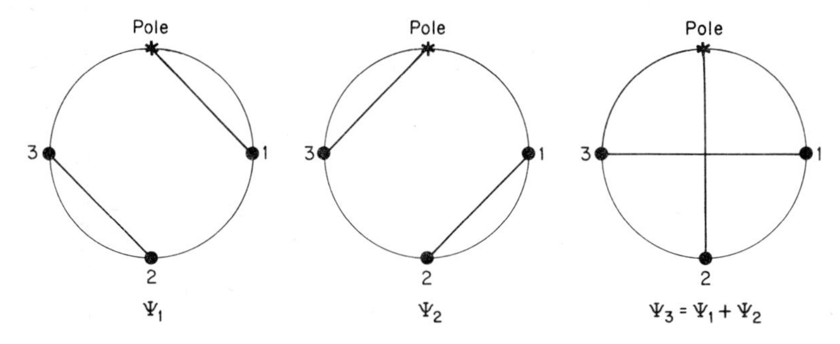

Fig. 6.2. Extended Rumer Diagrams $N = 3$, $S = \tfrac{1}{2}$.

First the unpaired orbitals, which are all assigned α spin, are linked directly to the pole and then the bonded orbitals are linked in pairs (Fig. 6.2).

Also shown in Fig. 6.2 is a third doublet function

$$\Psi_3(\tfrac{1}{2}, \tfrac{1}{2}) = \phi_1(\mathbf{r}_1)\phi_2(\mathbf{r}_2)\phi_3(\mathbf{r}_3) \cdot \alpha(\omega_2) \frac{\alpha(\omega_1)\beta(\omega_3) - \beta(\omega_1)\alpha(\omega_3)}{\sqrt{2}} \tag{6.104}$$

obtained by pairing ϕ_1 and ϕ_3. Clearly the function Ψ_3, which involves a crossing of links, is not independent of Ψ_1 and Ψ_2. In fact

$$\Psi_3 = \Psi_1 + \Psi_2. \tag{6.105}$$

This is a general feature of the method; any function whose diagram involves crossed links may be expressed in terms of diagrams without crossed links, and the diagrams without crossed links are linearly independent.

Using these rules we may immediately write down the independent diagram for any number, N, of singly occupied orbitals and any value S of the total spin. The spin eigenfunction corresponding to a typical diagram is of the form

$$\Psi_r^N(S, S) = \mathscr{A}\phi_1(\mathbf{r}_1)\ldots\phi_N(\mathbf{r}_N)\alpha(\omega_k)\alpha(\omega_l)\ldots$$

$$\frac{\alpha(\omega_u)\beta(\omega_v) - \beta(\omega_u)\alpha(\omega_v)}{\sqrt{2}} \times \frac{\alpha(\omega_x)\beta(\omega_y) - \beta(\omega_x)\alpha(\omega_y)}{\sqrt{2}} \cdots. \tag{6.106}$$

Equation (6.106) gives the spin eigenfunction corresponding to the diagram in which $2S$ points $k, l, \ldots$ are linked directly to the pole, while the remaining $N - 2S$ points $(u, v), (x, y) \ldots$ are paired. To specify the sign of $\Psi_r^N(S, S)$ uniquely we require $u < v$, $x < y, \ldots$.

Using the fact that the antisymmetrizer $\mathscr{A}$ and the spin operators commute, it is evident that

$$S_z\Psi_r^N(S, S) = S\Psi_r^N(S, S), \tag{6.107}$$

and

$$S_+\Psi_r^N(S, S) = 0. \tag{6.108}$$

The expression (6.76) for $\mathbf{S}^2$ then shows that $\Psi_r^N(S, S)$ is, indeed, a spin eigenfunction with the eigenvalues shown.

We have seen that the extended Rumer diagrams with no crossed links lead to independent spin eigenfunctions. To show that the set of functions obtained in this way is complete, that is, includes *all* spin eigenfunctions for a given N and S, we set up a one-to-one correspondence between paths in the branching diagram and extended Rumer diagrams (Fig. 6.3).

First we number the arrows on the path in the branching diagram, starting from the origin. From the starting point of each upward arrow we draw a horizontal line to the right. If this line crosses the end points of one or more downward arrows, we join the two points on the Rumer circle with the same numbers as the upward arrow and the first downward arrow encountered; if not we join the point on the Rumer circle with the same number as the upward arrow to the pole. This procedure leads from a given path on the branching diagram to a uniquely specified extended Rumer diagram.

By reversing the procedure we may go from a given extended Rumer diagram to a definite path on the branching diagram.

For a given value of N and S, therefore, there are as many extended Rumer diagrams as paths through the branching diagram, namely

$$f_S^N = \frac{(2S + 1)N!}{(\frac{1}{2}N + S + 1)!(\frac{1}{2}N - S)!} \tag{6.109}$$

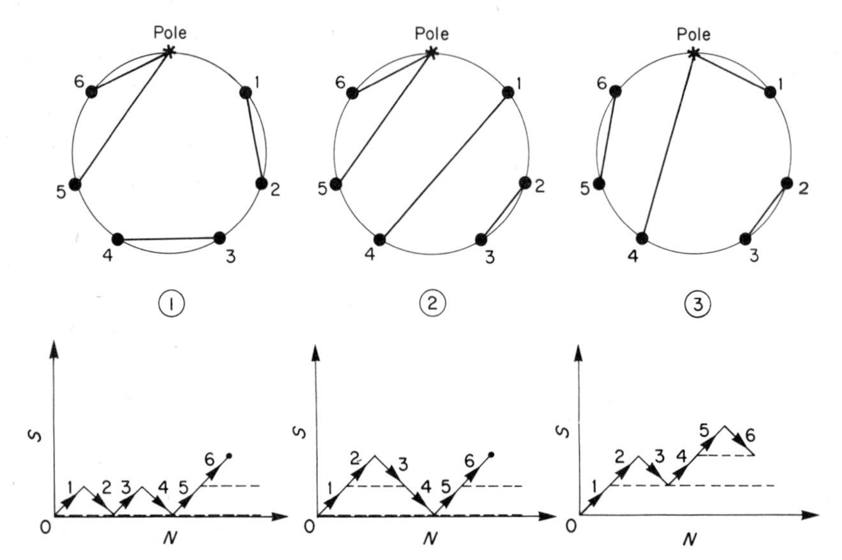

Fig. 6.3. Examples of correspondence between extended Rumer diagrams and paths through the branching diagram $N = 6$, $S = 1$.

and the spin eigenfunctions obtained from the extended Rumer diagrams with no crossed links form a complete set.

Doubly-occupied orbitals may be introduced into the formalism by including a factor

$$\phi_i(\mathbf{r}_i)\phi_i(\mathbf{r}_{i+1}) \ldots \alpha(\omega_i)\beta(\omega_{i+1}) \tag{6.110}$$

in equation (6.106) for each such orbital. This factor is the same as that obtained from equation (6.106) by identifying the two orbitals of a bonded pair, apart from a constant factor $2^{\frac{1}{2}}$. Doubly occupied orbitals may therefore be included in the extended Rumer diagrams provided two points are borne in mind;

(i) such orbitals must always be directly bonded,

(ii) there is a change in normalization by a factor $2^{\frac{1}{2}}$ for each doubly occupied orbital.

6.8. Evaluation of matrix elements

The spin eigenfunction (6.106) (with or without the inclusion of doubly occupied orbitals) may be expanded in terms of 2^{N-2S} Slater determinants, the first few terms in the expansion being as follows:

$$\begin{aligned}\Psi_r^N(S,S) = 2^{S-\frac{1}{2}N}\{ & |\phi_k(k)\phi_l(l)\ldots\phi_u(u)\bar{\phi}_v(v)\phi_x(x)\bar{\phi}_y(y)\ldots| \\ & -|\phi_k(k)\phi_l(l)\ldots\bar{\phi}_u(u)\phi_v(v)\phi_x(x)\bar{\phi}_y(y)\ldots| \\ & -|\phi_k(k)\phi_l(l)\ldots\phi_u(u)\bar{\phi}_v(v)\bar{\phi}_x(x)\phi_y(y)\ldots| \\ & +|\phi_k(k)\phi_l(l)\ldots\bar{\phi}_u(u)\phi_v(v)\bar{\phi}_x(x)\phi_y(y)\ldots| \\ & +\ldots\}. \qquad (6.111)\end{aligned}$$

We may, therefore, evaluate matrix elements of the Hamiltonian (and other one- and two-electron operators) between any pair of spin eigenfunctions using the results of Section 6.3.

Since the number of determinants involved in the evaluation of a single matrix element increases very rapidly with N, this method of evaluation soon becomes very tedious or impracticable if carried out by hand. However, the procedure is readily automated and, for a typical calculation on a small molecule by the method of configuration interaction, the computation time is much shorter than that for other aspects of the calculation; in particular the calculation of the basic one- and two-electron integrals.

Nevertheless, a much more efficient procedure is to devise general rules which give the matrix elements between spin eigenfunctions directly. These rules are an extension of those derived by Pauling(6) and others in early applications of the valence-bond method; they apply only to the case of orthonormal orbitals and this is an additional reason for using such orbitals whenever possible.

Two quite different derivations of the matrix element expressions in the most general case have been given by Reeves(8) and Sutcliffe(9) starting from the expansion (6.111) in terms of Slater determinants and by Cooper and McWeeny(10) starting from spin eigenfunctions of the form appearing in equation (6.106).

Cooper and McWeeny's rules are expressed in terms of superposition patterns obtained from two extended Rumer diagrams as follows:

(i) the pole and links connected to it are removed,

(ii) the link connecting a bonded pair (u, v) is replaced by an arrow directed from u to v,

(iii) the resulting patterns are superposed.

All superposition patterns are built up from two basic elements; islands and chains. An island is a closed sequence of arrows, including the limiting case of an arrow occurring in both spin eigenfunctions. A chain is an open sequence of arrows and includes the limiting case of an isolated point common to both functions. Chains naturally fall into two classes, O chains with an odd number of positions (and an even number of arrows) and E chains with an even number of positions (and an odd number of arrows).

As an example we show in Fig. 6.4 the superposition patterns obtained from the three extended Rumer diagrams of Fig. 6.3. The pattern (1) + (2) consists of one island (1, 2, 3, 4) and two O chains (5) and (6); pattern (1) + (3) two E chains (1, 2, 3, 4) and (5, 6); and pattern (2) + (3) one island (2, 3) and two E chains (1, 4) and (5, 6).

Reeves and Sutcliffe's derivation also involves the analysis of a set of diagrams which are essentially the duals of the superposition patterns. Both versions of the rules are rather complicated in the general case, and the reader is referred elsewhere for details.[8-10, 16]

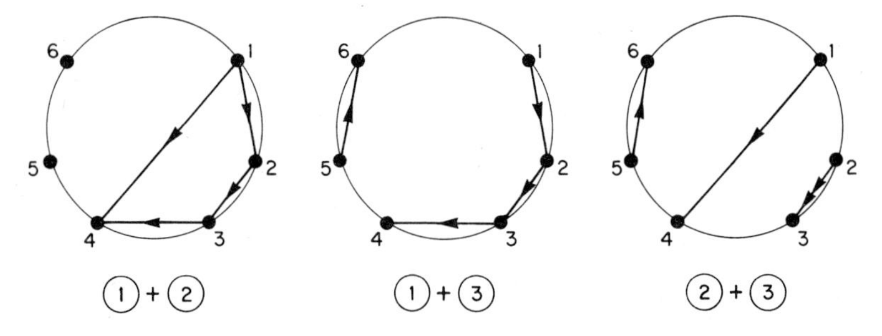

Fig. 6.4. Superposition patterns formed from the extended Rumer diagrams of Fig. 6.3.

6.9. Spin projection operators

Let Λ be any operator with a finite number of distinct eigenvalues $\lambda_1, \lambda_2, \ldots, \lambda_n$ and let Ψ be any wave function. Then, as Dirac[11] shows, Ψ may be expanded in terms of the eigenfunctions of Λ, that is

$$\Psi = \sum_{k=1}^{n} A_k \Psi_k \tag{6.112}$$

$$\Lambda \Psi_k = \lambda_k \Psi_k. \tag{6.113}$$

From equation (6.113) we see that, if the operator $\Lambda - \lambda_k$ is applied to the eigenfunctions Ψ_k and $\Psi_l (l \neq k)$, we obtain

$$(\Lambda - \lambda_k)\Psi_k = 0 \tag{6.114}$$

$$(\Lambda - \lambda_k)\Psi_l = (\lambda_l - \lambda_k)\Psi_l. \tag{6.115}$$

If we divide both these equations by $\lambda_l - \lambda_k$, we find that the operator

$$\frac{\Lambda - \lambda_k}{\lambda_l - \lambda_k} \tag{6.116}$$

annihilates the function Ψ_k and leaves Ψ_l unaltered.

Consider now the effect of applying the operator

$$\mathscr{P}_l = \prod_{k=1\ldots n}^{k\neq l} \frac{\Lambda - \lambda_k}{\lambda_l - \lambda_k} \tag{6.117}$$

to the expansion (6.112). Each term in the expansion with $k \neq l$ is annihilated by one of the factors in $\mathscr{P}_l$, whereas the term $A_l\Psi_l$ is unaltered by any of the factors. Consequently

$$\mathscr{P}_l\Psi = A_l\Psi_l \tag{6.118}$$

and $\mathscr{P}_l$ has the effect of projecting on to that component of Ψ which belongs to the eigenvalue λ_l of Λ.

Application of the operator $\mathscr{P}_l$ to equation (6.118) gives

$$\mathscr{P}_l^2\Psi = \mathscr{P}_l A_l\Psi_l = A_l\Psi_l = \mathscr{P}_l\Psi, \tag{6.119}$$

and, since Ψ is an arbitrary wave function, we conclude that $\mathscr{P}_l$ satisfies the relationship of idempotency

$$\mathscr{P}_l^2 = \mathscr{P}_l \tag{6.120}$$

characteristic of projection operators.

This relation leads to an important simplification when we come to evaluate matrix elements of the Hamiltonian, or other operators, between projected functions.

Suppose that $\mathscr{P}\Phi$ and $\mathscr{P}\Psi$ are two wave functions obtained from Φ and Ψ using a Hermitian projection operator $\mathscr{P}$ which commutes with H; for $\mathscr{P}_l$ above this implies that Λ is Hermitian and commutes with H. Then the matrix elements of H may be simplified by the *turn over rule*

$$\int (\mathscr{P}\Phi)^* H\mathscr{P}\Psi \, \mathrm{d}\tau = \int \Phi^* H\mathscr{P}\Psi \, \mathrm{d}\tau. \tag{6.121}$$

To establish this result we first rearrange the left-hand side to obtain

$$\int (\mathscr{P}\Phi)^* H\mathscr{P}\Psi \, \mathrm{d}\tau = \int (H\mathscr{P}\Psi)\mathscr{P}^*\Phi^* \, \mathrm{d}\tau. \tag{6.122}$$

The right-hand side of equation (6.122) is now transformed, using the facts that $\mathscr{P}$ is Hermitian, commutes with H and is idempotent ($\mathscr{P}^2 = \mathscr{P}$). Thus

$$\int (H\mathscr{P}\Psi)\mathscr{P}^*\Phi^* \, \mathrm{d}\tau = \int \Phi^*\mathscr{P}H\mathscr{P}\Psi \, \mathrm{d}\tau = \int \Phi^* H\mathscr{P}^2\Psi \, \mathrm{d}\tau = \int \Phi^* H\mathscr{P}\Psi \, \mathrm{d}\tau, \tag{6.123}$$

and the turn over rule (6.121) is proved.

In a typical application of the method of projection operators, Φ and Ψ are single determinants while $\mathscr{P}\Phi$ and $\mathscr{P}\Psi$ are sums of many (M say) determinants. The turn over rule (6.121) then reduces a sum containing M^2 terms to one with M terms.

As Löwdin[4] has pointed out the method of projection operators may be used to construct spin eigenfunctions from Slater determinants. As we have seen each determinant is already an eigenfunction of S_z and the possible eigenvalues of the operator $\mathbf{S}^2$ are given immediately by the branching diagram. Compared with the method based on shift operators (Section 6.5), the projection operator method has the advantage of yielding any desired spin eigenfunction directly; there is no need for a prior calculation of all spin eigenfunctions of higher multiplicity.

To illustrate the method we construct a spin eigenfunction $\Psi(\frac{1}{2}, \frac{1}{2})$ for the case of the three singly-occupied orbitals. Here $\mathbf{S}^2$ has the eigenvalue $\frac{1}{2}(\frac{1}{2}+1) = 3/4$ and, from the branching diagram (Fig. 6.1), the only other possible eigenvalue of $\mathbf{S}^2$ is $\frac{3}{2}(\frac{3}{2}+1) = 15/4$. The projection operator for doublet states is therefore

$$^2\mathscr{P} = \frac{\mathbf{S}^2 - 15/4}{3/4 - 15/4} = -\tfrac{1}{3}\mathbf{S}^2 + \tfrac{5}{4}. \tag{6.124}$$

Applying this operator to the determinant

$$\Delta_2 = |\phi_1 \phi_2 \bar{\phi}_3| \tag{6.125}$$

of equation (6.88), which is already an eigenfunction of S_z with $M_S = \frac{1}{2}$, and using the expression (6.76) for $\mathbf{S}^2$, we obtain

$$\begin{aligned}
\Psi(\tfrac{1}{2}, \tfrac{1}{2}) &= (\tfrac{5}{4} - \tfrac{1}{3}\mathbf{S}^2)\Delta_2 \\
&= (\tfrac{5}{4} - \tfrac{1}{3}\cdot\tfrac{3}{4})\Delta_2 - \tfrac{1}{3}S_- S_+ \Delta_2 \\
&= \Delta_2 - \tfrac{1}{3}(|\phi_1\phi_2\bar{\phi}_3| + |\phi_1\bar{\phi}_2\phi_3| + |\bar{\phi}_1\phi_2\phi_3|)
\end{aligned}$$

that is

$$\Psi(\tfrac{1}{2}, \tfrac{1}{2}) = \tfrac{1}{3}(2\Delta_2 - \Delta_3 - \Delta_4). \tag{6.126}$$

We note that $\Psi(\frac{1}{2}, \frac{1}{2})$, being the projection of a normalized function on a non-collinear axis, is not normalized.

A second doublet function, independent of $\Psi(\frac{1}{2}, \frac{1}{2})$ may be found by applying the projection operator $^2\mathscr{P}$ to either of the determinants Δ_3 or Δ_4. It is easily verified that these doublet functions are linear combinations of the pair found earlier using shift operators (equation (6.99)).

The turn over rule, equation (6.121), leads to the following expressions for the normalization integral and the expectation value of the Hamiltonian for the function $\Psi(\frac{1}{2}, \frac{1}{2})$ of equation (6.126);

$$\int \Psi^*(\tfrac{1}{2}, \tfrac{1}{2})\Psi(\tfrac{1}{2}, \tfrac{1}{2})\, d\tau = \tfrac{2}{3}S_{22} - \tfrac{1}{3}S_{23} - \tfrac{1}{3}S_{24}, \tag{6.127}$$

$$\int \Psi^*(\tfrac{1}{2}, \tfrac{1}{2})H\Psi(\tfrac{1}{2}, \tfrac{1}{2})\, d\tau = \tfrac{2}{3}H_{22} - \tfrac{1}{3}H_{23} - \tfrac{1}{3}H_{24}, \tag{6.128}$$

where

$$S_{ij} = \int \Delta_i^* \Delta_j \, \mathrm{d}\tau, \qquad H_{ij} = \int \Delta_i^* H \Delta_j \, \mathrm{d}\tau.$$

The expressions (6.127) and (6.128) are a lot simpler than those obtained by a direct expansion of the left-hand sides of these equations; in more complex situations the reductions effected by the turn over rule are even greater.

As in the case of valence-bond functions, we may derive general rules for the matrix elements of the Hamiltonian and other operators between spin projected Slater Determinants.[12–14] These rules are, however, rather complicated in the general case and the simplifications obtained with the help of the turn over rule are adequate for most practical applications.

It is clear that the method of projection operators is not restricted to problems of spin degeneracy; in atomic theory, similar operators may be constructed from the orbital angular momentum (**L**) and total angular momentum (**J**) operators. Slater[15] makes extensive use of these methods in his recent comprehensive study of atomic structure. The group theoretical operators discussed in Chapter 4 are also of similar type.

References

1. M. Kotani, A. Amemiya, E. Ishiguro and T. Kimura (1955), "Table of Molecular Integrals", Maruzen Co. Ltd., Tokyo.
2. J. C. Slater (1929), *Phys. Rev.* **34**, 1293.
3. J. C. Slater (1931), *Phys. Rev.* **38**, 1109.
4. P. O. Löwdin (1955), *Phys. Rev.* **97**, 1474, 1490, 1509.
5. M. Simonetta, E. Gianinetti and I. Vandoni (1968), *J. Chem. Phys.* **48**, 1579.
6. L. Pauling (1933), *J. Chem. Phys.* **1**, 280.
7. H. Eyring and G. E. Kimball (1933), *J. Chem. Phys.* **1**, 239, 626.
8. C. Reeves (1957), Thesis, University of Cambridge, England.
9. B. T. Sutcliffe (1966), *J. Chem. Phys.* **45**, 235.
10. I. L. Cooper and R. McWeeny (1966), *J. Chem. Phys.* **45**, 226.
11. P. A. M. Dirac (1958), "Quantum Mechanics", Clarendon Press, Oxford, 4th edition.
12. R. Pauncz, J. de Heer and P. O. Löwdin (1962), *J. Chem. Phys.* **36**, 2247.
13. R. Pauncz (1962), *J. Chem. Phys.* **37**, 2739.
14. F. E. Harris (1967), *J. Chem. Phys.* **46**, 2769.
15. J. C. Slater (1960), "Quantum Theory of Atomic Structure" Vols I and II, McGraw-Hill Book Company, Inc., London.
16. R. McWeeny and B. T. Sutcliffe (1969), "Methods of Molecular Quantum Mechanics", Academic Press, London and New York.

CHAPTER 7

Molecular Orbitals and the Hartree–Fock Method

The determinantal method enables us to treat larger diatomic molecules and polyatomic molecules with techniques which are very similar to those used for the hydrogen molecule in Chapter 5. The technique which has been most widely used, for both qualitative discussions and detailed calculations, is the molecular orbital (MO) method which is a natural extension of the orbital theory of atomic structure to molecules.

The basic framework of the MO method was built up soon after the development of quantum mechanics, the principal contributors being Hund,[1] Mulliken[2] and Lennard-Jones.[3] It enables us to interpret the nature of the ground and excited electronic states of molecules, both diatomic and polyatomic, and, in conjunction with spectroscopic observations, has led to a rather complete qualitative picture of the electronic energy levels of many molecular species (cf. Mulliken,[4] Herzberg[5]).

It is only comparatively recently, from about 1950 onwards, that the quantitative version of the MO method has been used to fill in this picture by the explicit construction of molecular electronic wave functions of adequate accuracy.

7.1. Qualitative molecular orbital theory

(a) THE ZEROTH APPROXIMATION

The physical picture underlying the MO method is that of each electron moving independently in a one-electron potential $v(\mathbf{r})$ produced by the nuclei and the averaged effect of the other electrons. In this zeroth approximation the total electronic Hamiltonian H_0 is separable

$$H_0(\mathbf{r}_1, \ldots, \mathbf{r}_N) = \sum_{n=1}^{N} h(\mathbf{r}_n) = \sum_{n=1}^{N} (-\tfrac{1}{2}\nabla_n^2 + v(\mathbf{r}_n)) \tag{7.1}$$

and the MO for each of the N electrons is determined by the one-electron Schrödinger equation

$$(-\tfrac{1}{2}\nabla^2 + v(\mathbf{r}))\phi_i(\mathbf{r}) = \epsilon_i\phi_i(\mathbf{r}). \tag{7.2}$$

The one-electron potential $v(\mathbf{r})$ is assumed to have the same symmetry as the nuclear framework. The MOs determined by equation (7.2) will, therefore, transform according to the irreducible representations of the molecular symmetry group (Chapter 4). The MOs belonging to one-dimensional representations will, in general, be non-degenerate with distinct energies; each such energy level can accommodate two electrons with opposite spins. In the case of an f-dimensional irreducible representation we will have f independent MOs, $\phi_{p+1}, \phi_{p+2}, \ldots, \phi_{p+f}$, say, having equal energies

$$\epsilon_{p+1} = \epsilon_{p+2} = \cdots = \epsilon_{p+f}. \tag{7.3}$$

This set of MOs belonging to the same one-electron energy level is said to form a *shell.* It can accommodate at most $2f$ electrons and in this case the shell is said to be *closed.*

A specific allocation of the available electrons to the various shells defines an electron *configuration* for the molecule. In particular the ground electronic configuration is obtained by allocating the available electrons to the MOs of lowest energy. Most stable molecules contain an even number $N = 2M$ of electrons and their ground electron configurations contain only closed shells. We then have a uniquely determined ground state whose total electronic wavefunction is the Slater determinant

$$\Psi_0 = |\phi_1\bar{\phi}_1\phi_2\bar{\phi}_2 \ldots \phi_M\bar{\phi}_M| \tag{7.4}$$

and whose total electronic energy in zeroth approximation is simply

$$E_0 = 2\sum_{i=1}^{M} \epsilon_i. \tag{7.5}$$

Here we have assumed that the MOs are labelled in order of increasing energy. Of course, if degenerate shells occur, some of the ϵ_i will be the same (equation (7.3)).

Ionized and excited configurations are obtained from the ground-state configuration by removing one or more electrons or by promoting them to shells of higher energy. Assuming that the MOs $i \leqslant M$ and $a > M$ are non-degenerate we can construct in this way two ionized determinants

$$\begin{aligned} \Psi_0(i\to\infty)_1 &= |\phi_1\bar{\phi}_1 \ldots \bar{\phi}_{i-1}\phi_i\phi_{i+1}\bar{\phi}_{i+1} \ldots \phi_M\bar{\phi}_M| \\ \Psi_0(i\to\infty)_2 &= |\phi_1\bar{\phi}_1 \ldots \phi_{i-1}\bar{\phi}_i\phi_{i+1}\bar{\phi}_{i+1} \ldots \phi_M\bar{\phi}_M| \end{aligned} \tag{7.6}$$

and four excited determinants

$$\begin{aligned}
\Psi_0(i\to a)_1 &= |\phi_1\bar{\phi}_1 \ldots \bar{\phi}_{i-1}\ \phi_i\phi_a\phi_{i+1} \ldots \phi_M\bar{\phi}_M| \\
\Psi_0(i\to a)_2 &= |\ldots \quad \ldots \quad \phi_i\bar{\phi}_a \ldots \quad \ldots| \\
\Psi_0(i\to a)_3 &= |\ldots \quad \ldots \quad \bar{\phi}_i\phi_a \ldots \quad \ldots| \\
\Psi_0(i\to a)_4 &= |\ldots \quad \ldots \quad \bar{\phi}_i\bar{\phi}_a \ldots \quad \ldots|.
\end{aligned} \tag{7.7}$$

In the zeroth approximation the two determinants (7.6) are degenerate as are the four determinants (7.7), with the energies

$$E_0(i\to\infty) = 2\sum_{k\neq i}^{M} \epsilon_k + \epsilon_i = E_0 - \epsilon_i \tag{7.8}$$

and

$$E_0(i\to a) = 2\sum_{k\neq i}^{M} \epsilon_k + \epsilon_i + \epsilon_a = E_0 + \epsilon_a - \epsilon_i \tag{7.9}$$

respectively.

From these equations we see that, in this simplest version of the MO method, the ionization potentials of a molecule are just the orbital energies ϵ_i (with reversed sign) and the promotion energy of an excited state is given by the differences between the energies of the MOs involved. The simple relationship between the MO energies and the ionization potentials is retained, to good approximation, in the more accurate Hartree–Fock theory (Section 7.2) but the expressions for excitation energies become more complicated.

Excited and ionized configurations of molecules usually contain one or more incompletely filled, or *open* shells so that there are more molecular spin orbitals (MSOs) $\phi_1, \bar{\phi}_1, \ldots$ available than there are electrons. Then, as in equations (7.6) and (7.7), we can form several independent Slater determinants corresponding to different allocations of the electrons to MSOs. In the zeroth approximation, when interactions between the electrons are neglected, these determinants are all degenerate. Electron interaction leads to a partial lifting of this degeneracy and we obtain several distinct states belonging to each configuration. The nature of this splitting is largely determined by the spin and spatial symmetry properties of the wave functions. The total electronic wave function for each state must have a definite spin multiplicity $(2S + 1)$ and must transform according to some irreducible representation Γ of the molecular symmetry group.

The sorting out of the states which arise from a given configuration is greatly simplified by the fact that a closed-shell wave function such as (7.4) is a singlet function and is totally symmetric, that is, belongs to the identity representation of the molecular symmetry group. The fact that (7.4) represents a singlet function was established in the previous chapter. To see that it is totally symmetric we consider the effect on Ψ_0 of a linear transformation of the orbitals $\phi_1 \ldots \phi_M$. Thus if

$$\Psi_0' = |\phi_1' \bar{\phi}_1' \phi_2' \bar{\phi}_2' \ldots \phi_M' \bar{\phi}_M'| \tag{7.10}$$

with

$$\phi_i' = \sum_{j=1}^{M} \phi_j T_{ji} \tag{7.11}$$

then, from the properties of determinants, we have

$$\Psi_0' = (\det T)^2 \Psi_0 \tag{7.12}$$

so that the total wavefunctions Ψ_0 and Ψ_0' are the same apart from a common factor. For this reason we may assume, without loss of generality, that the orbitals ϕ_i in equation (7.4) form an orthonormal set

$$\int \phi_i^* \phi_j \, dv = \delta_{ij}. \tag{7.13}$$

Consider now the effect of applying a symmetry operation O_R to Ψ_0. Because Ψ_0 represents a closed shell, the orbitals will transform among themselves, as in equation (7.11) without the introduction of any new functions. Furthermore the representation matrices T may be chosen as real and orthogonal with determinant ± 1. Hence from equation (7.12)

$$O_R \Psi_0 = \Psi_0 \tag{7.14}$$

for all elements R of the molecular symmetry group G, and Ψ_0 is totally symmetric. In sorting out the states from a given configuration we need consider only those electrons outside closed shells.

Let us see what these considerations imply for the degeneracies shown in equations (7.6) and (7.7). Since the MOs ϕ_i and ϕ_a have been assumed to be non-degenerate we need consider only spin degeneracy. If we drop the closed shells common to all the wave functions equations (7.6) and (7.7) simplify to

$$\left.\begin{aligned} \Psi_0(i \to \infty)_1 &= \phi_i \qquad (M_S = \tfrac{1}{2}) \\ \Psi_0(i \to \infty)_2 &= \bar{\phi}_i \qquad (M_S = -\tfrac{1}{2}) \end{aligned}\right\} \tag{7.15}$$

and

$$\left.\begin{aligned} \Psi_0(i \to a)_1 &= |\phi_i \phi_a| \qquad (M_S = 1) \\ \Psi_0(i \to a)_2 &= |\phi_i \bar{\phi}_a| \qquad (M_S = 0) \\ \Psi_0(i \to a)_3 &= |\bar{\phi}_i \phi_a| \qquad (M_S = 0) \\ \Psi_0(i \to a)_4 &= |\bar{\phi}_i \bar{\phi}_a| \qquad (M_S = -1). \end{aligned}\right\} \tag{7.16}$$

It is immediately clear that equation (7.15) represents the two components $M_S = \pm\frac{1}{2}$ of a doublet function ($S = \frac{1}{2}$); these functions will, therefore, remain degenerate when electron interaction is included.

The functions of equation (7.16) are easily sorted out into a singlet ($S = 0$) and the three components of a triplet ($S = 1$) using the methods of Section 6.5. Thus applying the operator $\mathbf{S}^2$ to the function $|\phi_i\phi_a|$ and using equation (6.76) we obtain

$$\mathbf{S}^2 |\phi_i\phi_a| = 1(1+1)|\phi_i\phi_a| \tag{7.17}$$

showing that this function is the $M_S = 1$ component of a triplet function ($S = 1$). We obtain the other two components of this triplet by applying the shift operator S_- (equation (6.74)); the singlet function is then found by orthogonalization to the $M_S = 0$ component of the triplet. In this way we find the triplet functions

$$\begin{aligned} {}^3\Psi_0(i \to a)(M_S = 1) &= |\phi_i\phi_a| \\ {}^3\Psi_0(i \to a)(M_S = 0) &= \frac{1}{\sqrt{2}}\{|\phi_i\bar{\phi}_a| + |\bar{\phi}_i\phi_a|\} \\ {}^3\Psi_0(i \to a)(M_S = -1) &= |\bar{\phi}_i\bar{\phi}_a| \end{aligned} \tag{7.18}$$

and the singlet function

$$ {}^1\Psi_0(i \to a) = \frac{1}{\sqrt{2}}\{|\phi_i\bar{\phi}_a| - |\bar{\phi}_i\phi_a|\}. \tag{7.19}$$

The first-order energies of the states from the open-shell configuration $(\phi_i\phi_a)$ are now given by the expectation values of the full many-electron Hamiltonian H with respect to the functions (7.18) and (7.19) (with the closed shells restored). Since H commutes with $\mathbf{S}^2$ and S_z the three triplet functions (7.18) remain degenerate when electron interaction is taken into account (equation (6.77)) and yield the common energy ${}^3E_1(i \to a)$, whereas a different energy ${}^1E_1(i \to a)$ is obtained for the singlet function (7.19). Indeed from equations (6.40)-(6.46) we find explicitly

$$ {}^3E_1(i \to a) = E_C + (\phi_i\phi_a|g|\phi_i\phi_a) - (\phi_i\phi_a|g|\phi_a\phi_i)$$

and

$$ {}^1E_1(i \to a) = E_C + (\phi_i\phi_a|g|\phi_i\phi_a) + (\phi_i\phi_a|g|\phi_a\phi_i)$$

where E_C represents terms common to the singlet and triplet states.

When one or both of the MOs i and a belongs to a degenerate shell ($f > 1$) there are additional degeneracies arising from the spatial symmetry of the molecule. The sorting out of the many-electron states is then more complicated. We return to this problem in Section 7.1(c) after considering the primary task of the qualitative

MO theory: the determination of the order and approximate values of the orbital energies ϵ_i and the qualitative form of the MOs ϕ_i.

(b) ORBITAL ENERGIES AND CORRELATION DIAGRAMS

The qualitative behaviour of the orbital energies $\epsilon_i(R)$ of a diatomic molecule, as a function of the nuclear separation R, may be inferred from the orbital energies of the atoms which are obtained in the two limiting cases $R \to 0$ (united atom) and $R \to \infty$ (separated atoms). Appropriate estimates of these atomic orbital energies may be found by averaging experimental atomic term values over the various states which arise from a given atomic configuration, making allowance for the statistical weight $(2L + 1)(2S + 1)$ of the atomic term ^{2S+1}L.

For example, the atomic configuration $(1s)^2 (2s)^2 (2p)^2$ gives terms ^3P, ^1D and ^1S and in this case we have

$$E_{\mathrm{AV}} [(1s)^2 (2s)^2 (2p)^2] = \tfrac{1}{15} [9E(^3P) + 5E(^1D) + E(^1S)].$$

In the zeroth approximation E_{AV} is just the sum of one-electron energies so that these one-electron energies may be obtained as differences of average-of-configuration energies. One electron energies obtained in this way by Slater[6] are given in Table 7.1 for atoms belonging to the first two rows of the periodic table.

Table 7.1. Orbital energies for atoms (au) (all entries negative)

	$1s$	$2s$	$2p$	$3s$	$3p$
H	0.50				
He	0.91				
Li	2.38	0.20			
Be	4.45	0.34			
B	7.25	0.57	0.21		
C	10.8	0.71	0.39		
N	15.0	0.94	0.48		
O	20.0	1.19	0.59		
F	25.6	1.48	0.68		
Ne	32.0	1.78	0.80		
Na	39.7	2.6	1.40	0.19	
Mg	48.2	3.5	2.05	0.28	
Al	57.6	4.5	2.9	0.42	0.22
Si	67.9	5.7	3.9	0.55	0.29
P	79.1	7.1	5.1	0.67	0.36
S	91.2	8.5	6.3	0.77	0.43
Cl	104.2	10.1	7.6	0.93	0.52
Ar	118.1	12.1	9.3	1.08	0.58

Table 7.1 provides a good deal of information concerning the chemical properties of the elements and the relationships between them embodied in the periodic table. The increase in the binding of the valence electrons $2s$ and $2p$ in the series Li to F reflects the steady increase in electro-negativity across the first row of the periodic table and this pattern is repeated with minor variations for the second row elements Na to Cl.

The following discussion is concerned largely with molecules formed from hydrogen and the first row elements but, using the values of Table 7.1, similar qualitative conclusions may be reached for molecules containing heavier atoms. A fuller discussion of these one-electron energies and an extension of Table 7.1 beyond the second row has been given by Slater.[6]

Consider the behaviour of the orbital energies $\epsilon_i(R)$ of a diatomic molecule AB as we vary the nuclear separation R. In the limiting case $R = 0$ the nuclei coalesce and equation (7.2) reduces to the zeroth-order equation for the orbital energies of the united atom with nuclear charge $Z_A + Z_B$ and $n_A + n_B$ electrons. In increasing order of energy the solutions of equation (7.2) are then the united atom orbitals (cf. Table 7.1)

$$1s, 2s, 2p, 3s, 3p, 3d, \ldots.$$

For small but finite nuclear separation R the symmetry of the system is reduced to $C_{\infty v}$ ($D_{\infty h}$ for a homonuclear molecule); that is, the orbital angular momentum l is no longer well defined but only its component λ along the internuclear axis. The non-degenerate united atom orbitals $1s$, $2s$, $3s$, . . . yield only σ MOs ($\lambda = 0$), while the degeneracy of the p, d, f . . . united atom orbitals is partially lifted to give σ, π ($\lambda = \pm 1$); σ, π, δ ($\lambda = \pm 2$); $\sigma, \pi, \delta, \phi$ ($\lambda = \pm 3$); . . . MOs respectively.

Analogy with the case of H_2^+ (Chapter 3) or a simple perturbation calculation leads us to expect that, for small R, the energies of MOs split from the same united atom orbital will increase with $|\lambda|$, that is $\epsilon_\sigma < \epsilon_\pi < \epsilon_\delta \ldots.$ For homonuclear diatomic molecules we have the additional symmetry requirement that all MOs from united-atom orbitals with l even ($s, d, \ldots$) are even under inversion (g-type), while those from united-atom orbitals with l odd ($p, f, \ldots$) are odd under inversion (u-type).

In the other limiting case, $R \to \infty$, equation (7.2) also simplifies. For very large R it is reasonable to take the effective one-electron potential $v(r)$ as the sum $v_A(r) + v_B(r)$ of two spherically symmetric potentials centred on the nuclei. As $R \to \infty$ the solutions of equation (7.2) then fall into two classes, the atomic orbitals of A, $1s_A$, $2s_A$, $2p_A$ . . . and the atomic orbitals of B, $1s_B$, $2s_B$, $2p_B$ . . .. The relative order of these two sets of orbitals will depend on the nature of the atoms A and B. If A and B are not too different we might expect

$$\epsilon(1s_A) < \epsilon(1s_B) \ll \epsilon(2s_A) < \epsilon(2s_B) < \epsilon(2p_A) < \epsilon(2p_B). \quad (7.20)$$

This order has been assumed in drawing Fig. 7.2 below. From the atomic energies listed in Table 7.1 we see that this diagram is appropriate for the almost homonuclear molecules CN, NO and CO, but that for molecules such as LiF and BeO with a larger difference in nuclear charge a different diagram, with the order of $\epsilon(2s_B)$ and $\epsilon(2p_A)$ reversed, is more appropriate.

In the homonuclear case (B = A) the individual separated atom orbitals $1s_A$, $1s_B$ . . . are not symmetry orbitals, that is, they do not transform as irreducible representations of the symmetry group $D_{\infty h}$. To obtain symmetry orbitals we must take sums and differences of the separated-atom orbitals. For large but finite R the degeneracy of p- and d-type orbitals is also partially lifted. In this way, just as for H_2^+ in Chapter 3, we obtain the bonding orbitals

$$\sigma_g(2p) \sim 2p_A^0 - 2p_B^0, \quad \sigma_g(1s) \sim 1s_A + 1s_B, \quad \pi_u(2p) \sim 2p_A^{\pm 1} + 2p_B^{\pm 1} \ldots \quad (7.21)$$

and the antibonding orbitals

$$\sigma_u(1s) \sim 1s_A - 1s_B, \quad \sigma_u(2p) \sim 2p_A^0 + 2p_B^0, \quad \pi_g(2p) \sim 2p_A^{\pm 1} - 2p_B^{\pm 1}. \quad (7.22)$$

In equations (7.21) and (7.22) the superscripts on the p atomic orbitals give the z component of the orbital angular momentum. The z axes on the two atoms are taken parallel to each other and along the internuclear axis. This is why we have a minus sign in $\sigma_g(2p)$ and a plus sign in $\sigma_u(2p)$. The alternative convention of inwardly directed z axes is frequently used.

To complete the picture for large values of R we need to know the energy order of the bonding and antibonding orbitals correlating with a degenerate $p, d, \ldots$ level of the separated atoms. We assume that, for very large R, all bonding orbitals (7.21) approach the limit from below, all antibonding orbitals from above and that this splitting decreases with increasing $|\lambda|$. The latter assumption may be justified by noting that, for large R, the atomic orbitals $2p_A^0$, $2p_B^0$ overlap more strongly than do $2p_A^{\pm}$ and $2p_B^{\pm}$. These assumptions lead to the ordering

$$\sigma_g(2p) < \pi_u(2p) < \pi_g(2p) < \sigma_u(2p) \qquad (\text{large } R), \quad (7.23)$$

shown in Figure 7.1.

We have now established the behaviour of the orbital energies in the two extreme regions of small and large R. The qualitative behaviour for intermediate nuclear separations can be inferred from the non-crossing rule which states that, in the absence of accidental degeneracy, two energy levels can cross only if they are of different symmetry type. To see this we consider the variation of two MOs $\phi_1(R)$, $\phi_2(R)$ and their energies $\epsilon_1(R)$, $\epsilon_2(R)$ with nuclear separation. These MOs which may be exact solutions of equation (7.2) or linear variational approximations, are chosen as real functions; for each value of R they span a two-dimensional function space. In each of these spaces we choose a real orthonormal basis $\chi_a(R)$, $\chi_b(R)$ which is arbitrary except for two requirements:

(i) the basis is symmetry adapted. For exact solutions of equation (7.2) this is always possible since, in the absence of accidental degeneracy, $\phi_1(R)$ and $\phi_2(R)$ are of pure symmetry types, which may be the same or different.

(ii) $\chi_a(R)$ and $\chi_b(R)$ vary smoothly with R.

The orbital energies $\epsilon_1(R)$, $\epsilon_2(R)$ now appear as the two roots of the determinantal equation

$$\begin{vmatrix} H_{aa}(R) - \epsilon, & H_{ab}(R) \\ H_{ba}(R), & H_{bb}(R) - \epsilon \end{vmatrix} = 0 \qquad (7.24)$$

where the matrix elements H_{aa}, $H_{ba} = H_{ab}$ and H_{bb} are real continuous functions of R.

In order that the two roots of equation (7.24) should coincide two independent conditions must be satisfied

$$H_{aa}(R) - H_{bb}(R) = 0 \qquad (7.25)$$

and

$$H_{ab}(R) = H_{ba}(R) = 0. \qquad (7.26)$$

Consider first the case when the MOs ϕ_1, ϕ_2 are of different symmetry types. The symmetry adapted basis functions must then coincide with the MOs, apart from unimportant phase factors, and the condition (7.26) is automatically satisfied for all values of R (cf. Theorem 6, p. 96). It is then quite possible for the remaining condition (7.25) to be satisfied for some value of R; if this happens the curves $\epsilon_1(R)$ and $\epsilon_2(R)$ cross at this value of R.

If, on the other hand, ϕ_1 and ϕ_2 are of the same symmetry type there is no *a priori* reason why either of the conditions (7.25) or (7.26) should be satisfied. Although each of the left hand sides may well vanish for certain values of R, it is infinitely improbable that both should vanish for the same R value. Excluding this case of accidental degeneracy we see that two curves of the same symmetry type will not cross as R is varied.

For any diatomic molecule we may use the non-crossing rule to correlate the short- and long-range behaviour of the orbital energies.

For each symmetry type the extreme segments of the curves are labelled in order of increasing energy and corresponding segments linked by smooth curves. The qualitative results of this procedure for all homonuclear diatomic molecules up to Ar_2 may be represented on a single correlation diagram (Fig. 7.1) in which the orbital energies are plotted against a suitably scaled or "effective" nuclear separation. The approximate positions on this scale of the equilibrium separations of stable molecules are also shown. For each symmetry type the orbitals are labelled in order of increasing energy. Using the diagram we may immediately translate these labels into the united-atom or separated-atom terminology. With the same notation as for H_2^+ in Chapter 3, we have for example

$$\begin{aligned} 2p\sigma_u &\leftarrow 1\sigma_u \rightarrow \sigma_u(1s) \\ 3d\pi_g &\leftarrow 1\pi_g \rightarrow \pi_g(2p). \end{aligned} \tag{7.27}$$

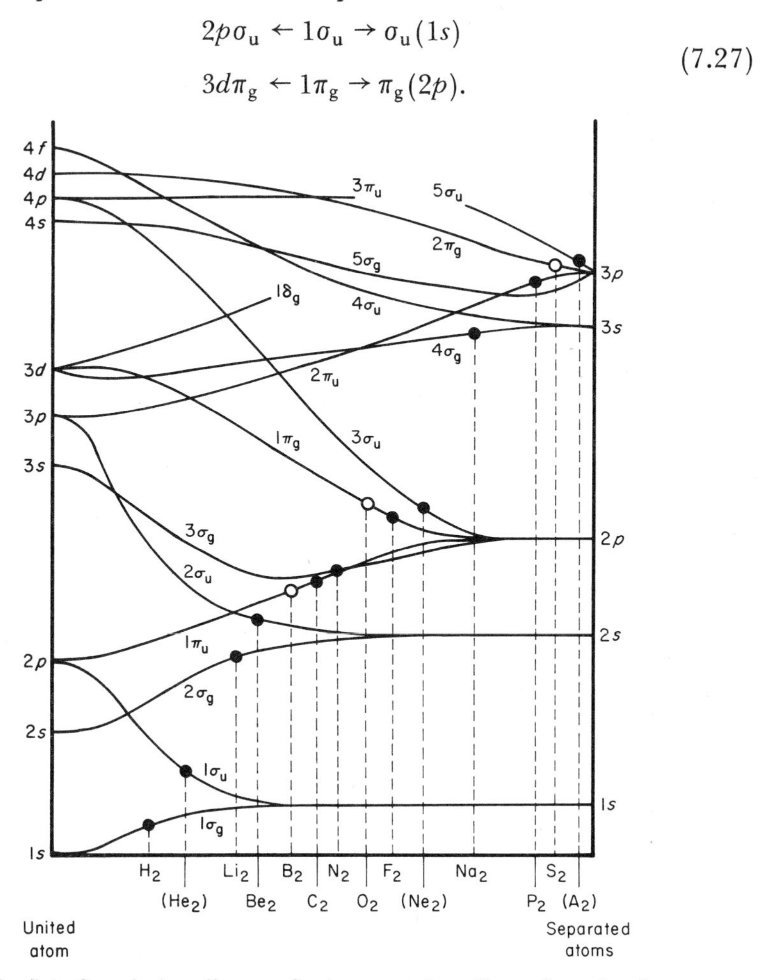

Fig. 7.1. Correlation diagram for homonuclear diatomic molecules.

In Fig. 7.1 the energies of the bonding MOs (7.21) fall steadily as we go from the separated atoms to the united atom, those of the antibonding MOs (7.22) rise steadily. The only exceptions to this behaviour are the orbital

$$3s\sigma_g \leftarrow 3\sigma_g \rightarrow \sigma_g(2p)$$

and its analogue $5\sigma_g$ for *M*-shell atoms. For large effective nuclear separations (O_2, F_2) $3\sigma_g$ behaves as a bonding orbital as discussed above. However, the non-crossing rule requires that $\sigma_g(2p)$ should correlate with the united atom orbital $3s\sigma_g$ with an increase of principal quantum number and hence of energy. For intermediate effective nuclear separations (C_2, N_2) $3\sigma_g$ behaves as a non-bonding or weakly bonding orbital.

The precise form of the curves in Fig. 7.1 incorporates a good deal of experimental and theoretical data on the states of homonuclear diatomic molecules. The basis of the figure is the correlation diagram of Mulliken's 1932 review article,[4] in which MO configurations were assigned to most of the then known states of diatomic molecules and a large number of states were successfully predicted. Figure 7.1 is essentially Mulliken's diagram subjected to a clarifying transformation and slightly modified in the light of more recent experimental and theoretical data. The transformation has the effect of making the bonding and antibonding character of the MOs immediately evident from the slopes of the lines.

From Fig. 7.1 we see that the MOs $1\sigma_g$, $1\sigma_u$ formed from $1s$ separated atom orbitals, which are strongly bonding and antibonding for H_2 and the unstable molecule He_2 (at $\sim 1 \times 10^{-8}$ cm) are but little split for Li_2, giving a very weak bonding and antibonding character. For all heavier homonuclear diatomics these orbitals are essentially degenerate and non-bonding. Similarly $2\sigma_g$ and $2\sigma_u$, strongly split for Li_2 are more weakly split for C_2, N_2 and essentially degenerate and non-bonding for F_2 and all heavier molecules. Again recent Hartree–Fock results for Na_2, and experimental ionization potentials show a small residual splitting of the MOs ($3\sigma_g$, $1\pi_u$, $1\pi_g$, $3\sigma_u$) the order of the levels being in accord with Fig. 7.1.

Figure 7.2 shows the correlation diagram for heteronuclear diatomic molecules whose atoms differ but little in atomic number. The main differences from Fig. 7.1 arise from the finite splitting of the bonding and antibonding orbitals even for infinite nuclear separation; this leads to a correlation of the bonding orbitals with atomic orbitals of the heavier atom (A) and the antibonding orbitals with those of B. The application of Fig. 7.2 is naturally restricted to molecules for which the atomic orbital levels fall in the order given by (7.20); it cannot be used for molecules containing atoms of widely differing atomic number such as diatomic hydrides.

In Table 7.2 we summarize the bonding and antibonding character

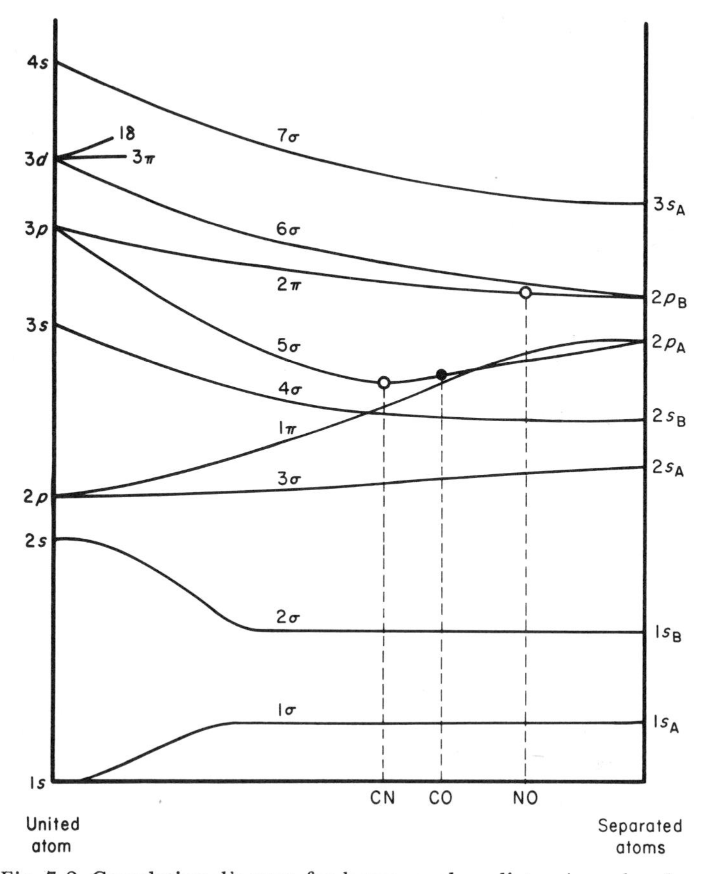

Fig. 7.2. Correlation diagram for heteronuclear diatomic molecules.

of the MOs for a typical heteronuclear diatomic molecule formed from first row atoms (e.g. CO), the approximate LCAO forms of these MOs appropriate for intermediate to large nuclear separations and the correspondence to the MOs of an iso-electronic homonuclear molecule.

The approximate forms in Table 7.2 are the simplest possible LCAO expressions Schmidt-orthogonalized in order of increasing energy (Appendix 1). As $R \to \infty$, $a, d, e, h, k, n \to 1$; $b, c, f, g, l, m \to 0$ and the orthogonalization terms involving the λs disappear. For values of R near equilibrium, however, these terms are important, especially the one orthogonalizing $5\sigma(3\sigma_g)$ to $3\sigma(2\sigma_g)$ which is largely responsible for the weakly bonding character of the 5σ orbital. At these nuclear separations there is an additional mixing (hybridization) of these two orbitals $3\sigma(2\sigma_g)$, $5\sigma(3\sigma_g)$ (and to a lesser extent all orbitals of the same symmetry type) just as in the case of the Dickinson

Table 7.2. MOs for a heteronuclear diatomic molecule BA

MO	Approximate LCAO form	Character	Corresponding homonuclear orbital
1σ	$1s_A$	Non-bonding	$1\sigma_g$
2σ	$1s_B - \lambda_{21} 1\sigma$	Non-bonding	$1\sigma_u$
3σ	$a2s_A + b2s_B - \sum_{n=1}^{2} \lambda_{3n} n\sigma$ $(ab > 0)$	Bonding	$2\sigma_g$
4σ	$c2s_A + d2s_B - \sum_{n=1}^{3} \lambda_{4n} n\sigma$ $(cd < 0)$	Antibonding	$2\sigma_u$
5σ	$e2p_A^0 + f2p_B^0 - \sum_{n=1}^{4} \lambda_{5n} n\sigma$ $(ef < 0)$	Weakly bonding	$3\sigma_g$
6σ	$g2p_A^0 + h2p_B^0 - \sum_{n=1}^{5} \lambda_{6n} n\sigma$ $(gh > 0)$	Antibonding	$3\sigma_u$
$1\pi^{\pm 1}$	$k2p_A^{\pm 1} + l2p_B^{\pm 1}$ $(kl > 0)$	Bonding	$1\pi_u$
$2\pi^{\pm 1}$	$m2p_A^{\pm 1} + n2p_B^{\pm 1} - \lambda 1\pi^{\pm 1}$ $(mn < 0)$	Antibonding	$1\pi_g$

function for H_2^+ (Section 3.2(d)). This too has the effect of increasing the bonding power of $3\sigma(2\sigma_g)$ at the expense of that of $5\sigma(3\sigma_g)$.

In Figs 7.1 and 7.2 we show the ground electronic configurations of a number of diatomic molecules; these are obtained simply by allocating the available electrons to the shells in increasing order of energy. In most cases a closed-shell configuration results, as indicated by a dot (●) on the curve for the highest filled shell; all these cases yield a unique $^1\Sigma_g^+$ ground state (Section 7.1(a)). However, for the molecules B_2, O_2 and, of course, for the odd-electron species CN and NO we obtain an incompletely filled or open uppermost shell indicated by an open circle (○) in Figs 7.1 and 7.2. Such open shells appear in almost all the excited configurations obtained by promoting one or more electrons from the ground configuration. We turn now to the problem of determining the number and symmetry type of the many-electron states which result from these open-shell configurations.

(c) THE SYMMETRY OF MO STATES

We consider first the case of a configuration with a single open shell of degeneracy f containing $n < 2f$ electrons. As we saw in Section 7.1(a) the remaining (closed) shells may be ignored in sorting out the electronic states. The case $n = 1$ is, therefore, trivial giving a doublet total electronic wave function of the same spatial symmetry as the singly-occupied orbital.

For two or more electrons in the same open shell (equivalent electrons) we may construct several Slater determinants, which will generate a reducible representation of the molecular symmetry group G. This representation appears as the direct sum of sub-representations, one for each value of M_S, the z-component of the total spin. The spatial symmetry types for a given value of the total spin S are then given by the formula

$$\chi_S = \chi(M_S = S) - \chi(M_S = S + 1) \tag{7.28}$$

relating the character of the representation spanned by functions with a given value of S to those spanned by functions with given values of M_S. Equation (7.28) follows from the argument used in constructing the branching diagram Fig. 6.1; the determinantal functions with $M_S = S$ include not only those with total spin S but also functions of higher total spin generated from functions with $M_S = S + 1$ by the application of S_-. Of course, when S is equal to the maximum value of M_S for the configuration there are no functions with $M_S = S + 1$ and the second term in equation (7.28) disappears. Consider the case of $n = 2$ equivalent electrons in a shell belonging to an irreducible representation $\Gamma(P)$ of dimension f. We then have f independent MOs $\phi_1 \ldots \phi_f$ which transform under a symmetry operation O_P of G as follows.

$$O_P\phi_i = \sum_{i=1}^{f} \Gamma_{i'i}\phi_{i'}. \tag{7.29}$$

For $M_S = 1$ we may construct $\frac{1}{2}f(f-1)$ linearly independent determinants

$$|\phi_i\alpha, \phi_j\alpha|, \qquad (i<j = 2, \ldots f) \tag{7.30}$$

which transform according to

$$O_P |\phi_i\alpha, \phi_j\alpha| = \sum_{i'=1}^{f} \sum_{j'=1}^{f} \Gamma_{i'i}\Gamma_{j'j} |\phi_{i'}\alpha, \phi_{j'}\alpha|, \tag{7.31}$$

and for $M_S = 0$ we have f^2 independent determinants

$$|\phi_i\alpha, \phi_j\beta|, \qquad (i, j = 1 \ldots f) \tag{7.32}$$

transforming according to

$$O_P |\phi_i\alpha, \phi_j\beta| = \sum_{i'=1}^{f} \sum_{j'=1}^{f} \Gamma_{i'i}\Gamma_{j'j} |\phi_{i'}\alpha, \phi_{j'}\beta|. \tag{7.33}$$

Now, whereas all the determinants in equation (7.33) are non-zero and linearly independent (indeed, orthonormal if the MOs are), the determinants in equation (7.31) with $i' = j'$ vanish identically and those with $i' \neq j'$ are equal in pairs

$$|\phi_{j'}\alpha, \phi_{i'}\alpha| = -|\phi_{i'}\alpha, \phi_{j'}\alpha|. \tag{7.34}$$

We, therefore, use equation (7.34) to re-express the sum in equation (7.31) in terms of linearly independent determinants

$$\begin{aligned} O_P|\phi_i\alpha, \phi_j\alpha| &= \sum_{i'<j'}^{f} (\Gamma_{i'i}\Gamma_{j'j} - \Gamma_{j'i}\Gamma_{i'j})|\phi_{i'}\alpha, \phi_{j'}\alpha| \\ &= \sum_{i'<j'}^{f} \begin{vmatrix} \Gamma_{i'i}, \Gamma_{i'j} \\ \Gamma_{j'i}, \Gamma_{j'j} \end{vmatrix} |\phi_{i'}\alpha, \phi_{j'}\alpha|. \end{aligned} \tag{7.35}$$

The character of the representation spanned by the functions (7.30) is now given by the trace of the transformation (7.35);

$$\begin{aligned} \chi(M_S = 1) &= \sum_{i<j}^{f} \begin{vmatrix} \Gamma_{ii}, \Gamma_{ij} \\ \Gamma_{ji}, \Gamma_{jj} \end{vmatrix} \\ &= \sum_{i<j}^{f} (\Gamma_{ii}\Gamma_{jj} - \Gamma_{ij}\Gamma_{ji}) \\ &= \tfrac{1}{2} \sum_{i=1}^{f}\sum_{j=1}^{f} \{\Gamma_{ii}(P)\Gamma_{jj}(P) - \Gamma_{ij}(P)\Gamma_{ji}(P)\} \\ &= \tfrac{1}{2}\left[\left\{\sum_{i=1}^{f} \Gamma_{ii}(P)\right\}^2 - \sum_{i=1}^{f} \Gamma(P^2)_{ii}\right] \\ &= \tfrac{1}{2}[\chi(P)^2 - \chi(P^2)], \end{aligned} \tag{7.36}$$

where $\chi(P)$ is the character of the representation (7.29) spanned by the MOs ϕ_i.

Similarly from equation (7.33)

$$\chi(M_S = 0) = \chi(P)^2. \tag{7.37}$$

From equations (7.28), (7.36) and (7.37) and the fact that there are no functions with $M_S = 2$, we obtain the characters of the singlet and triplet states

$$\chi_{S=1}(P) = \tfrac{1}{2}[\chi(P)^2 - \chi(P^2)] \tag{7.38}$$

and

$$\chi_{S=0}(P) = \tfrac{1}{2}[\chi(P)^2 + \chi(P^2)]. \tag{7.39}$$

The case of $n = 3$ equivalent electrons is similar but slightly more complicated. For $M_S = 3/2$ we have $f(f-1)(f-2)/6$ determinants

$$|\phi_i\alpha, \phi_j\alpha, \phi_k\alpha| \qquad (i<j<k = 3\ldots f) \tag{7.40}$$

which span a representation with character

$$\chi(M_S = \tfrac{3}{2}) = \sum_{i<j<k} \begin{vmatrix} \Gamma_{ii} & \Gamma_{ij} & \Gamma_{ik} \\ \Gamma_{ji} & \Gamma_{jj} & \Gamma_{jk} \\ \Gamma_{ki} & \Gamma_{kj} & \Gamma_{kk} \end{vmatrix}$$

$$= \tfrac{1}{6} \sum_{i=1}^{f} \sum_{j=1}^{f} \sum_{k=1}^{f} [\Gamma_{ii}\Gamma_{jj}\Gamma_{kk} - (\Gamma_{ii}\Gamma_{jk}\Gamma_{kj} + \Gamma_{ji}\Gamma_{ij}\Gamma_{kk} + \Gamma_{ki}\Gamma_{ik}\Gamma_{jj})$$

$$+ (\Gamma_{ij}\Gamma_{jk}\Gamma_{ki} + \Gamma_{ik}\Gamma_{ji}\Gamma_{kj})]$$

$$= \tfrac{1}{6}[\chi(P)^3 - 3\chi(P)\chi(P^2) + 2\chi(P^3)], \tag{7.41}$$

while for $M_S = \frac{1}{2}$ we have $\frac{1}{2}f^2(f-1)$ determinants

$$|\phi_i\alpha, \phi_j\alpha, \phi_k\beta| \qquad (i<j = 2\ldots f) \quad (k = 1\ldots f) \tag{7.42}$$

which generate a representation with character

$$\chi(M_S = \tfrac{1}{2}) = \sum_{i<j}^{f} \sum_{k=1}^{f} \begin{vmatrix} \Gamma_{ii} & \Gamma_{ij} \\ \Gamma_{ji} & \Gamma_{jj} \end{vmatrix} \Gamma_{kk}$$

$$= \tfrac{1}{2} \sum_{i=1}^{f} \sum_{j=1}^{f} \sum_{k=1}^{f} (\Gamma_{ii}\Gamma_{jj} - \Gamma_{ij}\Gamma_{ji})\Gamma_{kk}$$

$$= \tfrac{1}{2}[\chi(P)^3 - \chi(P)\chi(P^2)]. \tag{7.43}$$

Hence using (7.28) we obtain for the quartets

$$\chi_{S=\frac{3}{2}}(P) = \tfrac{1}{6}\chi(P)^3 - \tfrac{1}{2}\chi(P)\chi(P^2) + \tfrac{1}{3}\chi(P^3) \tag{7.44}$$

and for the doublets

$$\chi_{S=\frac{1}{2}}(P) = \tfrac{1}{3}\chi(P)^3 - \tfrac{1}{3}\chi(P^3). \tag{7.45}$$

It is possible to deal in this elementary way with the cases of four or more equivalent electrons, although the equations become increasingly cumbersome. An elegant solution of the general problem has been given by Kotani[8] in terms of the characters of the symmetric permutation group.

However, the simple results of equations (7.38), (7.39), (7.44) and (7.45) and the character tables of Appendix 2 are all that are needed for sorting out the states of MO configurations. This is because the highest degeneracy which occurs for molecular symmetry groups is $f = 3$ and, just as in atomic theory,[(6)] the states from an almost filled shell with $2f - n$ electrons are the same as those for n electrons in the shell. Table 7.3 shows the states obtained in this way for some frequently encountered states of simple molecules.

Table 7.3. Molecular electronic states from equivalent electrons

Molecule	Symmetry group	Electron configuration	Electronic states
A_2	$D_{\infty h}$	π_g^2, π_u^2	${}^3\Sigma_g^-, {}^1\Sigma_g^+, {}^1\Delta_g$
		δ_g^2, δ_u^2	${}^3\Sigma_g^-, {}^1\Sigma_g^+, {}^1\Gamma_g$
AB	$C_{\infty v}$	π^2	${}^3\Sigma^-, {}^1\Sigma^+, {}^1\Delta$
		δ^2	${}^3\Sigma^-, {}^1\Sigma^+, {}^1\Gamma$
NH_3	C_{3v}	e^2	${}^3A_2, {}^1A_1, {}^1E$
CH_4	T_d	e^2	${}^3A_2, {}^1A_1, {}^1E$
		$t_1^2, t_2^2, t_1^4, t_2^4$	${}^3T_2, {}^1A_1, {}^1E, {}^1T_1$
		t_1^3	${}^4A_1, {}^2E, {}^2T_1, {}^2T_2$
		t_2^3	${}^4A_2, {}^2E, {}^2T_1, {}^2T_2$
C_6H_6	D_{6h}	$e_{1g}^2, e_{1u}^2, e_{2g}^2, e_{2u}^2$	${}^3A_{2g}, {}^1A_{1g}, {}^1E_g$

Excited configurations of molecules often contain two or more incomplete shells. As for atoms the symmetries of the states in these cases are easily found by independent composition of the spins and spatial symmetries of the sub-states from each open shell. Two spins S_1 and S_2 combine according to the usual vector-coupling formula.

$$S_1 \times S_2 \to S_1 + S_2, S_1 + S_2 - 1, \ldots, |S_1 - S_2|, \qquad (7.46)$$

while the composition of two spatial symmetries (irreducible representations) Γ_1, Γ_2 is effected by reduction of the direct product representation $\Gamma_1 \times \Gamma_2$ with character $\chi_1(P)\chi_2(P)$. For the groups C_{3v}, T_d, $C_{\infty v}$ and $D_{\infty h}$ these reductions are given explicitly in Table 4.8; other cases are easily treated using the methods of Chapter 4 and the character tables of Appendix 2.

Two examples will illustrate the procedure:

(i) $D_{\infty h}$ configuration $(\pi_u)^3(\pi_g)^3$. Here each open shell gives rise to only one sub-state

$$(\pi_u)^3 \to {}^2\Pi_u, \qquad (\pi_g)^3 \to {}^2\Pi_g.$$

Coupling the spins of these two sub-states by (7.46) (with $S_1 = S_2 = \frac{1}{2}$) and the spatial symmetries using Table 4.8 we find

$$^2\Pi_u \times {}^2\Pi_g \rightarrow {}^{3,1}(\Sigma_u^+ + \Sigma_u^- + \Delta_u)$$
$$= {}^3\Sigma_u^+ + {}^3\Sigma_u^- + {}^3\Delta_u + {}^1\Sigma_u^+ + {}^1\Sigma_u^- + {}^1\Delta_u. \quad (7.47)$$

(ii) T_d configuration $(t_2)^4(e)^2$. Here we see from Table 7.3 that each open shell leads to several sub-states

$$(t_2)^4 \rightarrow {}^3T_2 + {}^1A_1 + {}^1E + {}^1T_1$$
$$(e)^2 \rightarrow {}^3A_2 + {}^1A_1 + {}^1E.$$

We find the array of states from the configuration $(t_2)^4(e)^2$ by forming the direct product of all pairs of sub-states; thus

$$\begin{aligned}
&(t_2)^4(e)^2 \rightarrow \\
&\quad {}^3T_2 \times ({}^3A_2 + {}^1A_1 + {}^1E) = \quad {}^5T_1 + {}^3T_1 + {}^1T_1 + {}^3T_2 + {}^3T_1 + {}^3T_2 \\
&+ {}^1A_1 \times ({}^3A_2 + {}^1A_1 + {}^1E) \quad + {}^3A_2 + {}^1A_1 + {}^1E \\
&+ {}^1E \times ({}^3A_2 + {}^1A_1 + {}^1E) \quad + {}^3E + {}^1E + {}^1A_1 + {}^1A_2 + {}^1E \\
&+ {}^1T_1 \times ({}^3A_2 + {}^1A_1 + {}^1E) \quad + {}^3T_2 + {}^1T_1 + {}^1T_1 + {}^1T_2.
\end{aligned} \quad (7.48)$$

It is a straightforward but somewhat tedious task to construct determinantal wave functions for the states arising from open-shell MO configurations. A general method which always works is to combine the group theoretical projection operators of Section 4.7 with the spin projection operators of Section 6.9. The other methods for constructing spin eigenfunctions described in Chapter 6 may also be used. A convenient combination of these techniques when a large number of configurations and states need to be considered has been described by Nesbet.(7) Once the symmetry adapted functions have been found as linear combinations of determinants, all matrix elements of the Hamiltonian and other operators may be obtained from the rules given in equations (6.41)-(6.43).

As a simple example we consider the open shell $(\pi_g)^2$ occurring in the ground configuration of O_2 (Fig. 7.1).

Here we have $\binom{4}{2} = 6$ determinants with the following M_L and M_S values.

$$\begin{aligned}
D_1 &= |g_+g_-| & M_L &= 0, & M_S &= 1 \\
D_2 &= |g_+\bar{g}_+| & M_L &= 2, & M_S &= 0 \\
D_3 &= |g_+\bar{g}_-| & M_L &= 0, & M_S &= 0 \\
D_4 &= |g_-\bar{g}_+| & M_L &= 0, & M_S &= 0 \\
D_5 &= |g_-\bar{g}_-| & M_L &= -2, & M_S &= 0 \\
D_6 &= |\bar{g}_+\bar{g}_-| & M_L &= 0, & M_S &= -1.
\end{aligned} \quad (7.49)$$

Here the MOs π_g^{+1}, π_g^{-1} are abbreviated to g_+, g_- respectively.

The determinants D_2 and D_5 are clearly singlets, and from the M_L values they are the $\Lambda = M_L = \pm 2$ components of a Δ state. Applying the inversion operator O_i we see that the state is ${}^1\Delta_g$.

$$
\begin{aligned}
{}^1\Delta_g:\ \Lambda &= 2,\ |g_+\bar{g}_+| \\
\Lambda &= -2,\ |g_-\bar{g}_-|.
\end{aligned}
\tag{7.50}
$$

As always, the determinant D_1 with the highest M_S value is an eigenfunction of S^2 with $S = M_S$ (= 1). Applying the operators for inversion (O_i) and reflexion in the internuclear axis (O_{σ_v}) we find

$$O_i\ |g_+g_-| = |g_+g_-| \tag{7.51}$$

$$O_{\sigma_v} |g_+g_-| = |g_-g_+| = -|g_+g_-|. \tag{7.52}$$

These equations and $\Lambda = M_L = 0$, show that the state is ${}^3\Sigma_g^-$. Applying the shift operator S_- to D_1 we obtain the $M_S = 0$ and $M_S = -1$ components of the triplet:

$$
\begin{aligned}
{}^3\Sigma_g^-: M_S &= 1,\ |g_+g_-| \\
M_S &= 0,\ \frac{1}{\sqrt{2}}\{|g_+\bar{g}_-| - |g_-\bar{g}_+|\} \\
M_S &= -1,\ |\bar{g}_+\bar{g}_-|.
\end{aligned}
\tag{7.53}
$$

Finally a singlet state is obtained as that linear combination of D_3 and D_4 which is orthogonal to the $M_S = 0$ component of ${}^3\Sigma_g^-$. From the M_L value and the effects of O_i and O_{σ_v} we see that the state is ${}^1\Sigma_g^+$.

$${}^1\Sigma_g^+: \frac{1}{\sqrt{2}}\{|g_+\bar{g}_-| + |g_-\bar{g}_+|\}. \tag{7.54}$$

In (7.50), (7.53) and (7.54) we have explicit wave functions for the ${}^3\Sigma_g^-$, ${}^1\Delta_g$ and ${}^1\Sigma_g^+$ states which Table 7.3 predicts for the configuration π_g^2. The first-order energies of these states are given by the diagonal matrix elements of the full many-electron Hamiltonian with respect to these functions (with all closed-shell functions included, e.g. $1\sigma_g^2\ 1\sigma_u^2\ 2\sigma_g^2\ 2\sigma_u^2\ 3\sigma_g^2\ 1\pi_u^4$ for O_2). The only terms in these energy expressions which differ among the three states are those from the $1/r_{12}$ operator involving the open-shell orbitals g_+ and g_-. These may be obtained using the simple two-electron wave functions (7.50), (7.53) and (7.54), and the matrix element rules of equations (6.41)-(6.43). In this way we find

$$E(^3\Sigma_g^-) = E_0 + \left(g_+g_-\left|\frac{1}{r_{12}}\right|g_+g_-\right) - \left(g_+g_-\left|\frac{1}{r_{12}}\right|g_-g_+\right)$$

$$= E_0 + J(g^+, g^-) - K(g^+, g^-)$$

$$E(^1\Delta_g) = E_0 + \left(g_+g_+\left|\frac{1}{r_{12}}\right|g_+g_+\right) = E_0 + J(g^+, g^-) \tag{7.55}$$

$$E(^1\Sigma_g^+) = E_0 + \left(g_+g_-\left|\frac{1}{r_{12}}\right|g_+g_-\right) + \left(g_+g_-\left|\frac{1}{r_{12}}\right|g_-g_+\right)$$

$$= E_0 + J(g^+, g^-) + K(g^+, g^-),$$

where E_0 represents all terms common to the three states.

Since the integrals J and K are inherently positive, equation (7.55) predicts the energy order

$$E(^3\Sigma_g^-) < E(^1\Delta_g) < E(^1\Sigma_g^+) \tag{7.56}$$

in accordance with Hund's rule that states from the same configuration increase in energy with decreasing multiplicity and, for equal multiplicities, states with higher values of Λ have lower energies. This simple rule holds for most MO configurations of diatomic molecules.

(d) STATES OF DIATOMIC MOLECULES AND IONS

Using these techniques and the correlation diagrams we may deduce the ground states and low-lying excited states of the molecules represented in Figs 7.1 and 7.2. For the homonuclear case these deductions are shown in Table 7.4 together with experimental data for the observed states. For simplicity the table is restricted to excited states for which T_e, the height of the minimum of the potential curve above that for the ground electronic state is less than about 4.5 eV ($\sim$36 000 cm^{-1}). Some of the molecules, notably O_2 and N_2, have a large number of additional excited states but, as expected from the correlation diagram, these lie at higher energies.

The experimental properties shown in Table 7.4 are in very good qualitative and semi-quantitative agreement with expectations based on the correlation diagram 7.1. In every case the symmetry of the ground state is given correctly, both for the neutral molecules and for the ions. The symmetry type, energy order and approximate location of the low lying excited states are also in good accord with estimates based on the separation of the MO energies. Thus when a sizeable orbital excitation is required to obtain an excited state as for N_2, O_2^+, F_2 there are no observed levels in the range (< 4.5 eV) covered in the table; similarly the only levels of O_2 in this range are the levels $X^3\Sigma_g^-$, $a^1\Delta_g$, $b^1\Sigma_g^+$ expected from the ground configuration.

Table 7.4. Ground states and low lying excited states of some homonuclear diatomic molecules and ions[a]

Species	Configurations from Fig. 7.1: $1\sigma_g^2 1\sigma_u^2$ + $2\sigma_g$	$2\sigma_u$	$3\sigma_g$	$1\pi_u$	$1\pi_g$	$3\sigma_u$	Expected states	Observed states	T_e (cm^{-1})	Bond order	R_e (Å)	D_e (eV)
Li_2	2						$^1\Sigma_g^+$	X	0	1	2.67	1.1
	1	1					$^3\Sigma_u^+$	Repulsive				
	1	1					$^1\Sigma_u^+$	A	14 068		3.11	
	1			1			$^3\Pi_u$					
	1			1			$^1\Pi_u$	B	20 439		2.94	
Be_2	2	2					$^1\Sigma^+{}_g$	Repulsive		0		
B_2	2	2		2			$^3\Sigma_g^-$	X	(0)	1	1.59	(3.6)
	2	2		2			$^1\Delta_g$					
	2	2		2			$^1\Sigma_g^+$					
	2	2	1	1			$^3\Pi_u$					
	2	2	1	1			$^1\Pi_u$					
	2	1	1	2			$^{5,3,3,1}\Sigma_u^-$	$A^3\Sigma_u^-$	30 573		1.62	
	2	1	1	2			$^{3,1}\Delta_u$					
	2	1	1	2			$^{3,1}\Sigma_u^+$					
C_2	2	2		4			$^1\Sigma_g^+$	a	0	2	1.24	6.36
	2	2	1	3			$^3\Pi_u$	X	716		1.31	
	2	2	1	3			$^1\Pi_u$	b	8392		1.32	
	2	2	2	2			$^3\Sigma_g^-$	A	6434		1.37	
	2	2	2	2			$^1\Delta_g$					
	2	2	2	2			$^1\Sigma_g^+$					
	2	1	1	4			$^{3,1}\Sigma_g^+$	$A''\ ^3\Sigma_u^+$	13 312		1.23	
N_2	2	2	2	4			$^1\Sigma_g^+$	X	0	3	1.09	9.90
N_2^+	2	2	1	4			$^2\Sigma_g^+$	X	0	$2\frac{1}{2}$	1.12	8.86

	2	2	2	3			$^2\Pi_u$	A	9300		1.18	
	2	1	2	4			$^2\Sigma_u^+$	B	25 462		1.07	
O_2	2	2	2	4	2		$^3\Sigma_g^-$	X	0	2	1.21	5.21
	2	2	2	4	2		$^1\Delta_g$	a	7918		1.22	
	2	2	2	4	2		$^1\Sigma_g^+$	b	13 195		1.23	
O_2^+	2	2	2	4	1		$^2\Pi_g$	X	0	$2\frac{1}{2}$	1.12	6.75
F_2	2	2	2	4	4		$^1\Sigma_g^+$	X	0	1	1.42	1.69
F_2^+	2	2	2	4	3		$^2\Pi_g$	X	0	$1\frac{1}{2}$	—	3.3
Ne_2	2	2	2	4	4	2	$^1\Sigma_g^+$	Repulsive		0		

[a] Experimental data from Ref. 5.

Table 7.5. Low-lying states of some heteronuclear diatomic molecules and ions[a]

Species	3σ	4σ	Configurations from Fig. 7.2: $1\sigma^2 2\sigma^2$ + 1π	5σ	2π	6σ	Expected states	Observed states T_e (cm^{-1})		Bond order	R_e (Å)	D_e (eV)
CN	2	2	4	1			$^2\Sigma^+$	X	0	$2\frac{1}{2}$	1.17	8.3
	2	2	3	2			$^2\Pi$	A	9242		1.23	
	2	1	4	2			$^2\Sigma^+$	B	25 952		1.15	
CO	2	2	4	2			$^1\Sigma^+$	X	0	3	1.13	11.24
CO^+	2	2	4	1			$^2\Sigma^+$	X	0	$2\frac{1}{2}$	1.12	
	2	2	3	2			$^2\Pi$	A	20 733		1.24	
	2	1	4	2			$^2\Sigma^+$	B	45 877		1.17	
NO	2	2	4	2	1		$^2\Pi$	X	0	$2\frac{1}{2}$	1.15	6.62
NO^+	2	2	4	2			$^1\Sigma^+$	X	0	3	1.06	10.7

[a] Experimental data from Ref. 5.

For C_2 on the other hand the correlation diagram shows a very low-lying unoccupied MO ($3\sigma_g$) and excitations to this MO produce a rich pattern of expected excited states which fit in well with the observed energy levels. A rich low energy spectrum is also predicted for B_2 from the open shell $(1\pi_u)^2$ and the neighbouring $3\sigma_g$ and $2\sigma_u$ orbitals. Here only one transition $A^3\Sigma_u^- \rightarrow X^3\Sigma_g^-$ has so far been observed. We should mention at this point that, although qualitative deductions about energy levels based on the correlation diagram are fairly reliable, quite elaborate wave functions involving extensive configuration interaction are needed if the correct pattern of energy levels is to be obtained in *ab initio* calculations; this is especially true in cases such as C_2 and B_2 where several MOs are nearly degenerate. The techniques and results of such calculations are discussed by Hurley.[69] In general, the assignments of the observed levels shown in Table 7.4 are confirmed but there are differences. For example, the ground state of B_2 is predicted to be the $^5\Sigma_u^-$ arising (principally) from the configuration $(2\sigma_g)^2(2\sigma_u)(3\sigma_g)(1\pi_u)^2$.

We may define a simple bond order for diatomic molecules and ions as half the difference between the number of electrons in bonding and antibonding MOs. The last three columns in Table 7.4 show that this quantity correlates well with nuclear separations R_e and dissociation energies D_e; a high bond order leads to a large dissociation energy and a small nuclear separation. Especially remarkable is the way in which changes in bond length on ionization or excitation mirror the bonding and antibonding character of the MOs shown by the slopes of the lines in Fig. 7.1. Firstly we see that excitation without change in the occupation of the MOs (e.g. O_2, $X^3\Sigma_g^-$, $a^1\Delta_g$, $b^1\Sigma_g^+$; C_2, $X^3\Pi_u$, $b^1\Pi_u$) has very little effect on R_e, that is states from the same configuration have similar nuclear separations. Again for N_2 the removal of a weakly bonding $3\sigma_g$ electron leads to an increase of about 0.03 in R_e, for the strongly bonding $1\pi_u$ the change in R_e is 0.09 and for the weakly antibonding $2\sigma_u$ the change is -0.02. For O_2 the removal of a strongly antibonding $1\pi_g$ electron decreases R_e by 0.09.

The effect of excitation on bond length is well illustrated by the states of C_2. Transfer of one electron from $1\pi_u$ to $3\sigma_g$ increases R_e by about 0.07 and the transfer of a second electron to form the $A^3\Sigma_g^-$ state leads to an additional increase of R_e of comparable magnitude. The slight decrease in R_e in going from a $^1\Sigma_g^+$ to $A''^3\Sigma_u^+$ is also consistent with the loss of a weakly antibonding electron ($2\sigma_u$) and the gain of a weakly bonding electron $3\sigma_g$. All these bond length regularities have a simple interpretation in terms of the virial and electrostatic theorems.[11]

Table 7.5 gives corresponding data for some heteronuclear molecules and ions. Here too the simple picture provided by qualitative MO theory and the correlation diagram is well borne out by the observations.

(e) SIMPLE POLYATOMIC MOLECULES

We have seen above that the qualitative form and energetic order of the MOs of diatomic molecules can be estimated from the behaviour of the orbitals in the two limiting cases $R \to 0$ (united atom) and $R \to \infty$ (separated atoms). Similar methods may be used for simple polyatomic molecules, but here the situation is more complicated since several parameters are needed to specify the nuclear positions and separate correlation diagrams must be drawn for each type of molecule.

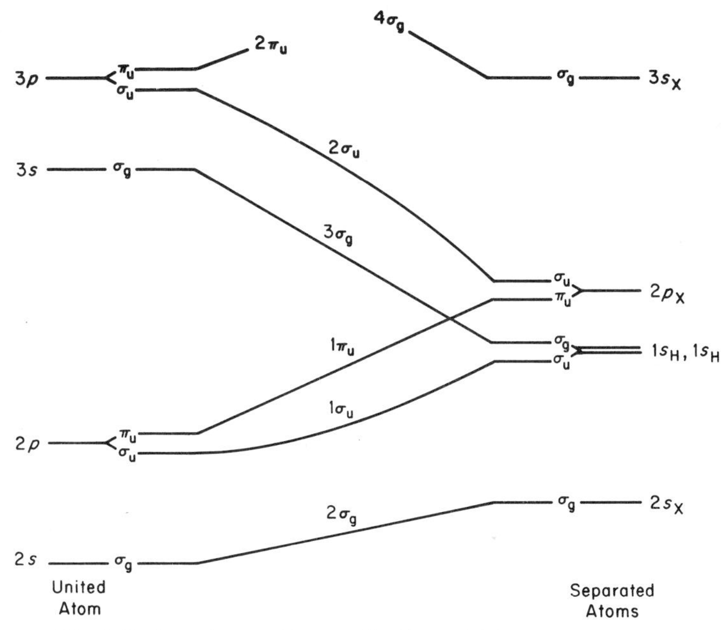

Fig. 7.3. Correlation diagram for linear XH_2 molecules.

Linear XH_2 molecules—If we assume that atom X always remains midway between the hydrogens, the geometry is specified by the single parameter R(XH) and we may construct a correlation diagram as in the diatomic case (Fig. 7.3).

Here the levels of the united atom are shown at the extreme left, those of the separated atoms at the extreme right, the hydrogen level being doubled since there are two hydrogens. This hydrogen level is shown between the $2s$ and $2p$ levels of X, as is appropriate for BeH_2, BH_2, CH_2, NH_2; for OH_2 the levels $1s_H$ and $2p_X$ should be reversed (Table 7.1). The inner-shell orbital $1s_X$, which gives $1\sigma_g$, is hardly affected by molecular formation and is omitted from the diagram but is counted in numbering the σ_g orbitals.

Next to the levels of the united atom on the left are shown the MOs which they give in the linear conformation and next to the extreme right the MO energies corresponding to the orbitals of the separated atoms. The possible symmetries of these MOs are easily determined from the character table of $D_{\infty h}$ (Appendix 2) or directly from their behaviour under the various symmetry operations. The energetic order of splitting of a united atom orbital ($\sigma < \pi < \delta \ldots$) may be established by a simple perturbation calculation. The order of splitting at the separated atom limit is chosen so that there are as few crossings as possible when the correlation diagram is completed by linking MOs of the same symmetry from the united and separated atom limits. Crossing of the curves for MOs of the same symmetry type is forbidden by the non-crossing rule.

Straight connecting lines are used except when two lines of the same symmetry type lie fairly close to each other. In such a case the lines interact and "repel" each other corresponding to the mixing or hybridization of the corresponding wave functions. Thus the bonding $1\sigma_u$ curve of Fig. 7.3 interacts with the antibonding $2\sigma_u$ and in so doing acquires a certain amount of $2p_X$ character.

The orbitals to be expected for a linear XH_2 molecule are now obtained by drawing a vertical line somewhere between the two limiting cases. For example, BeH_2 with 6 electrons is expected to

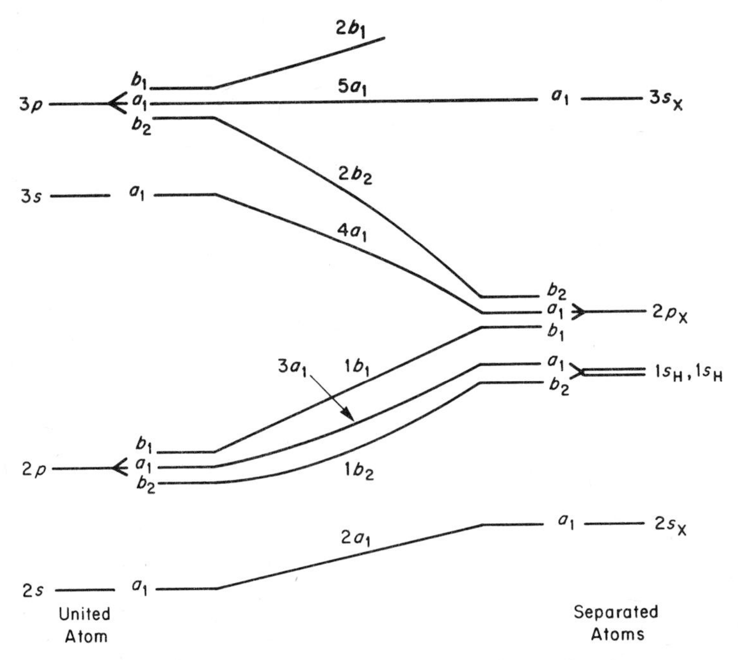

Fig. 7.4. Correlation diagram for bent XH_2 molecules.

have a ground-state configuration

$$BeH_2: (1\sigma_g)^2 (2\sigma_g)^2 (1\sigma_u)^2\ {}^1\Sigma_g^+.$$

Bent XH_2 molecules—The correlation of the MOs of bent XH_2 molecules between the united and separated atoms is shown in Fig. 7.4.

In drawing this figure we assume that the two X—H lengths are always equal and the bond angle remains constant at ~90°. Here the p atomic orbitals at both edges of the diagram split into three molecular orbitals a_1, b_1, b_2 corresponding to p_x, p_y and p_z (Appendix 2). The 1s orbitals of the two hydrogens form a_1 and b_2 MOs in this point group (C_{2v}) corresponding to $1s_H + 1s_{H'}$ and $1s_H - 1s_{H'}$.

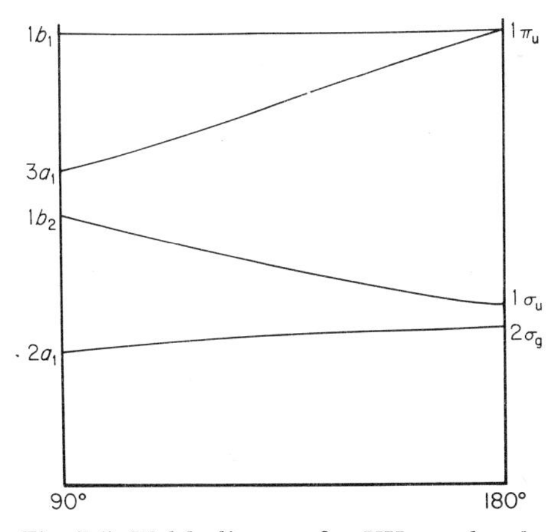

Fig. 7.5. Walsh diagram for XH_2 molecules.

Again the behaviour for intermediate bond lengths is obtained by linking united and separated atom with the same symmetry and the energetic order of the MOs in a bent XH_2 molecule may be estimated by drawing a vertical line at the appropriate place. Thus for NH_2 with 9 electrons we expect a ground configuration

$$NH_2: (1a_1)^2 (2a_1)^2 (1b_2)^2 (3a_1)^2 (1b_1):\ {}^2B_1.$$

The most striking difference between Figs 7.3 and 7.4 is in the behaviour of the levels correlating with the hydrogen orbitals. In Fig. 7.3 one of these levels is bonding ($1\sigma_u$), the other is antibonding ($3\sigma_g$) whereas in Fig. 7.4 both levels are bonding ($1b_2$, $3a_1$). Moreover, in the bent case there is a strong interaction and "repulsion" between the $3a_1$ and $4a_1$ levels. For these two reasons the $3a_1$ level falls well below the corresponding MO in the linear case (one of the degenerate $1\pi_u$ orbitals) and occupation of this orbital favours the bent conformation.

By combining the results of Figs 7.3 and 7.4 we obtain Fig. 7.5 which shows the variation of orbital energies as the HXH angle is varied from 180° to 90°. Such diagrams were introduced by Walsh[9] and are usually referred to as Walsh diagrams. Figure 7.5 differs somewhat from Walsh's original diagram in that the $2a_1$–$2\sigma_g$ curve rises slightly instead of falling as we go from 90° to 180° and the rise of the $3a_1$–$1\pi_u$ curve is somewhat less marked so that there is now no crossing with the $1b_2$–$1\sigma_u$ curve. These modifications have been made in the light of more recent calculations.[10]

From Fig. 7.5 we expect that XH_2 molecules with only 4 valence electrons should be linear in their ground states, since the net effect of 2 electrons in each of $2a_1$ and $1b_2$ is to favour the right-hand side of the figure ($\theta = 180°$). For 5 or 6 valence electrons, however, we expect a bent ground state since the additional electrons go into the $3a_1$ orbital which from Fig. 7.5, strongly favours bent conformations. Molecules with 7 or 8 valence electrons are also expected to be bent, the extra electrons being accommodated in the $1b_1$ orbital which shows little angular variation.

It is also possible to predict the geometry of some low-lying excited states of XH_2 molecules from Fig. 7.5. For example, the ground state of BH_2 with 5 valence electrons is expected to be bent with the ground configuration

$$BH_2: (1a_1)^2\,(2a_1)^2\,(1b_2)^2\,3a_1,\ {}^2A_1.$$

From Fig. 7.5 we expect a low-lying state in which an electron is excited from $3a_1$ to $1b_1$. Since the orbital $3a_1$ favouring bending is now vacant we expect (and find) a linear conformation for the excited state

$$BH_2: (1\sigma_g)^2\,(2\sigma_g)^2\,(1\sigma_u)^2\,(1\pi_u),\ {}^2\Pi_u.$$

Table 7.6 shows the geometry and electron configurations of the ground states and some low-lying excited states of XH_2 molecules. In general the qualitative expectations based on the Walsh diagram (Fig. 7.5) are well borne out.

Correlation diagrams and Walsh diagrams may be constructed in a similar way for other polyatomic species. As the molecules become more complicated, however, ambiguities arise as to the ordering of the levels and the slopes of the connecting lines. These ambiguities may only be resolved by extensive calculations or an appeal to spectroscopic data. Thus in these more complex situations the diagrams lose their predictive power but remain a simple, flexible, framework for correlating and summarizing detailed theoretical and experimental investigations. An excellent account of their application to the electronic spectra of polyatomic molecules has been given by Herzberg.[5]

Table 7.6. Some states of XH_2 molecules[a]

Species	Point group	$r_0(XH)$[b]	$H\hat{X}H$	Electron configuration	State	T_0[c]
BH_2	C_{2v}	1.18	131°	$\ldots(2a_1)^2(1b_2)^2 3a_1$	X^2A_1	0
	$D_{\infty h}$	1.17	180°	$\ldots(2\sigma_g)^2(1\sigma_u)^2 1\pi_u$	$A^2\Pi_u$	5150
CH_2	C_{2v}	1.11	102°	$\ldots(2a_1)^2(1b_2)^2(3a_1)^2$	a^1A_1	x
	$D_{\infty h}$	1.03	(180°)[d]	$\ldots(2\sigma_g)^2(1\sigma_u)^2(1\pi_u)^2$	$X^3\Sigma_g^-$	0
NH_2	C_{2v}	1.02	103°	$\ldots(2a_1)^2(1b_2)^2(3a_1)^2 1b_1$	X^2B_1	0
	C_{2v}	1.00	(144°)	$\ldots(2a_1)^2(1b_2)^2(3a_1)(1b_1)^2$	A^2A_1	10 249
H_2O	C_{2v}	0.956	105°	$\ldots(2a_1)^2(1b_2)^2(3a_1)^2(1b_1)^2$	X^1A_1	0

[a] Data from Ref. 5; [b] Units 10^{-8} cm; [c] Units cm^{-1}; [d] This state is probably bent ($H\hat{X}H \sim 140°$) (cf. Ref. 69).

7.2. Hartree–Fock theory

The Hartree–Fock method enables us to convert the qualitative discussion of the previous section into a quantitative theory. We use the same expression for the total electronic wave function as in the qualitative MO theory and determine the optimum form of the orbitals by the variation principle.

(a) THE UNRESTRICTED THEORY

The version of the theory which is formally the simplest is the so-called unrestricted Hartree–Fock approximation (UHF). Here we construct a wave function for an N-electron system as a Slater determinant

$$\Psi_0 = |\psi_1 \psi_2 \ldots \psi_N| \tag{7.57}$$

from N spin orbitals

$$\psi_i \qquad (i = 1 \ldots N) \tag{7.58}$$

and seek the absolute minimum of

$$E_0 = \frac{(\Psi_0|H|\Psi_0)}{(\Psi_0|\Psi_0)} \tag{7.59}$$

for arbitrary variations of the spin orbitals.

A necessary condition for this minimum is that

$$\delta(\Psi_0|H|\Psi_0) = 0 \tag{7.60}$$

for any variation of the ψ_i which preserves the normalization of Ψ_0

$$\delta(\Psi_0|\Psi_0) = 0. \tag{7.61}$$

Since the Hamiltonian H is Hermitian, we have

$$\begin{aligned}\delta(\Psi_0 | H | \Psi_0) &= (\delta\Psi_0 | H | \Psi_0) + (\Psi_0 | H | \delta\Psi_0) \\ &= (\delta\Psi_0 | H | \Psi_0) + (\delta\Psi_0 | H | \Psi_0)^*,\end{aligned} \tag{7.62}$$

so that the condition (7.60) is equivalent to the condition

$$\mathrm{Re}(\delta\Psi_0 | H | \Psi_0) = 0. \tag{7.63}$$

Since the total wave function (7.57) is unchanged, apart from a constant factor, by any linear transformation of the ψ_i among themselves, it is no restriction to assume that these occupied spin orbitals form an orthonormal set. Further spin orbitals, which together with the occupied spin orbitals form a complete orthonormal set, are denoted by

$$\psi_a (a = N + 1, \ldots, \infty). \tag{7.64}$$

We consider now the special variation

$$\delta\psi_i = \psi_a \, \mathrm{d}\alpha \tag{7.65}$$

where $\mathrm{d}\alpha$ is an infinitesimal constant. For the variation of the total wave function we obtain

$$\begin{aligned}\delta\Psi_0 &= | \psi_1 \ldots \delta\psi_i \ldots \psi_N | \\ &= | \psi_1 \ldots \psi_a \ldots \psi_N | \, \mathrm{d}\alpha \\ &= \Psi_i^a \, \mathrm{d}\alpha\end{aligned} \tag{7.66}$$

where Ψ_i^a denotes the Slater determinant obtained from Ψ_0 by replacing the occupied spin orbital ψ_i by the unoccupied spin orbital ψ_a. Since the spin orbitals are orthonormal, the special variation (7.65) preserves the normalization of $\Psi_0 (\delta(\Psi_0 | \Psi_0) = 0)$ and the condition (7.63) applies, that is

$$\mathrm{Re}\,(\Psi_i^a | H | \Psi_0) \, \mathrm{d}\alpha^* = 0. \tag{7.67}$$

The complex phase of $\mathrm{d}\alpha$ may be chosen so that $(\Psi_i^a | H | \Psi_0) \, \mathrm{d}\alpha^*$ is a real number. The condition (7.60) for the special variations (7.65) therefore implies the conditions

$$(\Psi_i^a | H | \Psi_0) = 0 \qquad (i = 1 \ldots N) \quad (a = N + 1, \ldots, \infty). \tag{7.68}$$

Conversely the conditions (7.68) imply (7.60) subject to (7.61); any linear variation of Ψ_0 which preserves normalization may be expressed as a linear combination of the special variations (7.65). This follows since the spin orbitals ψ_i and ψ_a form a complete set of functions and any variation of the form (7.65) with ψ_a replaced by an occupied spin orbital either changes the normalization $(a = i)$

or leads to zero variation of the total wave function Ψ_0 (equation (7.66) with $a = j \neq i$).

The matrix elements (7.68) for single substitutions in a Slater determinant were evaluated in Section 6.3. From equation (6.42) we obtain

$$
\begin{aligned}
(\Psi_i^a | H | \Psi_0) &= (\psi_a | f | \psi_i) + \sum_{\substack{j=1 \\ j \neq i}}^{n} [(\psi_a \psi_j | g | \psi_i \psi_j) - (\psi_a \psi_j | g | \psi_j \psi_i)] \\
&= (\psi_a | f | \psi_i) + \sum_{j=1}^{N} [(\psi_a \psi_j | g | \psi_i \psi_j) - (\psi_a \psi_j | g | \psi_j \psi_i)]
\end{aligned}
\tag{7.69}
$$

since the term added to the summation vanishes. In equation (7.69) f is a one-electron term

$$f = -\tfrac{1}{2}\nabla^2 - \sum_{\alpha} \frac{Z_\alpha}{r_\alpha} \tag{7.70}$$

from the Hamiltonian H and

$$g(1, 2) = \frac{1}{r_{12}}. \tag{7.71}$$

The right-hand side of equation (7.69) may be expressed in terms of an effective one-electron Hamiltonian. To this end we introduce Coulomb and exchange operators J_j and K_j whose effects on an arbitrary spin orbital ψ are defined by

$$J_j(1)\psi(1) = \left(\int \psi_j^*(2) \frac{1}{r_{12}} \psi_j(2)\, d\tau_2 \right) \psi(1)$$

and (7.72)

$$K_j(1)\psi(1) = \left(\int \psi_j^*(2) \frac{1}{r_{12}} \psi(2)\, d\tau_2 \right) \psi_j(1)$$

or, more briefly, by

$$J_j\psi = (\psi_j | g | \psi_j)\psi, \tag{7.73}$$

$$K_j\psi = (\psi_j | g | \psi)\psi_j. \tag{7.74}$$

Using these definitions and equation (7.69) we may now write the conditions (7.68) in the form

$$(\psi_a | F_0 | \psi_i) = 0 \qquad \begin{matrix} (i = 1 \ldots N) \\ (a = N+1, \ldots, \infty) \end{matrix} \tag{7.75}$$

where F_0, the Fock operator (or UHF Hamiltonian) is given by

$$F_0 = f + \sum_{j=1}^{N} (J_j - K_j). \tag{7.76}$$

It is readily verified that the operators J_j, K_j and F_0 are all Hermitian.

As indicated by the subscript zero, the Fock operator depends on the spin orbitals which are occupied in Ψ_0. However, like Ψ_0 itself, it is not affected by a unitary transformation

$$\psi_i' = \sum_{j=1}^{N} \psi_j U_{ji} \qquad (i = 1 \ldots N) \tag{7.77}$$

of these spin orbitals among themselves. This invariance follows readily from the invariance of the quantity

$$\gamma(1, 2) = \sum_{j=1}^{N} \psi_j(1)\psi_j^*(2) \tag{7.78}$$

for which we have

$$\begin{aligned} \sum_{j=1}^{N} \psi_j'(1)\psi_j'^*(2) &= \sum_{j=1}^{N} \sum_{k=1}^{N} \sum_{l=1}^{N} \psi_k(1)\psi_l^*(2) U_{kj} U_{lj}^* \\ &= \sum_{k=1}^{N} \sum_{l=1}^{N} \psi_k(1)\psi_l^*(2)\delta_{lk} \\ &= \sum_{k=1}^{N} \psi_k(1)\psi_k^*(2). \end{aligned} \tag{7.79}$$

We see that the conditions (7.75), known as the Hartree–Fock conditions, divide the total function space of a single electron into two sub-spaces:

(i) the occupied sub-space spanned by the spin orbitals $\psi_i (i = 1 \ldots N)$;

(ii) the unoccupied sub-space spanned by the remaining spin orbitals $\psi_a (a = N + 1, \ldots, \infty)$.

The UHF Hamiltonian F_0 has a zero matrix element between any two functions lying in different sub-spaces, but so far the choice of basis functions in each sub-space is quite arbitrary since the Hartree–Fock conditions (7.75) are invariant to unitary transformations of the occupied spin orbitals among themselves and of the unoccupied spin orbitals among themselves.

The Hartree–Fock conditions (7.75) may be expressed in a different form by expanding the N functions $F_0\psi_i$ in terms of the complete set of functions provided by the ψ_i and the ψ_a. Thus we obtain

$$F_0\psi_i = \sum_{j=1}^{N} \epsilon_{ij}\psi_j + \sum_{a=N+1}^{\infty} \epsilon_{ia}\psi_a \qquad (i = 1 \ldots N) \tag{7.80}$$

for appropriate values of the constants ϵ_{ij}, ϵ_{ia}. Multiplying equation (7.80) by ψ_a^* and integrating we find

$$\epsilon_{ia} = (\psi_a \mid F_0 \mid \psi_i) = 0 \tag{7.81}$$

from equation (7.75). The second summation in equation (7.80) therefore vanishes and an alternative form of the Hartree–Fock conditions is

$$F_0 \psi_i = \sum_{j=1}^{N} \epsilon_{ij} \psi_j \qquad (i = 1 \ldots N) \tag{7.82}$$

for suitable constants ϵ_{ij}. That is, the result of operating with F_0 on any function of the occupied sub-space is to give some other function lying wholly in the occupied sub-space.

Since F_0 is a Hermitian operator, the matrix of constants ϵ_{ij} (known as Lagrange multipliers) is also Hermitian.

$$\epsilon_{ij} = (\psi_j \mid F_0 \mid \psi_i) = (\psi_i \mid F_0 \mid \psi_j)^* = \epsilon_{ji}^*. \tag{7.83}$$

This property of the Lagrange multipliers permits a further simplification of the Hartree–Fock conditions (7.82). Since the matrix ϵ_{ij} is Hermitian it may be reduced to real diagonal form by a suitable unitary transformation (7.77) of the occupied spin orbitals. As we have seen such a transformation leaves F_0 invariant. The conditions (7.82) then reduce to the canonical Hartree–Fock equations

$$F_0 \psi_i = \epsilon_i \psi_i \qquad (i = 1 \ldots N) \tag{7.84}$$

where we have written $\epsilon_i = \epsilon_{ii}$ for the elements of the diagonalized matrix of Lagrange multipliers.

If all the eigenvalues ϵ_i are distinct the canonical form (7.84) leads to a unique orthonormal basis in the sub-space of occupied spin orbitals. The solutions of equations (7.84) will be referred to as canonical Hartree–Fock spin orbitals whenever it is necessary to distinguish them from the solutions of the more general equations (7.82), which may differ from the canonical spin orbitals by a unitary transformation.

In equation (7.84) the constants ϵ_i and the spin orbitals ψ_i appear as the eigenvalues and eigenfunctions of a one-electron effective Hamiltonian F_0; we appear to have a simple one-electron Schrödinger equation. However, from the definition of F_0 (equations (7.76), (7.73) and (7.74)) we see that this operator itself depends on the eigenfunctions of equation (7.84), so that these equations are in reality a complicated set of coupled integro-differential equations. Their solution is by no means unique, as is to be expected, since the equivalent conditions (7.84), (7.82), (7.75) and (7.68) are all necessary but not sufficient conditions for $(\Psi_0 \mid H \mid \Psi_0)/(\Psi_0 \mid \Psi_0)$ to assume its least possible value. Besides the absolute minimum, other extrema in the energy surface such as saddle-points, local minima or maxima may well arise. In the atomic case, for example, it is clear that, for N even, there will be a solution of equations (7.84) corresponding to the ^1S state from the configuration $(1s)^2(2s)^2(3s)^2$ . . .. Except for $N = 2$ or 4 it is clear that this solution does not lead to the absolute minimum of $(\Psi_0|H|\Psi_0)/(\Psi_0|\Psi_0)$. Such cases may be distinguished from true minima by investigating the second variations of the energy, that is, by studying the stability conditions of Hartree–Fock solutions.[22] However, the resulting theory is very complicated and one usually relies on physical intuition in picking appropriate exact or approximate solutions of equations (7.84).

We may use the Hartree–Fock equations (7.82) or (7.84) to derive several equivalent expressions for the total electronic energy E_0 corresponding to the wave function Ψ_0. From equations (6.40) and (6.41) with $\Delta_i = \Delta_{ij} = 1$, $k_i = l_i = \psi_i$ we obtain directly

$$E_0 = \sum_{i=1}^{N} (\psi_i|f|\psi_i) + \sum_{i<j}^{N} [(\psi_i\psi_j|g|\psi_i\psi_j) - (\psi_i\psi_j|g|\psi_j\psi_i)] \tag{7.85}$$

which may be written

$$E_0 = \sum_{i=1}^{N} (\psi_i|f|\psi_i) + \tfrac{1}{2} \sum_{i=1}^{N} \sum_{j=1}^{N} [(\psi_i\psi_j|g|\psi_i\psi_j) - (\psi_i\psi_j|g|\psi_j\psi_i)] \tag{7.86}$$

since the term $i = j$ added to the double summation vanishes.

The expression (7.86) may be simplified by introducing the definitions

$$f_i = (\psi_i|f|\psi_i) \tag{7.87}$$

$$J_{ij} = (\psi_i\psi_j|g|\psi_i\psi_j) = (\psi_i|J_j|\psi_i) = (\psi_j|J_i|\psi_j) \tag{7.88}$$

$$K_{ij} = (\psi_i\psi_j|g|\psi_j\psi_i) = (\psi_i|K_j|\psi_i) = (\psi_j|K_i|\psi_j) \tag{7.89}$$

$$G = \sum_{j=1}^{N} (J_j - K_j), \qquad G_i = (\psi_i|G|\psi_i). \tag{7.90}$$

We note that the total electron interaction operator G, like the UHF Hamiltonian F_0, depends on the spin orbitals occupied in Ψ_0 but is invariant under unitary transformations of these spin orbitals.

With these definitions we obtain

$$E_0 = \sum_{i=1}^{N} f_i + \tfrac{1}{2} \sum_{i=1}^{N} \sum_{j=1}^{N} (J_{ij} - K_{ij}) = \sum_{i=1}^{N} (f_i + \tfrac{1}{2}G_i). \tag{7.91}$$

Equation (7.91) is often referred to as the additive expression for the total electronic energy.

If we multiply the Hartree-Fock equations (7.82) by ψ_i^*, integrate, and sum over i we obtain the identity

$$\sum_{i=1}^{N} \epsilon_{ii} = \sum_{i=1}^{N} (\psi_i \,|\, F_0 \,|\, \psi_i) = \sum_{i=1}^{N} (\psi_i \,|\, f + G \,|\, \psi_i) = \sum_{i=1}^{N} (f_i + G_i). \tag{7.92}$$

Combining equation (7.92) with the additive expression for the total energy (7.91) we obtain two alternative expressions for E_0, namely

$$E_0 = \sum_{i=1}^{N} (\epsilon_{ii} - \tfrac{1}{2}G_i) = \sum_{i=1}^{N} \epsilon_{ii} - \tfrac{1}{2} \sum_{i=1}^{N} \sum_{i=1}^{N} (J_{ij} - K_{ij}) \tag{7.93}$$

and

$$E_0 = \tfrac{1}{2} \sum_{i=1}^{N} (\epsilon_{ii} + f_i). \tag{7.94}$$

We note that all these equations, in particular the alternative expressions (7.91), (7.93) and (7.94), are valid not only for the canonical spin orbitals which satisfy (7.84) but for any set of solutions of the more general equations (7.82).

The wave functions of UHF satisfy two general theorems which are important for applications and further developments.

Brillouin's theorem.[12] *The matrix element of the many-electron Hamiltonian H between the UHF determinant Ψ_0 and the determinant Ψ_i^x obtained from Ψ_0 by replacing any occupied spin orbital ψ_i by any spin orbital orthogonal to all occupied spin orbitals vanishes.*

Since any such spin orbital ψ_x may be expressed as a linear combination of the $\psi_a (a = N + 1, \ldots, \infty)$, Brillouin's theorem is an immediate consequence of the conditions (7.68), which we have shown to be equivalent to the Hartree–Fock equations (7.82). We note that Brillouin's theorem, like the conditions (7.82), applies to any representation of the UHF determinant not just to canonical spin-orbitals.

Corollary 1. If the UHF determinant Ψ_0 is taken as the zeroth order wavefunction in a perturbation calculation of the accurate ground-state wave function Ψ then corrections to the charge density function and the expectation value of any symmetric one-electron operator

$$P = \sum_{i=1}^{N} p_i \tag{7.95}$$

are all of second or higher order in the perturbation.

To establish this result we base our perturbation calculations on the complete set of many-electron functions

$$\Psi_0, \Psi_i^a, \Psi_{ij}^{ab}, \ldots, (1 \leqslant i < j \cdots \leqslant N < a < b \cdots \infty) \tag{7.96}$$

provided by Ψ_0 and the determinants obtained from Ψ_0 by substituting one, two, . . . occupied spin orbitals by unoccupied spin orbitals.

The zeroth-order wavefunction Ψ_0 is an eigenfunction of the zeroth-order Hamiltonian

$$H_0 = \sum_{i=1}^{N} F_0(i). \tag{7.97}$$

This follows from the Hartree–Fock equations (7.84) and the fact that for symmetric one-electron operators a Slater determinant behaves like a simple product (Chapter 6). The perturbing Hamiltonian H' is, therefore, given by

$$H' = H - H_0 \tag{7.98}$$

where H is the full many-electron Hamiltonian.

Now in standard perturbation theory, of either Rayleigh–Schrödinger or Brillouin–Wigner type, the only functions which appear in the expression for the first-order wave function $\Psi^{(1)}$ are those linked to Ψ_0 by non-zero matrix elements of H'. Because of the Hartree–Fock equations (7.82), Brillouin's theorem, the fact that H_0 is a one-electron operator and the general matrix element formulae (6.40)–(6.43), the only functions from the set (7.96) which satisfy this criterion are the doubly substituted determinants Ψ_{ij}^{ab}.

The first-order wave function $\Psi^{(1)}$ is, therefore, of the form

$$\Psi^{(1)} = \Psi_0 + \sum_{i<j}^{N} \sum_{a<b}^{\infty} C_{ij}^{ab} \Psi_{ij}^{ab} \tag{7.99}$$

where the constants C_{ij}^{ab} are of first order in the perturbation.

Abbreviating (7.99) by

$$\Psi^{(1)} = \Psi_0 + \sum_{\mu} C_\mu \Psi_\mu \tag{7.100}$$

and noting that, for any one-electron operator (7.95)

$$(\Psi_0 | P | \Psi_{ij}^{ab}) = (\Psi_0 | P | \Psi_\mu) = 0, \tag{7.101}$$

we obtain for the expectation value of P

$$\begin{aligned}\frac{(\Psi | P | \Psi)}{(\Psi | \Psi)} &= \frac{(\Psi^{(1)} | P | \Psi^{(1)})}{(\Psi^{(1)} | \Psi^{(1)})} + O(C_\mu^2) \\ &= P_{00} + \frac{\sum_{\mu} |C_\mu|^2 (P_{\mu\mu} - P_{00}) + \sum_{\nu} \sum_{\mu \neq \nu} C_\mu^* C_\nu P_{\mu\nu}}{1 + \sum_{\mu} |C_\mu|^2} + O(C_\mu^2) \\ &= P_{00} + O(C_\mu^2)\end{aligned} \tag{7.102}$$

as required. In equation (7.102) we have written

$$P_{\mu\nu} = (\Psi_\mu \,|P|\, \Psi_\nu) \qquad (\mu, \nu = 0, 1, \ldots). \tag{7.103}$$

In practical applications, Hartree–Fock expectation values (P_{00}) of many one-electron operators are even better than might be expected from the formula (7.102) and rough estimates of the larger *C*s. This is presumably because of extensive cancellation among the terms in the numerator, which are of indefinite sign.

That the charge density itself is correct to first order follows from applying equation (7.102) to the special one-electron operator

$$\rho(x) = \sum_{i=1}^{N} \delta(x - x_i). \tag{7.104}$$

Corollary 2. If the exact ground-state wavefunction Ψ is expanded in terms of the complete set (7.96)

$$\Psi = \Psi_0 + \sum_i \sum_a d_i^a \Psi_i^a + \sum_{i<j} \sum_{a<b} d_{ij}^{ab} \Psi_{ij}^{ab} + \ldots \tag{7.105}$$

then the exact ground-state energy E is given by

$$E = E_0 + \sum_{i<j} \sum_{a<b} d_{ij}^{ab} [(\psi_i\psi_j \,|g|\, \psi_a\psi_b) - (\psi_i\psi_j \,|g|\, \psi_b\psi_a)]. \tag{7.106}$$

In order to derive (7.106) we simply substitute the expression (7.105) in the many-electron Schrödinger equation

$$H\Psi = E\Psi, \tag{7.107}$$

multiply by the UHF wavefunction Ψ_0^* and integrate over all electronic coordinates. From Brillouin's theorem and the rules for evaluating matrix elements of H between Slater determinants, the only non-zero terms are those given explicitly in equation (7.106). Of course, this attractively simple formula is of no practical value unless means are available for estimating the coefficients d_{ij}^{ab}. (For a further discussion of this problem see Hurley.)[69]

Koopmans' theorem.[13] *Minus the eigenvalues ($-\epsilon_i$) of the* canonical *UHF equations (7.84) provide good estimates of the vertical ionization potentials of an atom or molecule. Furthermore the Slater determinant obtained by deleting the ith row and ith column of the canonical UHF determinant for an N-electron system provides a fair approximation to the UHF determinant for a state of N – 1 electrons in the same nuclear field.*

The discussion of this theorem is conveniently divided into two parts.

(i) If a state of the $N - 1$ electron system is represented by the $(N - 1) \times (N - 1)$ determinant

$$\Psi_k = |\psi_1 \psi_2 \ldots \psi_{k-1} \psi_{k+1} \ldots \psi_N| \tag{7.108}$$

where

$$\Psi_0 = |\psi_1 \psi_2 \ldots \psi_{k-1} \psi_k \psi_{k+1} \ldots \psi_N| \tag{7.109}$$

is *any* representation of the UHF determinant for the N-electron system, then

$$E_0 - (\Psi_k | H_{N-1} | \Psi_k) = \epsilon_{kk} = (\psi_k | F_0 | \psi_k). \tag{7.110}$$

Here H_{N-1} is the many-electron Hamiltonian for the $N - 1$ electron system and E_0, F_0 are the energy and UHF Hamiltonian for the N-electron system, respectively.

To derive (7.110) we subtract from (7.85) the corresponding expression for $(\Psi_k | H_{N-1} | \Psi_k)$. Since we are using the same spin orbitals for both systems most of the terms cancel leaving

$$\begin{aligned} E_0 - (\Psi_k | H_{N-1} | \Psi_k) &= (\psi_k | f | \psi_k) + \sum_{i=1}^{N} \{(\psi_i \psi_k | g | \psi_i \psi_k) \\ &\qquad - (\psi_i \psi_k | g | \psi_k \psi_i)\} \\ &= (\psi_k | f + \sum_{i=1}^{N} (J_i - K_i) | \psi_k) \\ &= (\psi_k | F_0 | \psi_k) = \epsilon_{kk} \end{aligned} \tag{7.111}$$

as required.

(ii) As we have seen the formal identity (7.110) holds for any representation of the UHF determinant (7.109). However, this identity has as yet no physical significance since we have not shown that Ψ_k is a reasonable approximation to a stationary state of the $N - 1$ electron system. We now show that if ψ_k is a *canonical* UHF orbital this is the case; for some stationary state of the $N - 1$ electron system Ψ_k is then the best single determinant representation that can be constructed within the occupied sub-space of Ψ_0, that is, using $\psi_1, \psi_2, \ldots, \psi_N$ as basis functions. We establish this result by finding the stationary points of the energy functional $(\Psi_k | H_{N-1} | \Psi_k)$ within this sub-space.

Equation (7.110) may be written

$$(\Psi_k | H_{N-1} | \Psi_k) = E_0 - (\psi'_k | F_0 | \psi'_k) \tag{7.112}$$

where we have used a dash to distinguish the UHF spin orbital ψ'_k, which may not be canonical, from the canonical spin orbitals $\psi_1, \psi_2, \ldots, \psi_N$.

Expanding ψ'_k in terms of the canonical spin orbitals

$$\psi'_k = \sum_{i=1}^{N} c_i \psi_i \tag{7.113}$$

we find

$$\begin{aligned}(\Psi_k | H_{N-1} | \Psi_k) &= E_0 - \sum_{i=1}^{N} \sum_{j=1}^{N} c_i^* c_j (\psi_i | F_0 | \psi_j) \\ &= E_0 - \sum_{i=1}^{N} \sum_{j=1}^{N} c_i^* c_j \delta_{ij} \epsilon_j \\ &= E_0 - \sum_{i=1}^{N} c_i^* c_i \epsilon_i \end{aligned} \tag{7.114}$$

where we have used the canonical UHF equations (7.84) and the orthonormality of the ψ_i.

We seek the stationary values of the functional (7.114) subject to the normalization condition

$$(\psi'_k | \psi'_k) = \sum_{i=1}^{N} c_i^* c_i = 1. \tag{7.115}$$

The linear variational method (Section 2.2) leads to the linear equations

$$\sum_{j=1}^{N} (-\delta_{ij}\epsilon_j - \lambda\delta_{ij}) c_j = 0 \qquad (i = 1, \ldots, N) \tag{7.116}$$

and the determinantal equation

$$\det\{-\delta_{ij}\epsilon_j - \lambda\delta_{ij}\} = 0, \tag{7.117}$$

that is

$$(\epsilon_1 + \lambda)(\epsilon_2 + \lambda) \ldots (\epsilon_N + \lambda) = 0. \tag{7.118}$$

Equation (7.118) has the obvious roots

$$-\epsilon_k \qquad (k = 1, 2, \ldots, N) \tag{7.119}$$

and, for the particular root $\lambda = -\epsilon_k$, the secular equations (7.116) reduce to

$$(\epsilon_k - \epsilon_i) c_i = 0 \qquad (i = 1, \ldots, N). \tag{7.120}$$

From equation (7.120) we see that, apart from an arbitrary normalization factor, $c_i = \delta_{ik}$ and equation (7.113) reduces to

$$\psi'_k = \psi_k. \tag{7.121}$$

Equations (7.119) and (7.121) complete the demonstration of Koopmans' theorem. Of course, the approximate UHF wavefunction for the $N-1$ electron system (Ψ_k of equation (7.108)) is unaltered by any unitary transformation of the remaining $N-1$ spin orbitals among themselves, but to obtain estimates of ionization potentials and reasonable wave functions for the $N-1$ electron system the spin orbital ψ_k omitted from Ψ_0 must be a canonical spin orbital.

In practice ionization potentials obtained from Koopmans' theorem may show errors of more than 1 eV even if accurate UHF spin orbitals are employed (Section 7.4). These errors arise from two sources:

(i) failure to allow for the relaxation of the orbitals after ionization. This error may be eliminated by using the true stationary values of $(\Psi \mid H_{N-1} \mid \Psi)$ in the full function space for $N-1$ electrons, that is, by separate UHF calculations on each ionized state. For the lowest ionization potential this refinement always reduces the calculated value;

(ii) correlation errors, that is, errors in the UHF approximation itself. Errors from this source may be of either sign (cf. Ref. 69).

Despite its formal simplicity the UHF theory does not, in itself, provide an appropriate quantitative version of the MO method of Section 7.1. This is because the UHF Hamiltonian F_0 does not, in general, display the full symmetry of the nuclear framework nor is it independent of the spin state (α or β) of the spin orbitals. In the general N-electron case we may have solutions of the UHF equations (7.84) which involve N distinct spatial functions none of which has simple symmetry properties. The UHF determinant Ψ_0 is then not a pure spin state and is highly degenerate, the determinants $O_R \Psi_0$ generated by the molecular symmetry operators all yielding the same expectation value for H. Although this degeneracy may be resolved by the use of projection operators, as discussed in Chapter 4, the resultant theory, the projected unrestricted Hartree–Fock (PUHF) theory,[14] is rather complicated and bears little relationship to the simple qualitative MO theory of Section 7.1.

A more widely used method is the restricted or traditional Hartree–Fock method based on equations originally derived by Fock[15] in the atomic case and extensively applied by Hartree[16] and others. Here the simple shell structure of the qualitative one-electron theory is retained for all atomic and molecular states. For certain states this may involve restricting the variations permitted in minimizing the electronic energy and/or introducing approximations in the solution of the Hartree–Fock equations.

For states containing only closed electronic shells, however, no restrictions or approximations are needed. Here we have the best of both worlds. On one hand the Hartree–Fock Hamiltonian F_0

has the full symmetry of the molecular framework and is independent of spin; this leads to paired MSOs of the form $\phi\alpha$, $\phi\beta$ and the usual simplifications and reductions arising from molecular symmetry. On the other hand these MSOs are accurate solutions of the *unrestricted* Hartree–Fock equations (7.84); the formal simplicity of the UHF theory is thereby retained and the useful theorems and corollaries established above may be used. Fortunately the ground states of nearly all stable molecules fall into the closed-shell category so that this simple case is widely applicable.

(b) THE CLOSED-SHELL CASE

We consider a molecule with an even number $N = 2M$ of electrons and seek solutions of the canonical UHF equations (7.84) of the form

$$\left.\begin{aligned} \psi_{2i-1} &= \phi_i\alpha = \phi_i \\ \psi_{2i} &= \phi_i\beta = \bar{\phi}_i \end{aligned}\right\} \quad (i = 1, 2, \ldots, M). \tag{7.122}$$

We assume that the spatial orbitals ϕ_i provide a basis for some representation of the molecular symmetry group G. That is, the result of applying any symmetry operator O_R to one of the functions ϕ_i may be expressed as a linear combination of the ϕ_i without introducing any new functions.

$$O_R\,\phi_i = \sum_{j=1}^{M} D_{ji}(R)\phi_j. \tag{7.123}$$

The representation of G provided by the matrices $\mathbf{D}(R)$ will, in general, be reducible.

The UHF determinant

$$\Psi_0 = |\,\phi_1\bar{\phi}_1 \ldots \phi_M\bar{\phi}_M\,| \tag{7.124}$$

is clearly a singlet function ($S = M_S = 0$). Since Ψ_0 is unaltered, apart from normalization, by any linear transformation of the orbitals ϕ_i among themselves we may assume that the ϕ_i form an orthonormal set

$$(\phi_i\,|\,\phi_j) = \int \phi_i^*\phi_j \, dv = \delta_{ij}. \tag{7.125}$$

The transformation (7.123) of the orbitals induced by a symmetry operation is, then, a unitary transformation of the ϕ_i among themselves. As we showed in the previous section (equation (7.79)) the quantity $\gamma(1, 2)$, its spin-free equivalent

$$\rho(1, 2) = 2\sum_{j=1}^{M} \phi_j(1)\phi_j^*(2), \tag{7.126}$$

and the UHF Hamiltonian F_0 are all invariant under such transformations. Since F_0 is also independent of spin, the spin functions α and β factor out from the UHF equations and the $N = 2M$ equations (7.82) reduce to M equations for the spatial orbitals

$$F_0\phi_i = \sum_{j=1}^{M} \epsilon_{ij}\phi_j,\ \epsilon_{ji} = \epsilon_{ij}^* \qquad (i = 1 \ldots M) \tag{7.127}$$

or, in canonical form,

$$F_0\phi_i = \epsilon_i\phi_i \qquad (i = 1 \ldots M). \tag{7.128}$$

Since F_0 is totally symmetric its eigenfunctions ϕ_i will be basis functions for irreducible representations of the molecular symmetry group. Our assumption (7.123) concerning the transformation properties of the orbitals ϕ_i is, therefore, self-consistent; if we start with approximate MOs satisfying (7.123), construct F_0, solve the equations (7.128) to obtain improved MOs and iterate these steps the transformation properties (7.123) will never be lost. Some such iterative procedure is involved in all techniques for obtaining Hartree–Fock solutions.

The solutions obtained in this way are by no means unique and may or may not yield the UHF determinant of lowest energy. We have already seen an example of this in the simple case of H_2 (Section 4.8). For small and moderate nuclear separations the closed-shell function $|\phi\bar{\phi}|$ obtained by solving (7.128) for $M = 1$ is, indeed, the lowest energy determinant but for large R the open-shell function $|a\bar{b}|$ is lower in energy. This open-shell function is also an approximate solution of the UHF equations (7.82) and, in fact, tends to an exact solution of these equations (and of the full two-electron Schrödinger equation) as $R \to \infty$.

In the closed-shell case all the operators and integrals which appear in the Hartree–Fock equations and the various expressions for the electronic energy may be written in terms of the MOs ϕ_i instead of the MSOs ψ_i. The bracket notation $(|\ |)$ then indicates integration over the spatial coordinates only. The most important definitions and expressions become

Core energies:

$$f_i = (\phi_i | f | \phi_i). \tag{7.129}$$

Coulomb and exchange operators and integrals:

$$\begin{aligned} &J_j\phi = (\phi_j | g | \phi_j)\phi, \qquad K_j\phi = (\phi_j | g | \phi)\phi_j \\ &J_{ji} = J_{ij} = (\phi_i\phi_j | g | \phi_i\phi_j), \qquad K_{ij} = K_{ji} = (\phi_i\phi_j | g | \phi_j\phi_i). \end{aligned} \tag{7.130}$$

Total electron interaction operator:

$$G = \sum_{j=1}^{M} (2J_j - K_j), \qquad G_i = (\phi_i \mid G \mid \phi_i). \tag{7.131}$$

Hartree–Fock Hamiltonian:

$$F_0 = f + G. \tag{7.132}$$

Total electronic energy:

$$E_0 = 2\sum_{i=1}^{M} f_i + \sum_{i=1}^{M}\sum_{j=1}^{M} (2J_{ij} - K_{ij}) \tag{7.133}$$

$$= \sum_{i=1}^{M} (2f_i + G_i) \tag{7.134}$$

$$= \sum_{i=1}^{M} (2\epsilon_{ii} - G_i) \tag{7.135}$$

$$= \sum_{i=1}^{M} (f_i + \epsilon_{ii}). \tag{7.136}$$

The difference in the coefficients multiplying J_j and K_j in equation (7.131) and J_{ij}, K_{ij} in equation (7.133) arises from the fact that the exchange operator K_j of equation (7.74) gives zero wherever the spin-orbitals ψ_j and ψ have different spin, whereas the Coulomb operator (7.73) always gives a non-zero result.

When the Coulomb and exchange operators J_j, K_j operate on the orbital ϕ_j itself the results are the same, namely (from the definitions (7.130))

$$K_j\phi_j = J_j\phi_j = (\phi_j \mid g \mid \phi_j)\phi_j. \tag{7.137}$$

When these operators act on any orbital ϕ_x orthogonal to ϕ_j, however, the results are very different. The Coulomb operator J_j still involves an integral over the normalized charge distribution $\phi_j^*(2)\phi_j(2)$, representing a complete "smeared" electron, and always gives a sizeable result, whereas for K_j the charge distribution involved is $\phi_j^*(2)\phi_x(2)$ which integrates to zero; there is extensive cancellation and the resulting function $K_j\phi_x$ is small everywhere.

These properties have an important bearing on the nature of the so-called "virtual orbitals" which are often used to describe excited molecular states. Such an orbital ϕ_a is defined as an eigenfunction of the Hartree–Fock Hamiltonian F_0 which is orthogonal to all the occupied orbitals.

$$F_0\phi_a = \epsilon_a\phi_a, \qquad (\phi_a \mid \phi_i) = 0 \qquad (i = 1, 2, \ldots, M). \tag{7.138}$$

If we expand F_0 according to the definitions (7.131), (7.132) and (7.70) we find that occupied (i) and virtual (a) orbitals are determined by the equations

$$\{-\tfrac{1}{2}\nabla^2 - \sum_{\alpha} \frac{Z_\alpha}{r_\alpha} + J_i + \sum_{\substack{j=1 \\ j\neq i}}^{M} (2J_j - K_j)\}\phi_i = \epsilon_i \phi_i \tag{7.139}$$

and

$$\{-\tfrac{1}{2}\nabla^2 - \sum_{\alpha} \frac{Z_\alpha}{r_\alpha} + \sum_{j=1}^{M} (2J_j - K_j)\}\phi_a = \epsilon_a \phi_a \tag{7.140}$$

respectively.

In equation (7.139) we have used the identity (7.137) to cancel one of the Coulomb operators J_i with K_i. Apart from the small effect of the other exchange operators, equation (7.139) has an obvious interpretation as the one-electron Schrödinger equation of electron i moving in the field of the nuclear framework and the averaged Coulomb field of the other $2M - 1$ electrons (one of which is also in orbital ϕ_i but with opposite spin). In other words the exchange operator K_i cancels the self-interaction of the electron i thus leading to an eminently reasonable physical picture. Indeed simple arguments based on just this picture first led Hartree to propose an equation identical with (7.139) except for the omission of the small exchange terms $K_j (j \neq i)$.

If equation (7.140) for the virtual orbital ϕ_a is analysed in a similar way we find that *all* the exchange operators K_j provide only small corrections and that (7.140) corresponds to the Schrödinger equation for an electron moving in the field of the nuclei and the averaged field of $2M$ other electrons. This physical picture suggests that it is unreasonable to expect the virtual orbitals ϕ_a (and their energies ϵ_a) to provide an adequate description of excited states of a neutral molecule with a closed-shell ground state; they are more relevant to a description of the bound states (if any) of the singly-charged negative ion. In fact, for many closed-shell atoms and molecules there are probably *no* accurate bound state solutions of equation (7.140); all eigenfunctions of F_0 orthogonal to the occupied orbitals may well lie in the continuum; this has been shown to be the case for the ground state $(1s)^2 (2s)^2\,{}^1S$ of Be by Kelly[23] in the course of a perturbation treatment of the correlation energy.

As already discussed in Chapter 5 these deficiencies of the "virtual orbitals" are often masked when crude approximations to Hartree–Fock solutions are obtained by the expansion method using a severely limited basis set (as, for example, in LCAO calculations).

We can now extend the discussion of the energies of the singly excited states

$$^3\Psi_0(i \to a) = \begin{cases} |\phi_i\phi_a| & M_S = 1 \\ \frac{1}{\sqrt{2}}\{|\phi_i\bar{\phi}_a| + |\bar{\phi}_i\phi_a|\} & M_S = 0 \\ |\bar{\phi}_i\bar{\phi}_a| & M_S = -1 \end{cases} \quad (7.141)$$

and

$$^1\Psi_0(i \to a) = \frac{1}{\sqrt{2}}\{|\phi_i\bar{\phi}_a| - |\bar{\phi}_i\phi_a|\} \quad (7.142)$$

given in Section 7.1(a). There we found that in zeroth order all four of the functions (7.141), (7.142) were degenerate, the excitation energy being simply the difference in one-electron energies.

$$E_0(i \to a) - E_0 = \epsilon_a - \epsilon_i. \quad (7.143)$$

The orbitals $\phi_j (j = 1 \ldots M)$ are taken as the occupied Hartree-Fock orbitals for the closed-shell ground state (equation (7.124)); ϕ_i is a non-degenerate ground-state orbital and ϕ_a is a non-degenerate excited orbital obtained either as a "virtual orbital" from the higher solutions of (7.140) or (preferably) in some other way.

The first-order energies of the states (7.141), (7.142) are now obtained as the diagonal matrix elements of the full many electron Hamiltonian between the wave functions (7.141), (7.142) (with the paired orbitals $\phi_1\bar{\phi}_1 \ldots \phi_{i-1}\bar{\phi}_{i-1}\phi_{i+1}\bar{\phi}_{i+1} \ldots \phi_M\bar{\phi}_M$ included in the Slater determinants). Using the rules of Section 6.3 we find for the three components of $^3\Psi_0(i \to a)$ the energy

$$\begin{aligned} ^3E_0(i \to a) &= E_0 + (f_a - f_i) + \sum_{j=1}^{M}(2J_{ja} - K_{ja}) - \sum_{j=1}^{M}(2J_{ji} - K_{ji}) - J_{ia} \\ &= E_0 + (\phi_a | F_0 | \phi_a) - (\phi_i | F_0 | \phi_i) - J_{ia} \end{aligned} \quad (7.144)$$

and for the singlet function $^1\Psi_0(i \to a)$ the energy

$$^1E_0(i \to a) = E_0 + (\phi_a | F_0 | \phi_a) - (\phi_i | F_0 | \phi_i) - J_{ia} + 2K_{ia}. \quad (7.145)$$

Here we have used equation (7.133) for the ground-state energy E_0 and the definition (7.131), (7.132) of the Fock operator F_0. We note that equations (7.144) and (7.145) are valid whether or not the orbitals ϕ_i and ϕ_a are solutions of the Fock equations (7.128), (7.138); if they are solutions the expectation values $(\phi_a | F_0 | \phi_a)$ and $(\phi_i | F_0 | \phi_i)$

may be replaced by the eigenvalues ϵ_a, ϵ_i respectively. The resulting expressions

$$\begin{aligned} {}^3E_0(i\to a) - E_0 &= \epsilon_a - \epsilon_i - J_{ia} \\ {}^1E_0(i\to a) - E_0 &= \epsilon_a - \epsilon_i - J_{ia} + 2K_{ia} \end{aligned} \tag{7.146}$$

differ from the zeroth order result of Section 7.1 (equation (7.9)) even if we identify F_0 with the effective one-electron Hamiltonian used there; both excited states are depressed from the zeroth order value $\epsilon_a - \epsilon_i$ by the Coulomb integral J_{ia} and they are split apart by twice the exchange integral K_{ia}. Since $K_{ia} > 0$ the calculated order of the excited states, triplet below singlet, is in accordance with Hund's rule.

(c) RESTRICTED HARTREE-FOCK THEORY FOR THE OPEN-SHELL CASE

Consider a wavefunction for an N-electron atom or molecule of the form

$$\Psi = |\phi_1\bar{\phi}_1'\phi_2\bar{\phi}_2' \ldots \phi_p\bar{\phi}_p'\phi_{p+1}\phi_{p+2} \ldots \phi_{p+q}| \qquad (N = 2p + q). \tag{7.147}$$

From the properties of the spin operators S^2, S_+ and S_- discussed in Chapter 6 it is clear that Ψ of equation (7.147) will be a spin eigenfunction (with eigenvalues $S = M_S = q/2$) only if

$$\phi_i \equiv \phi_i' \qquad (i = 1, 2, \ldots, p). \tag{7.148}$$

On the other hand if we substitute the spin orbitals of the wave function (7.147) into the UHF equations (7.84) and separate out the spin functions we obtain different equations for the dashed and undashed orbitals, namely

$$F\phi_i = \epsilon_i\phi_i \qquad (i = 1, \ldots, p + q) \tag{7.149}$$

and

$$F'\phi_i' = \epsilon_i'\phi_i' \qquad (i = 1, \ldots, p), \tag{7.150}$$

where, in an obvious notation

$$F = f + \sum_{j=1}^{p+q} (J_j - K_j) + \sum_{j=1}^{p} J_j' \tag{7.151}$$

and

$$F' = f + \sum_{j=1}^{p+q} J_j + \sum_{j=1}^{p} (J_j' - K_j'). \tag{7.152}$$

From these equations we see that, if the restrictions (7.148) are imposed, the equations (7.149) and (7.150) for $i = 1, \ldots, p$ become inconsistent, since the Fock operators F and F' contain different exchange contributions.

We may overcome this difficulty in several ways. In the (spin) unrestricted Hartree–Fock method the restrictions (7.148) are dropped and both sets of equations (7.149) and (7.150) are satisfied by using different sets of orbitals for electrons with α and β spin. This approach has been widely used in the discussions of spin densities of atomic states which, in the restricted theory, contain a half-filled open shell, for example, Li: $(1s)^2\ 2s$; N: $(1s)^2(2s)^2(2p)^3$. It has the advantage of retaining Brillouin's and Koopmans' theorems in their simplest forms; it is also possible to recover a spin eigenfunction at the end of the calculation by applying a spin projection operator to the unrestricted determinant (7.147). The main disadvantage is that the simple shell structure is lost and with it the relationship of the results to the qualitative MO theory of Section 7.1.

Nesbet(17) pointed out that, in many cases, the effect of the exchange operators in equations (7.149) and (7.150) is quite small and that a satisfactory approximation may be obtained by solving just one of the sets of equations, say (7.149), and using the solutions for both the dashed and undashed orbitals in (7.147). The simple shell structure of the qualitative MO theory is thereby retained.

More recently, efficient methods for solving the restricted Hartree–Fock equations have been developed enabling us to dispense with Nesbet's approximation and to still retain the simple shell structure of the qualitative MO theory. In this restricted Hartree–Fock (RHF) theory the restrictions (7.148) are imposed from the outset giving a total wave function

$$\Psi = |\phi_1\bar{\phi}_1 \cdots \phi_p\bar{\phi}_p\phi_{p+1} \cdots \phi_{p+q}|. \qquad (7.153)$$

The expectation value of the electronic Hamiltonian

$$E = \int \Psi^* H \Psi \, d\tau \qquad (7.154)$$

is then required to be stationary for variations of the orbitals $\phi_i (i = 1, \ldots, p + q)$ which preserve the orthonormality constraints

$$(\phi_i \mid \phi_j) = \int \phi_i^* \phi_j \, dv = \delta_{ij} \qquad (i, j = 1, \ldots, p + q). \qquad (7.155)$$

These constraints are incorporated by the method of Lagrange multipliers (Section 2.5). This leads to the condition

$$\delta\left\{E - \sum_{i=1}^{p+q} \sum_{j=1}^{p+q} 2\epsilon_{ij}(\phi_i \mid \phi_j)\right\} = 0, \qquad (7.156)$$

for arbitrary variations of the orbitals provided the constants ϵ_{ij} are suitably chosen.

Substituting from equation (7.153) into equation (7.154) and using the formulae of Section 6.3 we obtain an expression for the total electronic energy in terms of the unknown orbital functions ϕ_i

$$E = 2 \sum_{i=1}^{p} f_i + \sum_{i=p+1}^{p+q} f_i + \sum_{i=1}^{p} \sum_{j=1}^{p} (2J_{ij} - K_{ij}) + \sum_{i=1}^{p} \sum_{j=p+1}^{p+q} (2J_{ij} - K_{ij}) + \tfrac{1}{2} \sum_{i=p+1}^{p+q} \sum_{j=p+1}^{p+q} (J_{ij} - K_{ij}). \tag{7.157}$$

At this stage it is convenient to go over to the more general open-shell situation originally considered by Roothaan.[18] Here we have two sets of orbital functions, a set

$$\phi_i \qquad (i = 1, \ldots, p) \tag{7.158}$$

for the closed shell (or shells) of electrons and a set

$$\phi_i \qquad (i = p+1, \ldots, p+q) \tag{7.159}$$

for a single open shell. In cases of orbital degeneracy each of the sets (7.158) and (7.159) consists of complete sets of degenerate orbitals which form bases for representations of the appropriate symmetry group.

If the total number of electrons N lies between $2p$ and $2(p+q)$ we can construct from the orbitals (7.158) and (7.159) a total of $\binom{2q}{N-2p}$ Slater determinants, each of which contains all the closed-shell orbitals (doubly occupied) and a selection of the open-shell orbitals. Linear combinations of these determinants belonging to the different spectroscopic terms of the configuration may be obtained by the methods of Section 7.1(c). We suppose that this has been done and concentrate on one such term. We assume that the energy of the term may be written in the form

$$E = 2 \sum_{k} f_k + \sum_{k} \sum_{l} (2J_{kl} - K_{kl}) + \nu\Big[2 \sum_{m} f_m + \nu \sum_{m} \sum_{n} (2aJ_{mn} - bK_{mn}) + 2 \sum_{k} \sum_{m} (2J_{km} - K_{km})\Big], \tag{7.160}$$

where ν, a and b are numerical constants depending on the specific case. In equation (7.160) and subsequently we use subscripts $k, l (= 1, \ldots, p)$ for the closed-shell orbitals $m, n (= p+1, \ldots, p+q)$ for the open-shell orbitals and omit the limits of summations. Subscripts i, j will continue to be used for orbitals of either set.

The first two summations in equation (7.160) give the energy of the closed-shell core, the next two summations the energy of the open shell and the final summation the interaction energy of the closed and open shells. The constant ν represents the fractional occupation of the open shell, that is, the ratio of the number of occupied open-shell spin orbitals $(N-2p)$ to the total number $(2q)$ of available open-shell spin orbitals from the set (7.159) with α or β spin functions; clearly $0<\nu<1$. The constants a and b differ for different states of the same configuration. Even with this freedom of choice in a and b, not all states of atoms and molecules with a single open shell are covered by the energy formula (7.160). This limitation of the theory and further generalizations are considered below. Here we merely note that the energy expression (7.157) for the wave function (7.153) which represents a half-filled shell with all open-shell spins parallel, corresponds to the case $\nu=\frac{1}{2}$, $a=1$, $b=2$.

It is important to note that, despite the partial occupation of the open shell, the summations in equation (7.160) are over the *complete* sets of orbitals (7.158) or (7.159). In general, such an expression for the total energy will not be obtained by applying the usual formula

$$E_0=\int \Psi_0^* H \Psi_0 \, \mathrm{d}\tau \tag{7.161}$$

to a single total wavefunction Ψ_0 belonging to a degenerate set (spectroscopic term). To obtain (7.160) it may be necessary to average the different expressions obtained for the various members of the degenerate set. It is in this way that the fractional occupation number ν is introduced. Consider, for example, the simple atomic configuration s^2p with the single term ^2P. There are six degenerate wavefunctions belonging to this term each represented by a single determinant. None of the expressions obtained by substituting one of these determinants in equation (7.161) is of the form (7.160), but if the six expressions are averaged we obtain (using real orbitals s, p_x, p_y, p_z)

$$E_{\mathrm{AV}}=2f_s+(2J_{ss}-K_{ss})+\tfrac{1}{6}\left[2\sum_{m=x,y,z} f_m+2\sum_{m=x,y,z}(2J_{sm}-K_{sm})\right]. \tag{7.162}$$

This is a special case of (7.160) with $\nu=\frac{1}{6}$; $a=b=0$.

Because (7.160) contains only sums over the complete sets (7.158) and (7.159), it is invariant under any unitary transformation of the closed-shell orbitals among themselves and of the open-shell orbitals among themselves. Since, further, the sets (7.158) and (7.159) each comprise complete degenerate sets of orbitals, which form bases for representations of the appropriate symmetry group, the energy expression (7.160) is invariant under the full rotation group in the atomic case and under the appropriate point group in the molecular case. These invariance and symmetry properties are shared by the Fock Hamiltonians which we derive below by minimizing the expression (7.160) with respect to variations of the orbitals ϕ_i. Consequently, as in the closed-shell case, orbitals obtained as eigenfunctions of the Fock Hamiltonians will be symmetry orbitals, that is, will transform as partners in irreducible representations of the symmetry group. In short the fact that (7.160) contains only summations over *complete* degenerate sets ensures that our assumption of symmetry orbitals for both the closed and open shells is self-consistent.

Turning now to the variational problem we find that, for an arbitrary variation of the orbital functions, the variation of the total energy expression (7.160) may be expressed in the form

$$\delta E = 2 \sum_k [(\delta\phi_k \,|\, F_C \,|\, \phi_k) + (\delta\phi_k^* \,|\, F_C^* \,|\, \phi_k^*)] + 2\nu \sum_m [(\delta\phi_m \,|\, F_O \,|\, \phi_m) + (\delta\phi_m^* \,|\, F_O^* \,|\, \phi_m^*)] \tag{7.163}$$

where we have introduced different operators F_C and F_O for the closed-shell and open-shell orbitals. The explicit expressions for these operators are

$$F_C = f + \sum_k (2J_k - K_k) + \nu \sum_m (2J_m - K_m) \tag{7.164}$$

and

$$F_O = f + \sum_k (2J_k - K_k) + 2a\nu \sum_m J_m - b\nu \sum_m K_m \tag{7.165}$$

where J_j and K_j are the Coulomb and exchange operators of equation (7.130). We note that, like the J_j and K_j, the operators F_C and F_O are Hermitian and that this property of the operators must be used in reducing δE to the form (7.163).

For the remaining terms of equation (7.156) we require

$$\delta(\phi_i \,|\, \phi_j) = (\delta\phi_i \,|\, \phi_j) + (\phi_i \,|\, \delta\phi_j) = (\delta\phi_i \,|\, \phi_j) + (\delta\phi_j^* \,|\, \phi_i^*) \tag{7.166}$$

so that

$$\delta \sum_i \sum_j 2\epsilon_{ij}(\phi_i \,|\, \phi_j) = 2 \sum_i \sum_j [\epsilon_{ij}(\delta\phi_i \,|\, \phi_j) + \epsilon_{ji}(\delta\phi_i^* \,|\, \phi_j^*)]. \tag{7.167}$$

Here we have interchanged the roles of the summation variables in the second term.

Substituting the expressions (7.163) and (7.167) into equation (7.156) and remembering that the orbital variations $\delta\phi_i$ and $\delta\phi_i^*$ are now arbitrary we obtain

$$F_C \phi_k = \sum_j \epsilon_{kj}\phi_j = \sum_l \epsilon_{kl}\phi_l + \sum_n \epsilon_{kn}\phi_n \tag{7.168}$$

$$F_C^* \phi_k^* = \sum_j \epsilon_{jk} \phi_j^* \tag{7.169}$$

$$\nu F_O \phi_m = \sum_j \epsilon_{mj} \phi_j = \sum_l \epsilon_{ml} \phi_l + \sum_n \epsilon_{mn} \phi_n \tag{7.170}$$

$$\nu F_O^* \phi_m^* = \sum_j \epsilon_{jm} \phi_j^* . \tag{7.171}$$

Subtracting the complex conjugates of (7.169) and (7.171) from (7.168) and (7.170) respectively we obtain the conditions

$$\epsilon_{jk}^* = \epsilon_{kj}, \qquad \epsilon_{jm}^* = \epsilon_{mj}. \tag{7.172}$$

Recalling the ranges of the various indices we infer that the complete matrix of Lagrangian multipliers must be Hermitian

$$\epsilon_{ji} = \epsilon_{ij}^* \qquad (i, j = 1, \ldots, p + q). \tag{7.173}$$

Once the conditions (7.173) have been imposed, equations (7.169) and (7.171) become equivalent to (7.168) and (7.170) respectively and may, therefore, be dropped. Thus equations (7.168), (7.170) and (7.173) provide a complete set of conditions for the self-consistent determination of the optimum orbitals.

In the unrestricted Hartree–Fock theory and in the closed-shell case of the restricted theory we could use the invariance of the total wave function, the expression for the total energy and the Fock Hamiltonian under unitary mixing of the occupied orbitals to eliminate all the off-diagonal Lagrange multipliers $\epsilon_{ij} (i \neq j)$ and so reduce the Fock equations to the canonical form of pseudo-eigenvalue equations. The complete elimination of the $\epsilon_{ij} (i \neq j)$ by this simple method is not possible in the open-shell case, since the total energy expression (7.160) and the Fock Hamiltonians (7.164), (7.165) are *not* invariant under unitary mixing of all the occupied orbitals ϕ_j. The best we can do here by this simple approach is to use the invariance of (7.160), (7.164) and (7.165) under separate unitary transformations of the ϕ_k among themselves and of the ϕ_m among themselves to eliminate all off-diagonal ϵ_{ij} except those linking closed-shell orbitals with open-shell orbitals. The Fock equations then assume the form

$$F_C \phi_k = \epsilon_k \phi_k + \sum_m \epsilon_{km} \phi_m \tag{7.174}$$

and

$$\nu F_O \phi_m = \epsilon_m \phi_m + \sum_k \epsilon_{mk} \phi_k \tag{7.175}$$

with ϵ_k, ϵ_m real and

$$\epsilon_{mk}^* = \epsilon_{km}. \tag{7.176}$$

For atomic systems, where the spherical symmetry reduces equations (7.174) and (7.175) to coupled one-dimensional integro-differential equations, the most efficient method of solution is probably an adaptation of Hartree's[16] original numerical integration technique. Computer programs for this purpose have been developed by Froese.[24] The off-diagonal Lagrange multipliers ϵ_{km} cause no particular difficulty in these calculations since estimates derived from earlier iterations may be used.

For molecules, on the other hand, the reduction to one-dimensional equations is not possible and approximate solutions of (7.174) and (7.175) are usually sought by the expansion technique introduced by Hall[19] and Roothaan.[20] In these calculations the off-diagonal Lagrange multipliers are more troublesome and most early treatments of open-shell systems neglected or approximated these terms. As Roothaan[18] first showed, however, it is possible to avoid these approximations by using projection operators to transform equations (7.174) and (7.175) into pseudo-eigenvalue form. The essence of this transformation is already contained in the simple variational calculation of Section 2.5.

If we multiply equation (7.174) on the left by ϕ_m^* and integrate we obtain

$$\epsilon_{km} = (\phi_m \mid F_{\text{C}} \mid \phi_k). \tag{7.177}$$

An alternative expression for ϵ_{km} may be obtained using equations (7.175), (7.176) and the Hermitian character of the operator F_{O}; thus

$$\epsilon_{km} = \epsilon_{mk}^* = (\phi_k \mid \nu F_{\text{O}} \mid \phi_m)^* = (\phi_m \mid \nu F_{\text{O}} \mid \phi_k). \tag{7.178}$$

A more general expression for ϵ_{km} is obtained from a weighted mean of (7.177) and (7.178). This gives

$$\epsilon_{km} = (\phi_m \mid F_{\text{A}} \mid \phi_k) \tag{7.179}$$

with

$$F_{\text{A}} = xF_{\text{C}} + (1 - x)\nu F_{\text{O}}. \tag{7.180}$$

For each orbital ϕ_j we now introduce a projection operator $\mathscr{P}_j$ which, as in Section 2.5, is defined by its action on an arbitrary orbital function ϕ

$$\mathscr{P}_j\phi = \phi_j(\phi_j \mid \phi). \tag{7.181}$$

If we sum equation (7.181) over the set of closed-shell orbitals we obtain a projection operator Q_{C} for the closed-shell sub-space or manifold

$$Q_{\text{C}}\phi = \sum_k \mathscr{P}_k\phi = \sum_k \phi_k(\phi_k \mid \phi). \tag{7.182}$$

As in Section 2.5, it is readily shown that this operator is Hermitian ($Q_C^\dagger = Q_C$) and idempotent

$$Q_C^2 = Q_C. \tag{7.183}$$

Similarly, a projection operator for the open-shell manifold is defined by

$$Q_O = \sum_m \mathscr{P}_m \tag{7.184}$$

and has the properties

$$Q_O^\dagger = Q_O \tag{7.185}$$

$$Q_O^2 = Q_O. \tag{7.186}$$

The effect of the projection operators Q_C and Q_O on the occupied orbitals ϕ_i follows immediately from the orthogonality conditions (7.155); we find

$$Q_C \phi_k = \phi_k, \qquad Q_C \phi_m = 0 \tag{7.187}$$

$$Q_O \phi_k = 0, \qquad Q_O \phi_m = \phi_m. \tag{7.188}$$

These properties of the operators Q_C and Q_O and the expression (7.179) for ϵ_{km} enable us to express equation (7.174) in the desired form. Thus we find

$$\begin{aligned} F_C \phi_k &= \epsilon_k \phi_k + \sum_m \phi_m (\phi_m | F_A | \phi_k) \\ &= \epsilon_k \phi_k + Q_O F_A \phi_k \end{aligned} \tag{7.189}$$

that is

$$(F_C - Q_O F_A)\phi_k = \epsilon_k \phi_k. \tag{7.190}$$

Now although the operators F_C, Q_O and F_A are all Hermitian the product $Q_O F_A$ may not be. However, the symmetrized product $Q_O F_A + F_A Q_O$ is necessarily Hermitian and, furthermore, from equation (7.188)

$$F_A Q_O \phi_k = 0. \tag{7.191}$$

Subtracting equation (7.191) from equation (7.190) we obtain for the closed-shell orbitals ϕ_k the pseudo-eigenvalue equation

$$H_C \phi_k = \epsilon_k \phi_k \tag{7.192}$$

with an Hermitian effective Hamiltonian

$$H_C = F_C - (Q_O F_A + F_A Q_O). \tag{7.193}$$

In a similar way equation (7.175) for the open-shell orbitals may be reduced to

$$(\nu F_{\mathrm{O}} - Q_{\mathrm{C}} F_{\mathrm{A}} - F_{\mathrm{A}} Q_{\mathrm{C}})\phi_m = \epsilon_m \phi_m \tag{7.194}$$

which is, again, equivalent to a pseudo-eigenvalue equation

$$H_{\mathrm{O}} \phi_m = \eta_m \phi_m \tag{7.195}$$

with $\eta_m = (1/\nu)\epsilon_m$ and the effective Hamiltonian

$$H_{\mathrm{O}} = F_{\mathrm{O}} - \frac{1}{\nu} (Q_{\mathrm{C}} F_{\mathrm{A}} + F_{\mathrm{A}} Q_{\mathrm{C}}). \tag{7.196}$$

In equations (7.192) and (7.195) we have the required pseudo-eigenvalue form of the Fock equations (7.174) and (7.175).

The introduction of projection operators into Hartree–Fock theory results in some arbitrariness in the definition of the effective Hamiltonians H_{O} and H_{C}. This is apparent in the above derivation, since the parameter x of equation (7.180) is still at our disposal. Furthermore, if R is any Hermitian one-electron operator which, like F_{O} and F_{C}, is invariant under unitary mixing of the closed-shell orbitals, we have from equation (7.187).

$$\begin{aligned}\{R - (Q_{\mathrm{C}} R + R Q_{\mathrm{C}})\} \phi_k &= R\phi_k - Q_{\mathrm{C}} R \phi_k - R\phi_k \\ &= -Q_{\mathrm{C}} R \phi_k = \sum_l \phi_l \zeta_{kl}\end{aligned} \tag{7.197}$$

with

$$\zeta_{kl} = -(\phi_l \,|\, R \,|\, \phi_k). \tag{7.198}$$

Addition of (7.192) and (7.197) then leads to the equation

$$\{H_{\mathrm{C}} + R - (Q_{\mathrm{C}} R + R Q_{\mathrm{C}})\} \phi_k = \sum_l (\zeta_{kl} + \delta_{kl} \epsilon_k) \phi_l. \tag{7.199}$$

Since the operator appearing on the left-hand side of equation (7.199) is Hermitian and invariant under a unitary transformation of the closed-shell orbitals, we may use such a transformation to diagonalize the Hermitian matrix $\zeta_{kl} + \delta_{kl}\epsilon_k$ and so reduce equation (7.199) to pseudo-eigenvalue form. Hence the addition of the term $R - (Q_{\mathrm{C}} R + R Q_{\mathrm{C}})$ to H_{C} simply results in a unitary mixing of the closed-shell orbitals.

In just the same way the addition of a term $S - (Q_{\mathrm{O}} S + S Q_{\mathrm{O}})$ to H_{O} results in a unitary mixing of the open-shell orbitals.

Because of the arbitrariness in the effective Hamiltonians H_{C} and H_{O} there is no unique canonical form for the Hartree–Fock equations and orbitals as there is in the closed-shell case. The individual orbitals ϕ_i are not uniquely determined but only the manifolds (linear sub-spaces) spanned by the closed-shell orbitals ϕ_k and by the open-shell orbitals ϕ_m.

As Roothaan[18] shows it is easy to choose R and S so that the closed- and open-shell effective Hamiltonians become identical. Although such a procedure has some formal advantages, the technique which has proved most successful in practical applications is based on the distinct effective Hamiltonians H_C and H_O of equations (7.193) and (7.196) and a choice of the parameter x which causes many of the terms in F_A to cancel. From equations (7.164), (7.165) and (7.180) we see that the choice $x + (1 - x)\nu = 0$, that is

$$x = -\frac{\nu}{1-\nu}, \tag{7.200}$$

causes all terms in F_A to cancel except those involving the Coulomb and exchange operators J_m, K_m for the open-shell orbitals.

With this choice of x and the definitions

$$\begin{aligned} J_C &= \sum_k J_k, \quad K_C = \sum_k K_k, \quad & J_O &= \nu \sum_m J_m, \quad K_O = \nu \sum_m K_m \\ L_C &= Q_C J_O + J_O Q_C, & M_C &= Q_C K_O + K_O Q_C, \\ L_O &= \nu(Q_O J_O + J_O Q_O), & M_O &= \nu(Q_O K_O + K_O Q_O), \end{aligned} \tag{7.201}$$

the effective Hamiltonians (7.193) and (7.196) reduce to the form originally given by Roothaan,[18] namely

$$H_C = f + 2J_C - K_C + 2J_O - K_O + 2\alpha L_O - \beta M_O \tag{7.202}$$

and

$$H_O = f + 2J_C - K_C + 2aJ_O - bK_0 + 2\alpha L_C - \beta M_C \tag{7.203}$$

with

$$\alpha = \frac{1-a}{1-\nu}, \qquad \beta = \frac{1-b}{1-\nu}. \tag{7.204}$$

With this choice of x the influence of the coupling terms L, M of equations (7.202), (7.203) in the Hartree–Fock equations

$$H_C \phi_k = \epsilon_k \phi_k \tag{7.205}$$

$$H_O \phi_m = \eta_m \phi_m \tag{7.206}$$

is usually small. Consequently, although Koopmans' theorem is no longer valid in the open-shell case, it is often a fair approximation to use the orbitals derived from (7.205) and (7.206) to describe singly ionized states. If this approximation is made it is a simple matter to determine, in any particular case, the correction terms which must be added to $-\epsilon_k$ and $-\eta_m$ to obtain estimates of the vertical ionization potentials.

Multiplying (7.205) and (7.206) by ϕ_k^*, ϕ_m^* respectively and integrating, the coupling terms disappear and we obtain

$$\epsilon_k = f_k + \sum_l (2J_{kl} - K_{kl}) + \nu \sum_m (2J_{km} - K_{km}) \qquad (7.207)$$

and

$$\eta_m = f_m + \sum_k (2J_{km} - K_{km}) + \nu \sum_n (2aJ_{mn} - bK_{mn}). \qquad (7.208)$$

Comparing (7.207) and (7.208) with the expression (7.160) for the total electronic energy we see that

$$E = \sum_k (f_k + \epsilon_k) + \nu \sum_m (f_m + \eta_m) \qquad (7.209)$$

which is completely analogous to equation (7.136) for the closed-shell case.

When solving the equations (7.205) and (7.206) we must be careful to select appropriate solutions at each iteration. Let us suppose that some trial open- and closed-shell orbitals which approximately satisfy the equations are available. From these we construct the operators H_O and H_C and determine their eigenvalues and eigenfunctions. The appropriate eigenfunctions of H_C are usually those belonging to the p lowest eigenvalues. Since there is little difference between the effective Hamiltonians H_C and H_O, eigenfunctions of H_O belonging to its p lowest eigenvalues are expected to be very similar to the closed-shell orbitals and clearly must be rejected as open-shell orbitals. Instead we must use the eigenfunctions of H_O belonging to the q next lowest eigenfunctions as is suggested by the labelling of the orbitals in equations (7.205) and (7.206).

The open-shell theory developed here is applicable only if the total energy of the state under consideration may be expressed in the form (7.160). This formula covers the following three important classes of states:

(i) The half-closed shell. The open shell consists of any number of singly occupied, complete degenerate sets of orbitals and all the spins are parallel. The total wavefunction may be expressed as a single Slater determinant (7.153) and we find $\nu = \frac{1}{2}$, $a = 1$, $b = 2$, $\alpha = 0$, $\beta = -2$. This case, together with the closed-shell case, already includes the ground states of almost all molecules and radicals. The lowest excited triplet state of a closed-shell molecule is also included provided that the excitation is from a non-degenerate orbital to a non-degenerate orbital. Examples of atomic states in this class are C: $(1s)^2(2s)(2p)^3$, ^5S; Cr: $(1s)^2(2s)^2(2p)^6(3s)^2(4s)(3d)^5$, ^7S.

(ii) All states arising from the configurations π^N, δ^N, . . .; $1 \leqslant N \leqslant 3$ of a linear molecule. The appropriate values of the constants are given in Table 7.7.

(iii) All states of an atom arising from a partially filled p shell. Table 7.8 lists the constants applicable to each state.

It is a straightforward matter to extend the theory to cover all cases where there is at most one open shell of each symmetry type.[21]

Table 7.7. Coefficients for the configurations π^N, δ^N of linear molecules

N	1	2			3
ν	1/4		1/2		3/4
π^N	$^2\Pi$	$^3\Sigma^-$	$^1\Delta$	$^1\Sigma^+$	$^2\Pi$
δ^N	$^2\Delta$	$^3\Sigma^-$	$^1\Gamma$	$^1\Sigma^+$	$^2\Delta$
a	0	1	1/2	0	8/9
b	0	2	0	−2	8/9
α	4/3	0	1	2	4/9
β	4/3	−2	2	6	4/9

Table 7.8. Coefficients for the configurations p^N of atoms

N	1	2			3			4			5
ν	1/6		1/3			1/2			2/3		5/6
	^2P	^3P	^1D	^1S	^4S	^2D	^2P	^3P	^1D	^1S	^2P
a	0	3/4	9/20	0	1	4/5	2/3	15/16	69/80	3/4	24/25
b	0	3/2	−3/10	−3	2	4/5	0	9/8	27/40	0	24/25
α	6/5	3/8	33/40	3/2	0	2/5	2/3	3/16	33/80	3/4	6/25
β	6/5	−3/4	39/20	6	−2	2/5	2	−3/8	39/40	3	6/25

(d) APPROXIMATE HARTREE-FOCK ORBITALS BY THE EXPANSION METHOD

The direct numerical solution of the Hartree–Fock equations is, at present, only feasible for atoms where the spherical symmetry of the system permits reduction to a system of coupled one-dimensional equations.

For molecules, approximate solutions of the equations may be obtained by expanding all functions ϕ which arise in terms of some set of fixed basis functions χ_r.

$$\phi = \sum_r \chi_r c_r. \tag{7.210}$$

If the basis functions $\chi_r (r = 1, \ldots, \infty)$ constitute a complete set, then any function may be expressed in the form (7.210) with negligible error by including sufficient terms in the expansion. Results obtained by the expansion method then tend to exact solutions of the Hartree-Fock equations and are, of course, independent of the nature of the basis functions. In practice it is often not feasible to include sufficient terms in the expansion for convergence to accurate Hartree-Fock solutions and the quality of the approximation (7.210) is then critically dependent on the choice of basis functions.

The expansion (7.210) reduces the Hartree-Fock equations of this section to sets of matrix eigenvalue equations. The details of the reduction are very similar in the various cases and will be given only for the restricted Hartree-Fock theory in the open-shell case (Section 7.2(c)). The closed-shell theory of Section 7.2(b) is, of course, a special case of this open-shell theory with $q = 0$.

With all functions represented in the form (7.210) any one-electron operator M is represented by a matrix $\mathbf{M}$ with elements

$$M_{rs} = \int \chi_r^* M \chi_s \, dv. \tag{7.211}$$

If the operator M is Hermitian, so is the matrix $\mathbf{M}$, that is

$$M_{rs}^\dagger = M_{sr}^* = M_{rs}.$$

To derive the matrix form of equations (7.205) and (7.206) we expand both the closed-shell orbitals and the open-shell orbitals in the form (7.210)

$$\phi_i = \sum_r \chi_r c_{ri}. \tag{7.212}$$

The orthogonality constraints (7.155) now appear as

$$(\phi_i \mid \phi_j) = \sum_r \sum_s c_{ri}^* S_{rs} c_{sj} = \delta_{ij} \tag{7.213}$$

where the $S_{rs} = \int \chi_r^* \chi_s \, dv$ are elements of the overlap matrix between the basis functions.

An arbitrary variation of the orbital ϕ_i is now represented by a variation in the coefficients in the expansion (7.212)

$$\delta\phi_i = \sum_r \chi_r \delta c_{ri}.$$

Applying such a variation to equation (7.213) we obtain

$$\begin{aligned}\delta(\phi_i \mid \phi_j) &= \sum_r \sum_s (\delta c_{ri}^* S_{rs} c_{sj} + c_{ri}^* S_{rs} \delta c_{sj}) \\ &= \sum_r \sum_s (\delta c_{ri}^* S_{rs} c_{sj} + \delta c_{sj} S_{sr}^* c_{ri}^*)\end{aligned} \tag{7.214}$$

where we have used the Hermitian property of $\mathbf{S}$ to transform the second term.

At this stage it is convenient to use explicit matrix notation with the conventions introduced in Section 2.5. Equations (7.213), (7.214) then become

$$\mathbf{c}_i^\dagger \mathbf{S} \mathbf{c}_j = \delta_{ij} \tag{7.215}$$

$$\delta(\phi_i \mid \phi_j) = \delta \mathbf{c}_i^\dagger \mathbf{S} \mathbf{c}_j + \delta \mathbf{c}_j^{*\dagger} \mathbf{S}^* \mathbf{c}_i^* \tag{7.216}$$

where $\delta\mathbf{c}^\dagger$, $\delta\mathbf{c}^{*\dagger}$ are row matrices and $\mathbf{c}$, $\mathbf{c}^*$ are column matrices. The matrix $\delta\mathbf{c}^{*\dagger}$ is, of course, just the transpose of the column matrix $\delta\mathbf{c}$. It is written in this form to ensure that all row matrices are distinguished by an explicit $\dagger$.

We note that for any 1 x 1 matrix the operations Hermitian conjugate ($\dagger$) and complex conjugate ($*$) are the same. For any scalar product we have, therefore

$$\mathbf{c}^\dagger \mathbf{d} = (\mathbf{c}^\dagger \mathbf{d})^{*\dagger} = \mathbf{d}^{*\dagger} \mathbf{c}^* \tag{7.217}$$

and for a Hermitian matrix $\mathbf{M}$

$$\mathbf{c}^\dagger \mathbf{M} \mathbf{d} = (\mathbf{c}^\dagger \mathbf{M} \mathbf{d})^{*\dagger} = \mathbf{d}^{\dagger *} \mathbf{M}^* \mathbf{c}^*. \tag{7.218}$$

When the expansions (7.212) are substituted in the total energy expression (7.157) (or the more general expression (7.160)) all terms may be expressed in this matrix notation. In this way we obtain

$$f_i = \mathbf{c}_i^\dagger \mathbf{f} \mathbf{c}_i \tag{7.219}$$

$$J_{ij} = \mathbf{c}_i^\dagger \mathbf{J}_j \mathbf{c}_i = \mathbf{c}_j^\dagger \mathbf{J}_i \mathbf{c}_j \tag{7.220}$$

$$K_{ij} = \mathbf{c}_i^\dagger \mathbf{K}_j \mathbf{c}_i = \mathbf{c}_j^\dagger \mathbf{K}_i \mathbf{c}_j \tag{7.221}$$

with the Hermitian matrices $\mathbf{J}_i$, $\mathbf{K}_i$ defined by

$$J_{i,rs} = \sum_t \sum_u c_{ti}^*(tr \mid us) c_{ui} = \sum_t \sum_u c_{ti}^* C_{ui} [tu \mid rs] \tag{7.222}$$

$$K_{i,rs} = \sum_t \sum_u c_{ti}^*(tr \mid su) c_{ui} = \sum_t \sum_u c_{ti}^* c_{ui} [ts \mid ru]. \tag{7.223}$$

Here the bracket notation for the two-electron integrals over the basis functions follows the conventions introduced in Section 5.3, namely

$$(rs \mid tu) = [rt \mid su] = \int \frac{\chi_r^*(1)\chi_s^*(2)\chi_t(1)\chi_u(2)}{r_{12}} \, dv_1 \, dv_2. \tag{7.224}$$

Using the alternative expressions for J_{ij} and K_{ij} given in (7.220) and (7.221) the variations of all terms in the energy expression (7.160) are readily expressed in a form analogous to (7.163).

$$\delta f_i = \delta \mathbf{c}_i^\dagger \mathbf{f} \mathbf{c}_i + \mathbf{c}_i^\dagger \mathbf{f} \delta \mathbf{c}_i = \delta \mathbf{c}_i^\dagger \mathbf{f} \mathbf{c}_i + \delta \mathbf{c}_i^{*\dagger} \mathbf{f}^* \mathbf{c}_i^* \tag{7.225}$$

$$\delta J_{ij} = \delta \mathbf{c}_i^\dagger \mathbf{J}_j \mathbf{c}_i + \delta \mathbf{c}_i^{*\dagger} \mathbf{J}_j^* \mathbf{c}_i^* + \delta \mathbf{c}_j^\dagger \mathbf{J}_i \mathbf{c}_j + \delta \mathbf{c}_j^{*\dagger} \mathbf{J}_i^* \mathbf{c}_j^* \tag{7.226}$$

$$\delta K_{ij} = \delta \mathbf{c}_i^\dagger \mathbf{K}_j \mathbf{c}_i + \delta \mathbf{c}_i^{*\dagger} \mathbf{K}_j^* \mathbf{c}_i^* + \delta \mathbf{c}_j^\dagger \mathbf{K}_i \mathbf{c}_j + \delta \mathbf{c}_j^{*\dagger} \mathbf{K}_i^* \mathbf{c}_j^* . \tag{7.227}$$

Assembling these results we obtain an expression for δE which is precisely analogous to (7.163), namely

$$\delta E = 2 \sum_k [\delta \mathbf{c}_k^\dagger \mathbf{F}_C \mathbf{c}_k + \delta \mathbf{c}_k^{*\dagger} \mathbf{F}_C^* \mathbf{c}_k^*] + 2\nu \sum_m [\delta \mathbf{c}_m^\dagger \mathbf{F}_O \mathbf{c}_m + \delta \mathbf{c}_m^{*\dagger} \mathbf{F}_O^* \mathbf{c}_m^*] \tag{7.228}$$

with

$$\mathbf{F}_C = \mathbf{f} + \sum_k (2\mathbf{J}_k - \mathbf{K}_k) + \nu \sum_m (2\mathbf{J}_m - \mathbf{K}_m) \tag{7.229}$$

$$\mathbf{F}_O = \mathbf{f} + \sum_k (2\mathbf{J}_k - \mathbf{K}_k) + 2a\nu \sum_m \mathbf{J}_m - b\nu \sum_m \mathbf{K}_m . \tag{7.230}$$

Substituting the expressions (7.228) and (7.216) into equation (7.156) and equating to zero the coefficients of the arbitrary variations $\delta \mathbf{c}_i$, $\delta \mathbf{c}_i^*$ we obtain the matrix equivalents of the Hartree–Fock equations (7.168)-(7.171)

$$\mathbf{F}_C \mathbf{c}_k = \sum_j \epsilon_{kj} \mathbf{S} \mathbf{c}_j \tag{7.231}$$

$$\mathbf{F}_C^* \mathbf{c}_k^* = \sum_j \epsilon_{jk} \mathbf{S}^* \mathbf{c}_j^* \tag{7.232}$$

$$\nu \mathbf{F}_O \mathbf{c}_m = \sum_j \epsilon_{mj} \mathbf{S} \mathbf{c}_j \tag{7.233}$$

$$\nu \mathbf{F}_O^* \mathbf{c}_m^* = \sum_j \epsilon_{jm} \mathbf{S}^* \mathbf{c}_j^* . \tag{7.234}$$

Again consistency of the equations requires that the matrix of Lagrange multipliers be Hermitian

$$\epsilon_{ji}^* = \epsilon_{ij} \tag{7.235}$$

and, once this condition has been imposed equations (7.232) and (7.234) become redundant.

The elimination of the off-diagonal Lagrange multipliers from equations (7.231) and (7.233) follows the same lines as in Section 7.2(c), although the appearance of the formulae is modified by the presence of the overlap matrix **S** in the orthogonality conditions (7.213) (cf. the simpler problem analysed in detail in Section 2.5).

Thus all off-diagonal multipliers, except those linking closed-shell orbitals with open-shell orbitals, are removed first by separate unitary transformations of the closed- and open-shell manifolds. For the remaining multipliers we have, using the orthogonality conditions (7.213)

$$\epsilon_{km} = \mathbf{c}_m^\dagger \mathbf{F}_\mathrm{C} \mathbf{c}_k = \mathbf{c}_m^\dagger \nu \mathbf{F}_\mathrm{O} \mathbf{c}_k = \mathbf{c}_m^\dagger \mathbf{F}_\mathrm{A} \mathbf{c}_k \tag{7.236}$$

with

$$\mathbf{F}_\mathrm{A} = x\mathbf{F}_\mathrm{C} + (1-x)\nu\mathbf{F}_\mathrm{O} \tag{7.237}$$

so that equation (7.231) for the closed-shell orbitals may be written

$$\begin{aligned} \mathbf{F}_\mathrm{C} \mathbf{c}_k &= \epsilon_k \mathbf{S}\mathbf{c}_k + \sum_m \mathbf{S}\mathbf{c}_m \mathbf{c}_m^\dagger \mathbf{F}_\mathrm{A} \mathbf{c}_k \\ &= \epsilon_k \mathbf{S}\mathbf{c}_k + \sum_m (\mathbf{S}\mathbf{c}_m \mathbf{c}_m^\dagger \mathbf{F}_\mathrm{A} \mathbf{c}_k + \mathbf{F}_\mathrm{A} \mathbf{c}_m \mathbf{c}_m^\dagger \mathbf{S}\mathbf{c}_k), \end{aligned} \tag{7.238}$$

that is

$$\mathbf{H}_\mathrm{C} \mathbf{c}_k = \epsilon_k \mathbf{S}\mathbf{c}_k \tag{7.239}$$

with the effective Hamiltonian matrix $\mathbf{H}_\mathrm{C}$ given by

$$\mathbf{H}_\mathrm{C} = \mathbf{F}_\mathrm{C} - (\mathbf{S}\mathbf{D}_\mathrm{O} \mathbf{F}_\mathrm{A} + \mathbf{F}_\mathrm{A} \mathbf{D}_\mathrm{O} \mathbf{S}) \tag{7.240}$$

where

$$\mathbf{D}_\mathrm{O} = \sum_m \mathbf{c}_m \mathbf{c}_m^\dagger. \tag{7.241}$$

Similarly we obtain

$$\mathbf{H}_\mathrm{O} \mathbf{c}_m = \eta_m \mathbf{S}\mathbf{c}_m \tag{7.242}$$

with

$$\mathbf{H}_\mathrm{O} = \mathbf{F}_\mathrm{O} - \frac{1}{\nu}(\mathbf{S}\mathbf{D}_\mathrm{C} \mathbf{F}_\mathrm{A} + \mathbf{F}_\mathrm{A} \mathbf{D}_\mathrm{C} \mathbf{S}) \tag{7.243}$$

$$\mathbf{D}_\mathrm{C} = \sum_k \mathbf{c}_k \mathbf{c}_k^\dagger \tag{7.244}$$

and

$$\eta_m = \frac{1}{\nu}\epsilon_m. \tag{7.245}$$

The matrices $\mathbf{D}_\mathrm{O}$ and $\mathbf{D}_\mathrm{C}$ defined by equations (7.241) and (7.244) are usually referred to as the open- and closed-shell "density matrices". However, for a non-orthogonal basis set they are *not* matrix representations of the quantum mechanical first-order density matrix (cf. Ref. 69).

Again the choice $x = -\nu/(1-\nu)$ in equation (7.237) eliminates all terms from F_A except those arising from the Coulomb and exchange operators for the open-shell orbitals and the effective Hamiltonian matrices of equations (7.239) and (7.242) reduce to the matrix versions of (7.202) and (7.203), namely

$$\mathbf{H}_C = \mathbf{f} + 2\mathbf{J}_C - \mathbf{K}_C + 2\mathbf{J}_O - \mathbf{K}_O + 2\alpha\mathbf{L}_O - \beta\mathbf{M}_O \tag{7.246}$$

and

$$\mathbf{H}_O = \mathbf{f} + 2\mathbf{J}_C - \mathbf{K}_C + 2a\mathbf{J}_O - b\mathbf{K}_O + 2\alpha\mathbf{L}_C - \beta\mathbf{M}_C \tag{7.247}$$

with

$$\mathbf{J}_C = \sum_k \mathbf{J}_k, \qquad \mathbf{K}_C = \sum_k \mathbf{K}_k, \qquad \mathbf{J}_O = \nu \sum_m \mathbf{J}_m, \qquad \mathbf{K}_O = \nu \sum_m \mathbf{K}_m \tag{7.248}$$

$$\mathbf{L}_C = \mathbf{S}\mathbf{D}_C\mathbf{J}_O + \mathbf{J}_O\mathbf{D}_C\mathbf{S}, \qquad \mathbf{M}_C = \mathbf{S}\mathbf{D}_C\mathbf{K}_O + \mathbf{K}_O\mathbf{D}_C\mathbf{S} \tag{7.249}$$

$$\mathbf{L}_O = \nu(\mathbf{S}\mathbf{D}_O\mathbf{J}_O + \mathbf{J}_O\mathbf{D}_O\mathbf{S}), \qquad \mathbf{M}_O = \nu(\mathbf{S}\mathbf{D}_O\mathbf{K}_O + \mathbf{K}_O\mathbf{D}_O\mathbf{S}) \tag{7.250}$$

and

$$\alpha = \frac{1-a}{1-\nu}, \qquad \beta = \frac{1-b}{1-\nu}.$$

In the closed-shell case $q = 0$, there are no open-shell orbitals and equation (7.239) reduces to the simpler pseudo-eigenvalue equation

$$\mathbf{F}\mathbf{c}_i = \epsilon_i \mathbf{S}\mathbf{c}_i \tag{7.251}$$

with

$$\mathbf{F} = \mathbf{f} + \sum_i (2\mathbf{J}_i - \mathbf{K}_i). \tag{7.252}$$

We note that the matrices **J**, **K**, and **D** are all quadratic functions of the unknown coefficients c_{ri}; consequently the coupling matrices **L**, **M** are of the fourth degree, and the pseudo-eigenvalue equations (7.239) and (7.242) of the open-shell theory are quintic equations in the unknown coefficients. They are, therefore, inherently more complicated than the cubic equations (7.251) of closed-shell theory. They may, however, be solved by an iterative procedure which is very similar to that used in the simpler closed-shell case. Starting from an assumed set of approximate eigenvectors $\mathbf{c}_k^{(0)}, \mathbf{c}_m^{(0)}$ a basic iterative step involves the following:

(i) Calculation of the matrices $\mathbf{H}_C^{(0)}, \mathbf{H}_O^{(0)}$ from $\mathbf{c}_k^{(0)}, \mathbf{c}_m^{(0)}$ via equations (7.222), (7.223), (7.241), (7.244) and (7.246)–(7.250).

(ii) Solution of the pseudo-eigenvalue equations (7.239) and (7.242).

(iii) Selection of the appropriate solutions $\epsilon_k^{(1)}$, $\mathbf{c}_k^{(1)}$, $\epsilon_m^{(1)}$, $\mathbf{c}_m^{(1)}$ and comparison with the input vectors $\mathbf{c}_k^{(0)}$, $\mathbf{c}_m^{(0)}$.

Successive iterative steps are carried out until input and output vectors and eigenvalues differ by less than some preset measure of convergence. In practice various special devices may be used to accelerate convergence and even to obtain solutions in some cases where the basic iterative step is inherently divergent.[21]

For symmetrical molecules and atoms the efficiency of matrix Hartree–Fock calculations is greatly increased by using a fully symmetry adapted basis. An arbitrary set of basis functions may be symmetry adapted using the group theoretical projection operators discussed in Section 4.7. Each basis function $\chi_{r\lambda\alpha}$ now bears three subscripts: λ which specifies an irreducible representation of the molecular symmetry group, α which specifies one of the partners in the representation λ and a running index r.

A typical molecular orbital $\phi_{i\lambda\alpha}$ also bears three subscripts and its expansion in terms of the basis functions contains only functions of the same symmetry species (λ) and sub-species (α). Thus equation (7.210) assumes the form

$$\phi_{i\lambda\alpha} = \sum_r \chi_{r\lambda\alpha} c_{\lambda ri}, \tag{7.253}$$

where, as indicated by the notation, the expansion coefficients are independent of the sub-species. These coefficients are also taken to be real.

We may now use the general results of Section 4.7 (in particular Theorem 6 and the subsequent discussion) to simplify the Hartree–Fock equations (7.239) and (7.242). Since the operator f is totally symmetric, that is, commutes with all operators of the symmetry group, its matrix $\mathbf{f}$ and the overlap matrix $\mathbf{S}$, split up into smaller matrices each of which links basis functions of the same species (λ) and sub-species (α). Furthermore, the matrices for different sub-species of the same species are identical. Explicitly we have

$$f_{r\lambda\alpha,\, s\mu\beta} = \delta_{\lambda\mu}\delta_{\alpha\beta} f_{\lambda rs} \tag{7.254}$$

with

$$f_{\lambda rs} = \frac{1}{d_\lambda}\sum_{\alpha=1}^{d_\lambda} f_{r\lambda\alpha,\, s\lambda\alpha} \tag{7.255}$$

and similar equations for the matrix $\mathbf{S}$.

In equation (7.255) d_λ is the dimensionality of the irreducible representation λ and all terms in the summation are in fact equal.

Now the matrices $\mathbf{J}_C, \mathbf{K}_C, \mathbf{J}_O, \mathbf{K}_O, \mathbf{D}_C$ and $\mathbf{D}_O$ appearing in the Hartree–Fock equations are all defined by summations over *complete* degenerate sets of orbitals, the partial occupancy of the open-shell orbitals being accounted for by the coefficients ν, a and b in the total energy expression (7.160). The underlying operators are, therefore, totally symmetric and each of these matrices has the structure shown in equations (7.254) and (7.255). The matrix Hartree–Fock equations (7.239) and (7.242) thus separate into independent sets, one for each symmetry species

$$\mathbf{H}_{C\lambda}\mathbf{c}_{\lambda k} = \epsilon_{\lambda k}\mathbf{S}_{\lambda}\mathbf{c}_{\lambda k} \tag{7.256}$$

$$\mathbf{H}_{O\lambda}\mathbf{c}_{\lambda m} = \eta_{\lambda m}\mathbf{S}_{\lambda}\mathbf{c}_{\lambda m}, \tag{7.257}$$

with $\mathbf{H}_{C\lambda}$ and $\mathbf{H}_{O\lambda}$ given by equations (7.246)-(7.250) with a subscript λ attached to every matrix.

When equations (7.256) and (7.257) are solved by an iterative procedure one of the most time consuming steps is the construction of the matrices $\mathbf{J}_{C\lambda}$, $\mathbf{K}_{C\lambda}$, $\mathbf{J}_{O\lambda}$ and $K_{O\lambda}$ from the eigenvectors of the previous iteration. The efficiency of this step can be improved by a suitable reorganization of the summations. Consider for example the calculation of the matrix $2\mathbf{J}_C - \mathbf{K}_C$, which is the only one needed in the closed-shell case. From equations (7.254) and (7.255) we have, in this case

$$(2\mathbf{J}_C - \mathbf{K}_C)_{r\lambda\alpha, s\mu\beta} = \delta_{\lambda\mu}\delta_{\alpha\beta}P_{\lambda rs} \tag{7.258}$$

where

$$P_{\lambda rs} = \frac{1}{d_\lambda}\sum_{\alpha=1}^{d_\lambda}(2\mathbf{J}_C - \mathbf{K}_C)_{r\lambda\alpha, s\lambda\alpha}. \tag{7.259}$$

We expand the expression (7.259) in terms of the definitions (7.248), (7.222) and (7.223) bearing in mind that each index $k, r, \ldots$ in these equations now stands for a triple of indices $(k\lambda\alpha)$, $(r\mu\beta), \ldots$. Making use of the fact that the expansion coefficients are real we obtain

$$P_{\lambda rs} = \sum_{\mu tu}\mathscr{P}_{\lambda rs, \mu tu}D_{\mu tu} \tag{7.260}$$

where

$$\mathscr{P}_{\lambda rs, \mu tu} = \frac{1}{d_\lambda d_\mu}\sum_{\alpha=1}^{d_\lambda}\sum_{\beta=1}^{d_\mu}[2(\chi_{r\lambda\alpha}\chi_{t\mu\beta}|\chi_{s\lambda\alpha}\chi_{u\mu\beta}) - (\chi_{r\lambda\alpha}\chi_{t\mu\beta}|\chi_{u\mu\beta}\chi_{s\lambda\alpha})] \tag{7.261}$$

and

$$D_{\mu tu} = d_\mu\sum_k c_{\mu tk}c_{\mu uk}. \tag{7.262}$$

Comparing equations (7.262) and (7.244) we see that, apart from the factor d_μ, the square matrix $\mathbf{D}_\mu$ is just the μ component of the density matrix for the closed-shell manifold

$$\mathbf{D}_\mu = d_\mu \mathbf{D}_{C\mu}. \qquad (7.263)^\ddagger$$

The expression (7.260) for $P_{\lambda rs}$ is well adapted to an iterative calculation. First the basis functions χ are chosen and the array $\mathscr{P}$ of integrals is calculated; this array remains fixed through all iterations. Current values of the eigenvectors $\mathbf{c}_{\mu k}$ are substituted in equation (7.262) to obtain the array of coefficients D. The matrix $\mathbf{P}_\lambda$ appearing in the Hartree–Fock Hamiltonian

$$\mathbf{F}_\lambda = \mathbf{f}_\lambda + \mathbf{P}_\lambda \qquad (7.264)$$

is then obtained by carrying out the triple summation in equation (7.260).

It is clear from equation (7.260) that, if P and D are regarded as linear arrays indexed in dictionary order with respect to λrs and μtu and $\mathscr{P}$ is regarded as a square matrix with rows and columns similarly ordered, the set of equations for all λ values appears as a simple matrix product

$$P = \mathscr{P}D. \qquad (7.265)$$

For this reason arrays such as $\mathscr{P}$ are often referred to as "super-matrices" and arrays such as P and D as "(column) super-vectors". This is in contrast to the form which the array P assumes in the Hartree–Fock Hamiltonians, namely a collection of square matrices, one for each value of λ (equation (7.264)).

Further economies in the storage of the arrays $\mathscr{P}, D, P, \ldots$ and in the execution of the sums (7.260) may be effected by using the symmetry of the arrays. The arrays D and P are real and symmetric and may be stored in lower diagonal form; the array $\mathscr{P}$ has the symmetry properties

$$\mathscr{P}_{\lambda rs,\,\mu tu} = \mathscr{P}_{\mu tu,\,\lambda rs} = \mathscr{P}^*_{\mu ut,\,\lambda sr} = \mathscr{P}^*_{\lambda sr,\,\mu ut}. \qquad (7.266)$$

A detailed description of a matrix Hartree–Fock computer program for atoms utilizing these economies is given by Roothaan and Bagus.[21] In the same article the open-shell theory described here is extended to all those configurations (atomic or molecular) which contain at most one open shell for each symmetry species.

‡ In Roothaan's original development[18] the factor d_μ is avoided by a renormalization of the eigenvectors $\mathbf{c}_{\mu k}$. Roothaan and Bagus,[21] on the other hand, introduce occupation numbers (here $2d_\mu$) in the definition of the "density matrices". Their $\mathbf{D}_{C\lambda}$ is therefore, not just the λ component of the matrix $\mathbf{D}_C$ given by equation (7.244).

(e) USE OF ORTHOGONALIZED BASES

Although the Roothaan scheme described above is probably the most efficient matrix Hartree–Fock technique various other schemes have been developed and used.[25–27] Some of these are applicable to cases involving several open shells of the same symmetry species.

Perhaps the most straightforward approach which is applicable to these more general cases is to use basis functions for each shell which are explicitly orthogonalized to the occupied orbitals of the other shells. At each stage of the calculation one is then faced with a standard eigenvalue–eigenvector problem, just as in the much simpler problem analysed in Section 2.5(b). In the present context the occupied and unoccupied orbitals available from previous iterations may be used as orthogonalized bases. Here we outline such a treatment for the two-shell case in which equation (7.163) for the first-order variation of the energy involves two distinct Fock operators F_{A} and F_{B}.

Let

$$\{\phi_k\} \qquad (k = 1, \ldots, n_a) \tag{7.267}$$

$$\{\phi_m\} \qquad (m = n_a + 1, \ldots, n_a + n_b) \tag{7.268}$$

denote the occupied Hartree–Fock orbitals of the two shells and let

$$\{\phi_u\} \qquad (u = n_a + n_b + 1, \ldots, n_a + n_b + n_c = n) \tag{7.269}$$

denote any set of functions which, together with the sets $\{\phi_k\}$ and $\{\phi_m\}$ form an orthonormal basis for the space spanned by the original fixed basis functions

$$\{\chi_r\} \qquad (r = 1, \ldots, n). \tag{7.270}$$

We assume that approximate estimates $\{\phi_k^{(0)}\}$, $\{\phi_m^{(0)}\}$ and $\{\phi_u^{(0)}\}$ of the functions (7.267), (7.268) and (7.269) are available, either as an initial guess or as the result of previous iterations. These estimates are all expressed in terms of the fixed basis functions

$$\phi_k^{(0)} = \sum_r \chi_r c_{kr}^{(0)} \tag{7.271}$$

$$\phi_m^{(0)} = \sum_r \chi_r c_{mr}^{(0)} \tag{7.272}$$

$$\phi_u^{(0)} = \sum_r \chi_r c_{ur}^{(0)} \tag{7.273}$$

and are assumed to be accurately orthonormal.

The first step in the iterative procedure is to use the functions $\{\phi_k^{(0)}, \phi_u^{(0)}\}$ as basis functions in a matrix Hartree–Fock calculation of the A-shell orbitals. Since these basis functions are all orthogonal to the current estimates $\{\phi_m^{(0)}\}$ of the B-shell orbitals no Lagrange multipliers are required and we have a standard (pseudo-) eigenvalue–eigenvector problem (cf. Section 2.5(b)). Since, furthermore, the $n_a + n_c$ basis functions are orthonormal the eigenvector equations are of the form

$$\mathbf{F}_{\mathrm{A}}^{(0)}\mathbf{d} = \epsilon\mathbf{d}. \tag{7.274}$$

The n_a eigenvectors of equation (7.274) belonging to the n_a lowest eigenvalues provide us with improved estimates $\{\phi_k^{(1)}\}$ of the A-shell orbitals; the remaining n_c eigenvectors provide a revised set $\{\phi_u^{(1)}\}$ of unoccupied orbitals. We note that the set of orbitals $\{\phi_k^{(1)}, \phi_m^{(0)}, \phi_u^{(1)}\}$ is accurately orthonormal.

The second step is to use the basis set $\{\phi_m^{(0)}, \phi_u^{(1)}\}$ for a calculation of improved B shell orbitals $\{\phi_m^{(1)}\}$ and revised unoccupied orbitals $\{\phi_u^{(2)}\}$. Again no Lagrange multipliers are required and the resulting set of orbitals $\{\phi_k^{(1)}, \phi_m^{(1)}, \phi_u^{(2)}\}$ is orthonormal.

The final step in each iterative cycle is to minimize the total energy with respect to mixing of the two sets of orbitals $\{\phi_k^{(1)}\}$ and $\{\phi_m^{(1)}\}$. This again leads to a standard eigenvalue–eigenvector problem, but with a Fock matrix $\mathbf{F}_{\mathrm{AB}}^{(0)}$ which is, in general, distinct from both $\mathbf{F}_{\mathrm{A}}^{(0)}$ and $\mathbf{F}_{\mathrm{B}}^{(0)}$. The resulting orbital sets $\{\phi_k^{(2)}\}$ and $\{\phi_m^{(2)}\}$ are again orthonormal and orthogonal to each other and to the unoccupied orbitals $\{\phi_u^{(2)}\}$.

The procedure outlined above, which is essentially due to Lefebvre,[25] is rather inefficient if the one- and two-electron integrals are all transformed to the new basis at each intermediate step. However, as Hunt *et al.*[26,27] point out, it is possible to delay the complete recalculation of the Fock matrices $\mathbf{F}_{\mathrm{A}}$, $\mathbf{F}_{\mathrm{B}}$ and $\mathbf{F}_{\mathrm{AB}}$, in the fixed basis $\{\chi_r\}$ until the end of a complete iterative cycle when improved estimates $\{\phi_k^{(2)}\}$, $\{\phi_m^{(2)}\}$ of all the occupied orbitals are available. Used in this way the technique may be only slightly less efficient than the Roothaan–Bagus procedure. Unlike the latter it is readily extended to cases with more than one open shell of a given symmetry type, although the complexity of the iterative cycle increases rapidly. Thus in the three shell case we have, in general, six distinct Fock matrices $\mathbf{F}_{\mathrm{A}}$, $\mathbf{F}_{\mathrm{B}}$, $\mathbf{F}_{\mathrm{C}}$, $\mathbf{F}_{\mathrm{AB}}$, $\mathbf{F}_{\mathrm{AC}}$, $\mathbf{F}_{\mathrm{BC}}$ and six elementary steps in each iterative cycle.

Hunt *et al.*[27] have applied the technique to the state

$$\text{He: } (1s)(2s),\ {}^1S \tag{7.275}$$

of the helium atom and the states

$$\text{H}_2\text{O: } (1a_1)^2(2a_1)^2(1b_1)^2(1b_2)^2(3a_1)^2,\ {}^1A_1 \tag{7.276}$$

$$(1a_1)^2(2a_1)^2(1b_1)^2(1b_2)^2(3a_1)(4a_1)\begin{cases} {}^3A_1 & (7.277) \\ {}^1A_1 & (7.278) \end{cases}$$

$$\text{H}_2\text{O}^+\text{: } (1a_1)^2(2a_1)^2(1b_1)^2(1b_2)^2\,3a_1,\ {}^2A_1 \tag{7.279}$$

of the water molecule and its positive ion, assuming ground-state geometry for the nuclear framework (symmetry C_{2v}). In all cases they find good convergence for the iterative scheme outlined above. In an earlier calculation(26) the same authors obtained incorrect results for the H_2O and H_2O^+ due to a non-optimal mixing of the occupied orbitals (in these earlier calculations the final step involving the Fock matrix $\mathbf{F}_{AB}^{(0)}$ was omitted from the iterative scheme).

The ground states of H_2O and H_2O^+ (7.276) and (7.279) are, of course, amenable to the standard Roothaan–Bagus approach, as is the excited $^3A_1$ state (7.277), but the excited $^1A_1$ state (7.278) and the excited ^1S state of He (7.275) are not, since these states contain two open shells of the same symmetry type.

We note also that the Hartree–Fock energy for a state such as (7.275) or (7.278), which is not the lowest of its symmetry type, is not an upper bound to the energy obtained from an exact solution of the Schrödinger equation.

This loss of the upper bound property is characteristic of any non-linear variational calculation, such as the Hartree–Fock method, and occurs because the approximate wave function for the excited state is not orthogonal to the accurate wave function for some lower state. It is only in the linear variational method that the appropriate root of the secular equation retains the upper bound property for states other than the lowest (Section 2.2).

(f) NUMERICAL INTEGRATION AND ACCUMULATIVE ACCURACY

So far we have assumed that all quantities appearing in the matrix Hartree–Fock equations are evaluated via one- and two-electron integrals over a fixed set of n basis functions. In such a calculation the number of two-electron integrals required increases as n^4, for large n, leading to a very rapid increase in computational effort with increasing accuracy. Boys and Rajagopal(28) have shown that this rapid rise may be avoided if the techniques of the previous subsection are combined with a direct numerical evaluation of the Fock matrix. For simplicity we outline their method for the closed-shell case.

We suppose then that we are seeking the Hartree–Fock determinant

$$\Psi = |\phi_1\bar{\phi}_1 \cdots \phi_N\bar{\phi}_N| \tag{7.280}$$

for a closed-shell state and that some approximate Hartree–Fock orbitals $\phi_1^{(0)}, \ldots, \phi_N^{(0)}$ are available. These approximate orbitals are used both to construct the Fock operator $F(\phi_i^{(0)})$ (equation (7.132)) and as the first N functions

$$\chi_i = \phi_i^{(0)} \qquad (i = 1, \ldots, N) \tag{7.281}$$

of the basis for the expansion of the next approximation

$$\phi_i^{(1)} = \sum_{j=1}^{n} \chi_j c_{ji} \qquad (n > N) \tag{7.282}$$

to the Hartree–Fock orbitals ϕ_i. The remaining $n - N$ functions in the expansion (7.282) are functions which are estimated to be the most effective in improving the $\phi_i^{(0)}$. The basic step in the iterative procedure is the solution of the eigenvalue problem

$$\sum_{t=1}^{n} \{(\chi_s \,|\, F(\phi_j^{(0)}) \,|\, \chi_t) - \epsilon_i (\chi_s \,|\, \chi_t)\} c_{ti} = 0. \tag{7.283}$$

The lowest N solutions of equation (7.283) now provide improved estimates $\phi_1^{(1)}, \ldots, \phi_N^{(1)}$ of the Hartree–Fock functions, which are used in place of $\phi_1^{(0)}, \ldots, \phi_N^{(0)}$ in the next iterative step. Iterations are continued until no choice of improvement functions has any appreciable effect on the orbitals or the total energy.

In the whole procedure there are two iterative processes; one is the usual self-consistent field (SCF) convergence using a limited basis set and the other arises from the need to introduce new types of improvement function. If the basis set $\chi_i (i = 1, \ldots, n)$ is fixed but the improvement functions are chosen in rotation until the results converge we obtain the usual SCF wave function with the fixed basis set. But if we introduce new improvement functions before this SCF convergence is reached, the adjustment of $F(\phi_i)$ proceeds in the background while other relaxations are being effected.

There is one inconvenient feature of this approach. The total energy corresponding to the wave function found in the pth iteration is only evaluated in the $(p + 1)$th iteration. Thus from equation (7.136) the total energy corresponding to the starting orbitals $\phi_i^{(0)}$ is given by

$$E(\phi_i^{(0)}) = \sum_{i=1}^{N} (\phi_i^{(0)} \,|\, F(\phi_j^{(0)}) \,|\, \phi_i^{(0)}) + \sum_{i=1}^{N} (\phi_i^{(0)} \,|\, f \,|\, \phi_i^{(0)}) \tag{7.284}$$

The first term in equation (7.284) is just the sum of the first N diagonal terms of the Fock matrix appearing in equation (7.283) and the second term is also required in setting up this Fock matrix. The quantity $E(\phi_i^{(0)})$ appears, therefore as a by-product of the setting up of the Fock matrix of equation (7.283) whose first N eigenvectors provide the next set of orbitals $\phi_i^{(1)} (i = 1, \ldots, N)$.

Although all integrals appearing in equation (7.283) are evaluated numerically using suitable three- and six-dimensional grids of points it is important to have analytic expressions for the orbitals and improvement functions. The effect of the differential operators representing the kinetic energy terms in the Hamiltonian may then be evaluated analytically prior to the numerical integration.

The system of functions employed by Boys and Rajagopal are Cartesian Slater functions centred on the nuclei. For such functions they estimate that, if the computational effort of the usual integral expansion method is $Cn^{4.5}$, then the corresponding effort in their numerical integration procedure is approximately $C(18)^3n^{1.5}$. This would indicate that, on the count of computation alone, their technique is preferable to the integral expansion method when the number of basis functions exceeds 18. However, these estimates are based on integration grids which are not quite adequate to ensure an accuracy of 0.001 au ($\sim$1 kcal mole^{-1}) in the computed total energy. Such a "chemical" level of accuracy is desirable in such calculations but, as we shall see below, very difficult to achieve in practice.

Another feature of this numerical approach is that of accumulative accuracy. This implies that at any stage it is easy to restart from the best results previously obtained, possibly by other methods. This is also a characteristic feature of Hartree's numerical integration technique for atomic SCF calculations.

7.3. Localized molecular orbitals and chemical bonds

Consider a molecule with a closed-shell ground state. In the Hartree–Fock approximation the total electronic wave function appears as a Slater determinant.

$$\Psi = |\phi_1\bar{\phi}_1 \cdots \phi_M\bar{\phi}_M|. \tag{7.285}$$

We assume that the MOs of equation (7.285) are all real; as Roothaan[20] has shown such a choice is always possible in the closed-shell case. The total wave function (7.285) is unchanged if the MOs are subjected to any unitary transformation; that is, if

$$\lambda_i = \sum_{j=1}^{M} \phi_j T_{ji}, \qquad (i = 1 \cdots M) \tag{7.286}$$

with T unitary, then

$$|\lambda_1\bar{\lambda}_1 \cdots \lambda_M\bar{\lambda}_M| = C|\phi_1\bar{\phi}_1 \cdots \phi_M\bar{\phi}_M| \tag{7.287}$$

where C is some constant phase factor of modulus unity. If, like the ϕ_i, the functions λ_i are restricted to be real, C will be unity and the transformation matrix of equation (7.286) will be real orthogonal with

$$\sum_{j=1}^{M} T_{ji}T_{jk} = \sum_{j=1}^{M} T_{ij}T_{kj} = \delta_{ik} \qquad (i, k = 1 \cdots M). \tag{7.288}$$

A transformation of the form (7.286) has already been used in Section 7.2 to reduce the Fock equations to the canonical form of pseudo-eigenvalue equations. This canonical form is the most convenient to use for the exact or approximate calculation of the MOs. It leads to canonical MOs (CMOs) which transform irreducibly under the molecular symmetry group and hence to considerable economy in the expansions required in the matrix method of solution. For a polyatomic molecule the CMOs of the valence shell are typically delocalized and, in the expansion method, will each contain appreciable contributions from functions centred on all or most of the nuclei.

In the derivation of Koopmans' theorem we saw that, if the ionization of a molecule is to be interpreted as a one-electron process, it is essential to use these delocalized CMOs. Similar considerations apply to the excitation energies associated with the promotion of an electron from an occupied MO i to a virtual orbital a. Equation (7.146) above expresses these excitation energies as a difference in orbital energies corrected by simple interaction terms. Thus there are special advantages in using CMOs for the discussion of ionization and excitation processes. The difference in energy between many-electron states is largely accounted for by the transfer of a single electron from one level to another, just as in the qualitative MO theory of Section 7.1.

However, for some other properties these delocalized CMOs are inconvenient. They provide no framework for a discussion of the classical chemical concepts of electron-pair bonds linking two neighbouring atoms or of chemically active lone-pair electrons localized on one atom. Empirically it is found that much of the chemistry of polyatomic molecules may be understood in terms of properties inherent in such localized electron groups, the influence of the rest of the molecule on the localized groups being of secondary importance.

These localized properties of polyatomic molecules (and even diatomic molecules) are better described in terms of localized molecular orbitals (LMOs) which are derived by applying an orthogonal transformation of the form (7.286) to the CMOs. Such a transformation was first used for this purpose by Coulson,[(29)] who showed that the usual CMOs of the methane molecule could be replaced by an inner-shell orbital localized near the carbon nucleus and four equivalent bond orbitals, each of which was concentrated in the region of a CH bond. For this simple, highly symmetrical molecule, the required transformation matrix T of equation (7.286) is completely determined by the assumed geometric equivalence of the four CH bonds, provided that the canonical inner-shell orbital, which is already well localized about the carbon nucleus, is not allowed to mix with the other orbitals.

Localized equivalent orbitals of a similar type were introduced for a number of simple molecules by Lennard-Jones,[(30)] who showed that such orbitals could, indeed, be associated with the bonds, lone-pairs and inner shells of simple chemical formulae. The original definition of these localized orbitals was in terms of geometric equivalence, and there was a measure of arbitrariness in the determination of the transformation matrix T for molecules less symmetric than methane. It is clearly more satisfactory to have some intrinsic criterion of localization which specifies the orbitals uniquely and implies the property of geometric equivalence in appropriate cases.

In order to derive such a criterion we express the total electronic energy (equation (7.133)) in two different but equivalent forms, namely

$$E_0 = 2\sum_{i=1}^{M} f_i + 2\sum_{i=1}^{M}\sum_{j=1}^{M} J_{ij} - \sum_{i=1}^{M}\sum_{j=1}^{M} K_{ij} \tag{7.289}$$

and

$$E_0 = 2\sum_{i=1}^{M} f_i + \sum_{i=1}^{M} J_{ii} + 4\sum_{i<j}^{M} J_{ij} - 2\sum_{i<j}^{M} K_{ij}. \tag{7.290}$$

The equivalence of these forms follows from the equality of the diagonal elements of the Coulomb and exchange matrices

$$J_{ii} = K_{ii} \qquad (i = 1 \ldots M). \tag{7.291}$$

Now the quantity

$$\rho(1, 2) = 2\sum_{j=1}^{M} \phi_j(1)\phi_j(2) \tag{7.292}$$

is invariant under the orthogonal transformation (7.286) (cf. equation (7.126)). Consequently each summation in the expression (7.289) for the total energy is invariant. However, the same does not apply to the individual summations in the expression (7.290) and we may use this fact to derive an intrinsic definition of localization. To see this we consider the physical interpretation of the different terms in equation (7.290).

The first summation represents the kinetic energy of the electrons and the energy of their Coulomb attraction for the nuclei; it is invariant. Each term in the second summation gives the Coulomb repulsion energy of a pair of electrons in the same space orbital. This summation is not invariant and the maximization of these intra-orbital repulsions is our inherent criterion of localization. Because of the invariance of the summations in equation (7.289), this maximization automatically leads to a minimization of the third summation in (7.290) (the inter-orbital Coulomb repulsions) and of the fourth summation (the inter-orbital exchange interactions).

All the interaction terms in equation (7.290) except the exchange terms have a simple physical interpretation in terms of the classical electrostatic interaction of smeared-out charge distributions. When we minimize the exchange terms in the transformation to localized bonding and lone-pair orbitals, we end up with an energy expression which is dominated by the classical electrostatic repulsions between bonds and lone pairs. The geometrical shapes adopted by many molecules can be accounted for very simply from this point of view.

The localization criterion, namely that

$$D(\lambda) = \sum_{i=1}^{M} J_{ii} = \sum_{i=1}^{M} [\lambda_i\lambda_i \mid \lambda_i\lambda_i] \tag{7.293}$$

should be maximum, can be put in an explicit form by considering infinitesimal orthogonal transformations of the form

$$\lambda_i' = \sum_{j=1}^{M} \lambda_j T_{ji} \tag{7.294}$$

with

$$T_{ji} = \delta_{ji} + t_{ji}, \tag{7.295}$$

where δ is the Kronecker delta symbol and t_{ji} is an infinitesimal.

Substituting the expressions (7.295) into the orthogonality conditions (7.288) we see that T will be orthogonal to first order in the parameter t provided that

$$t_{ij} = -t_{ji}. \tag{7.296}$$

A necessary condition that $D(\lambda)$ should be maximum is that its first-order variation should vanish for all infinitesimal orthogonal transformations, that is

$$\begin{aligned} 0 = \delta D(\lambda) &= 4 \sum_{i=1}^{M} [\delta\lambda_i\lambda_i \mid \lambda_i\lambda_i] \\ &= 4 \sum_{i=1}^{M} \sum_{j=1}^{M} t_{ji}[\lambda_j\lambda_i \mid \lambda_i\lambda_i] \\ &= 4 \sum_{i<j}^{M} t_{ji}\{[\lambda_j\lambda_i \mid \lambda_i\lambda_i] - [\lambda_i\lambda_j \mid \lambda_j\lambda_j]\} \end{aligned} \tag{7.297}$$

where we have used equation (7.296) and the fact that all the orbitals are real functions.

Now for an arbitrary infinitesimal orthogonal matrix the elements $t_{ji}(i<j)$ are independent, so that equation (7.297) implies

$$[\lambda_j\lambda_i \mid \lambda_i\lambda_i] = [\lambda_i\lambda_j \mid \lambda_j\lambda_j] \qquad (i<j, = 2 \cdots M). \tag{7.298}$$

We note that the $M(M+1)/2$ orthogonality conditions (7.288) and the $M(M-1)/2$ conditions (7.298) are just sufficient to fix the M^2 elements of the matrix T which effects the transformation from the CMOs $\phi_1, \ldots, \phi_M$ to the localized molecular orbitals (LMOs) $\lambda_1, \ldots, \lambda_M$. It is obvious that any pair of geometrically equivalent orbitals satisfy (7.298) so that the intrinsic definition of LMOs includes equivalent orbitals as a special case.

When they are expressed in terms of the matrix elements T_{ij} of the transformation (7.286) the conditions (7.288) and (7.298) becomes simultaneous algebraic equations of degree two and higher. As well as the solution corresponding to the LMOs these equations will, in general, possess several other solutions; there will always be one other solution corresponding to the least possible value of $D(\lambda)$ and perhaps further solutions corresponding to saddle-points and local extrema of $D(\lambda)$. The minimum solution will often be equal or close to the original CMOs. These orbitals, being basis functions for irreducible representations satisfy many of the conditions (7.298) automatically.

The most widely used technique for determining LMOs is based, not on solving the equations (7.288) and (7.298), but on a direct maximization procedure introduced by Edmiston and Ruedenberg.[31] Here, starting with the CMOs, the orbitals are transformed by a succession of 2 x 2 rotations of the form

$$\begin{aligned} \phi_i &\to \phi_i \cos\theta - \phi_j \sin\theta \\ \phi_j &\to \phi_i \sin\theta + \phi_j \cos\theta \\ \phi_k &\to \phi_k \qquad (k \neq i, j). \end{aligned} \tag{7.299}$$

It is a simple matter to express the sum of intra-orbital repulsions (7.293) as an explicit function of θ, $D(\theta)$, and to maximize $D(\theta)$ for each elementary rotation. All possible 2 x 2 rotations are considered successively and convergence of the orbitals to LMOs is usually quite satisfactory. An advantage of this direct procedure is that it does not depend on *a priori* assumptions as to the qualitative form of the LMOs (such as geometrical equivalence); indeed in some cases rather unexpected LMOs are obtained.[32]

Another technique for finding LMOs is to determine most of the parameters in the transformation T by the property of equivalence and the remainder by a direct application of the conditions (7.298). As an example we consider the ammonia molecule NH_3 in its equilibrium nuclear configuration of symmetry C_{3v}. We choose a cartesian coordinate system and label the hydrogen atoms as in Fig. 4.1 (p. 68).

The ground electronic state of NH_3 is the totally symmetric closed-shell singlet

$$NH_3: (1a_1)^2 (2a_1)^2 (3a_1)^2 (1e)^4, \; ^1A, \tag{7.300}$$

where the symmetry labels are from Table 4.4 (p. 79).

We assume that the LMOs comprise a nitrogen inner-shell orbital

$$N(i) \equiv a_1(i) \equiv i(= 1a_1)$$

which is identical to the CMO $1a_1$, a nitrogen lone-pair orbital

$$N(l) \equiv a_1(l) \equiv l,$$

which is also totally symmetric and three equivalent NH bond orbitals

$$NH_a \equiv b_a$$

$$NH_b \equiv b_b$$

$$NH_c \equiv b_c$$

which are permuted by the symmetry operations of C_{3v}.

First we check that these assumptions are consistent with the occupied CMOs shown in (7.300). The effects of the operators of C_{3v} on the bond orbitals and hence the character χ_b of the representation which they span are easily found from Fig. 4.1. We obtain

$$\chi_b(E) = 3, \qquad \chi_b\left\{C\left(\pm\frac{2\pi}{3}\right)\right\} = 0, \qquad \chi_b\{\sigma_a, \sigma_b, \sigma_c\} = 1. \tag{7.301}$$

Hence from Table 4.4

$$\chi_b = \chi(a_1) + \chi(e). \tag{7.302}$$

The relation (7.302) shows that the three equivalent bond orbitals may be expressed in terms of basis functions for the a_1 and e representations. The totally symmetric LMOs i and l require two further a_1 basis functions. We therefore need a total of three a_1 basis functions and one set of e basis functions. Since these requirements just match the available CMOs (7.300), our assumed LMOs are consistent with this ground-state wave function for NH_3.

The a_1 and e basis functions appearing implicitly in equation (7.302) may be generated explicitly from the bond orbitals using the group theoretical projection operators of Section 4.7. Indeed substituting b_a for ψ in equation (4.57) using the matrices of Table 4.4 and normalizing we obtain

$$a_1(b) = N\left\{E + C\left(\frac{2\pi}{3}\right) + C\left(\frac{2\pi}{3}\right) + \sigma_a + \sigma_b + \sigma_c\right\} b_a$$

$$= \frac{1}{\sqrt{3}}(b_a + b_b + b_c), \tag{7.303}$$

$$e_1(b) \equiv e_x(b) = N'\left\{E - \tfrac{1}{2}C\left(\frac{2\pi}{3}\right) - \tfrac{1}{2}C\left(-\frac{2\pi}{3}\right) + \sigma_a - \tfrac{1}{2}\sigma_b - \tfrac{1}{2}\sigma_c\right\} b_a$$

$$= \frac{1}{\sqrt{6}}(2b_a - b_b - b_c), \tag{7.304}$$

and

$$e_2(b) \equiv e_y(b) = N'\left\{0 \cdot \mathrm{E} + \frac{\sqrt{3}}{2} C\left(\frac{2\pi}{3}\right) - \frac{\sqrt{3}}{2} C\left(-\frac{2\pi}{3}\right) + 0 \cdot \sigma_a - \frac{\sqrt{3}}{2} \sigma_b + \frac{\sqrt{3}}{2} \sigma_c\right\} b_a = \frac{1}{\sqrt{2}} (b_b - b_c). \qquad (7.305)$$

To complete the specification of the orthogonal transformation from CMOs to LMOs we must relate the basis functions on the left-hand side of equations (7.303), (7.304) and (7.305) together with $a_1(i)$ and $a_1(l)$ to the CMOs of (7.300). Since we assume that the inner-shell orbital is unaffected by the transformation, and since basis functions of different irreducible symmetries do not mix (Theorem 6, Section 4.7) we obtain

$$\begin{aligned} 1a_1 &= a_1(i) \\ 2a_1 &= a_1(l) \cos\theta - a_1(b) \sin\theta \\ 3a_1 &= a_1(l) \sin\theta + a_1(b) \cos\theta \\ 1e_x &= e_x(b) \\ 1e_y &= e_y(b), \end{aligned} \qquad (7.306)$$

where the angle θ is yet to be determined.

Combining the transformations (7.303)-(7.305) and (7.306) we obtain the matrix equation

$$(1a_1, 2a_1, 3a_1, 1e_x, 1e_y) = (i, l, b_a, b_b, b_c)\,\mathbf{U} \qquad (7.307)$$

with

$$\mathbf{U} = \begin{bmatrix} 1, & 0, & 0, & 0, & 0 \\ 0, & \cos\theta, & \sin\theta, & 0, & 0 \\ 0, & -\frac{\sin\theta}{\sqrt{3}}, & \frac{\cos\theta}{\sqrt{3}}, & \sqrt{\frac{2}{3}}, & 0 \\ 0, & -\frac{\sin\theta}{\sqrt{3}}, & \frac{\cos\theta}{\sqrt{3}}, & -\frac{1}{\sqrt{6}}, & \frac{1}{\sqrt{2}} \\ 0, & -\frac{\sin\theta}{\sqrt{3}}, & \frac{\cos\theta}{\sqrt{3}}, & -\frac{1}{\sqrt{6}}, & -\frac{1}{\sqrt{2}} \end{bmatrix}. \qquad (7.308)$$

The transformation T which expresses the LMOs in terms of the CMOs may now be written down immediately, since this inverse of the real orthogonal matrix U is just the transposed matrix

$$\mathbf{T} = \mathbf{U}' = \mathbf{U}^{-1} \qquad (7.309)$$

obtained by interchanging rows and columns of **U**. Like **U** the matrix **T** depends on the single parameter θ. To find the value of θ which maximizes the sum of intra-orbital repulsions we use the single independent condition (7.298) which is not satisfied by virtue of symmetry, namely

$$[b_a l \mid ll] = [lb_a \mid b_a b_a]. \tag{7.310}$$

Equations (7.308), (7.309) and (7.310) now determine the LMOs completely.

It is a straightforward matter to generalize the above analysis to allow for a small difference between the inner shell orbital $i \equiv a_1(i)$ and the lowest energy CMO $1a_1$. The first three equations (7.306) are now replaced by a general 3 x 3 rotation

$$(1a_1, 2a_1, 3a_1) = (a_1(i), a_1(l), a_1(b))\mathbf{R}(\theta, \phi, \psi), \tag{7.311}$$

which mixes all three a_1 orbitals. As indicated in equation (7.311) such a general rotation may be expressed in terms of three independent parameters (e.g. Euler angles). To fix the values of these three parameters we need two independent conditions from the set (7.298) in addition to (7.310). These are obviously the conditions linking the inner-shell orbital i with the lone pair and bond orbitals, namely

$$[il \mid ll] = [li \mid ii]$$

and (7.312)

$$[b_a i \mid ii] = [ib_a \mid b_a b_a].$$

The group theoretical method we have used to obtain the LMOs for NH_3 is much more efficient for simple symmetrical molecules than the direct maximization method of Edmiston and Ruedenberg. However, this gain in efficiency is of marginal importance since the Edmiston–Ruedenberg method itself is a much simpler computational problem than the original determination of the CMOs, say by the expansion method. The direct maximization method is also indispensable for investigating orbital following, that is changes in the LMOs consequent on distortions of the nuclear framework from the equilibrium geometry.

On the other hand group theoretical methods do enable us to determine very simply the qualitative features of the sets of equivalent LMOs which can be formed from a given set of CMOs. For example, Hall[33] has shown that the six π electrons of the ground state of benzene (C_6H_6) can not be transformed into an equivalent set with respect to the full symmetry (C_{6v}) of the basic set of π orbitals. One of the C_{3v} sub-groups of C_{6v} must be used leading, qualitatively, to one or other of the two Kekulé structures

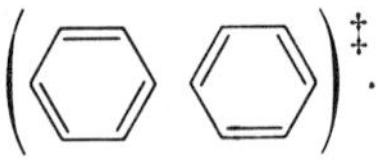

$$\left(\text{Kekulé structure I} \quad \text{Kekulé structure II}\right)^{\ddagger}.$$

Furthermore, the resulting π-bonds of benzene are not well localized; each contains appreciable contributions from at least three atomic orbitals. The implications of this delocalization of the maximally localized π orbitals of conjugated molecules has been studied by England and Ruedenberg.[34]

For planar molecules of this type, which are traditionally described in terms of σ MOs (symmetrical under reflexion in the plane of symmetry) and π MOs (antisymmetric under such reflexion), there is, in addition, an ambiguity in the set of LMOs. One may localize the σ MOs and the π MOs separately to obtain a description in terms of σ-bonds (each localized between two atoms) and π-bonds, each of which may be well localized between two atoms (non-conjugated molecules such as ethylene C_2H_4) or may be more diffuse (conjugated molecules such as benzene, naphthalene, etc). It is easy to see that all the extremal conditions (7.298) may be satisfied in such a description.

Alternatively, one may combine each π-bond of the first description with the corresponding σ-bond to obtain a pair of bent bonds which are equivalent under reflection in the symmetry plane. Again all the extremal conditions may be satisfied. Both descriptions correspond to stationary values of $D(\lambda)$ (equation (7.293)) and a decision between them cannot be made on the basis of qualitative symmetry arguments. Detailed calculations by the Edmiston–Ruedenberg technique show that, for ethylene, the description in terms of bent equivalent bonds ("banana" double bonds) between the two carbon atoms, σ-type CH bonds and carbon inner-shells corresponds to the absolute maximum of $D(\lambda)$. Similarly, by this criterion, maximum localization of the MOs in N_2 and C_2H_2 leads to equivalent "banana" triple bonds rather than to the traditional picture in terms of one σ-bond and two π-bonds.

The LMOs considered so far have been defined by the maximization of the quantity $D(\lambda)$ of equation (7.293) with respect to unitary transformations of the CMOs. As we have seen $D(\lambda)$ represents the sum of Coulomb repulsion energies of the paired electrons in each space orbital and thus provides a natural localization criterion. However, it is quite possible to define localized orbitals by the maximization of some other quantity which depends on the MOs and is not invariant under unitary transformations. For example, the

‡ There are also other possible "equivalent sets" with the same value of $D(\lambda)$.

exclusive orbitals of Foster and Boys[35] are defined to maximize the quantity

$$\prod_{i<j} |\mathbf{R}_i \quad \mathbf{R}_j|^2 \tag{7.313}$$

where

$$\mathbf{R}_i = \int \phi_i^* \mathbf{r} \phi_i \, dv \tag{7.314}$$

is the centroid of the charge distribution of the MO ϕ_i. By maximizing (7.313) these centroids are kept as far apart as possible, which again is a natural criterion for localization. In many cases the criterion (7.313) leads to exclusive orbitals which are very similar to the LMOs defined by the criterion (7.293); it is a good deal easier to apply since the integrals (7.314) are much simpler than those in (7.293) and although it fails for some cases (e.g. $1s$–$2s$ mixing in atoms) this failure is not serious since the CMOs involved are themselves well localized.

Gilbert[36] has investigated a wide variety of localization criteria, including both (7.293) and (7.313) and has shown how the Hartree-Fock equations may be modified to yield LMOs directly instead of CMOs. His scheme is probably less efficient than the indirect methods for atoms and small molecules, but may prove essential for really large molecules and crystals.

7.4. Results of some molecular orbital calculations

There have been a very large number of explicit numerical calculations by the methods of this chapter, especially since about 1960 when high speed digital computers became widely available.

Most of these calculations have used the expansion technique of Section 7.2(d) and are usually referred to as Hartree–Fock–Roothaan (HFR) or self-consistent-field (SCF) calculations. We shall use the latter term for all MO calculations by the expansion technique and reserve the terms HF (near HF) for calculations which have converged (nearly converged) to accurate numerical solutions of the Hartree–Fock equations. As a practical convergence criterion to this Hartree–Fock limit we allow an error of not more than 0.001 au per atom (other than H) in the total electronic energy; this criterion of "chemical accuracy" is comparable with that suggested by Boys and Rajagopal,[28] namely an error of 1 kcal mole^{-1} or better for simple reactions involving the formation and/or breaking of a single chemical bond.

All the calculations we consider are non-empirical (or *ab initio*) in that the complete many-electron Hamiltonian is used and all intermediate quantities, such as integrals over the basis functions are evaluated to adequate accuracy. We concentrate on HF (or near

HF) results. These are currently available only for rather small molecules ($\leqslant C_2H_6$), most of which contain only H and first row elements.

(a) BASIS FUNCTIONS

Two types of basis functions, STOs (Slater-type-orbitals) and GTOs (Gaussian-type-orbitals) have assumed a dominant position in recent SCF calculations on diatomic and polyatomic molecules. Both sets of basis functions are usually centred on the atomic nuclei and, in a local spherical coordinate system, assume the form

$$\chi_{nlm}(r) = NR_n(r)Y_{lm}(\theta, \phi) \tag{7.315}$$

where N is an overall normalization constant and $Y_{lm}(\theta, \phi)$ are the normalized surface harmonics of equation (3.51).

For STOs the radial factor has the form

$$R_n(r) = r^{n-1}\,e^{-\zeta r} \tag{7.316}$$

and the integral parameters n, l, m are restricted to the same ranges as for hydrogenic orbitals, namely

$$l \geqslant 0, \qquad n \geqslant l + 1, \qquad -l \leqslant m \leqslant l. \tag{7.317}$$

We may therefore refer to $1s$, $2s$, $2p_0$, $3d_1$, . . . Slater-type-orbitals; the subscript giving the angular momentum component is frequently omitted. Such a designation specifies the STO completely apart from the value of the orbital exponent ζ.

Slater[37] first showed that a single term of the form (7.315) and (7.316) provides a useful rough approximation to atomic SCF orbitals and derived simple rules for fixing the ζ-value for a given orbital in a given atom or ion. At least for the first two rows of the periodic table, Slater's rules provide ζ-values quite close to the optimum values obtained by minimizing the total electronic energy.

The molecular basis set consisting of one STO for each occupied atomic orbital is often referred to as a minimal or single $-\zeta$\{SZ\} basis. Another commonly used basis is the double-zeta \{DZ\} basis containing two STOs with different ζ-values for each atomic orbital. It provides a useful half-way house between a minimal STO basis and the more extended STO bases needed for an accurate representation of atomic Hartree–Fock orbitals.

A basis of STOs is quite efficient for atomic Hartree–Fock calculations, as we shall see below; if the ζ-values are suitably chosen the short-range behaviour (a cusp at $r = 0$) and the usual long-range decay (exponential) of accurate Hartree–Fock orbitals are each reproduced by a single term. Consequently quite short STO expansions can provide accurate atomic orbitals. The biggest obstacle to using such a basis in a molecular SCF calculation arises from the evaluation of the many difficult integrals involved, especially the two-electron repulsion integrals involving basis functions on several different centres. Much effort in recent years has gone into the development of efficient, general, computer programs[38] for this purpose. This effort has succeeded to the extent that an STO basis, or a closely related basis of elliptical functions (cf. equation (5.104)), is probably the most efficient choice for an SCF calculation on a diatomic or small linear polyatomic molecule.

Most recent calculations on non-linear molecules, however, have used the basis of Gaussian-type-orbitals (GTOs) suggested originally by Boys.[39] In local spherical coordinates these functions are of the form (7.315) with

$$R_n(r) = r^{n-1}\, \mathrm{e}^{-\zeta r^2}. \tag{7.318}$$

Boys showed that all the necessary integrals involving these Gaussian orbitals could be easily evaluated provided that the conditions (7.317) were supplemented by the condition

$$n - l = \text{odd integer.} \tag{7.319}$$

Thus allowed GTOs may be designated

$$(1s)_G, (3s)_G, \ldots; \qquad (2p)_G, (4p)_G, \ldots; \qquad (3d)_G, (5d)_G \ldots; \tag{7.320}$$

the subscript G serving to distinguish the Gaussian orbitals from STOs or other types.

The reason for the restriction (7.319) is most readily apparent if a local cartesian coordinate system is used instead of spherical coordinates. Simple linear combinations of STOs or GTOs of the form (7.315) may then be expressed as follows

$$\text{STOs } \chi_{pqr}(\mathrm{r}) = \begin{cases} x^p y^q z^r \exp(-\zeta r) & ((n-l) \text{ odd}) \quad (7.321) \\ x^p y^q z^r \cdot r \exp(-\zeta r) & ((n-l) \text{ even}) \quad (7.322) \end{cases}$$

$$\text{GTOs } \chi_{pqr}(\mathrm{r}) = \begin{cases} x^p y^q z^r \exp(-\zeta(x^2+y^2+z^2)) & ((n-l) \text{ odd}) \quad (7.323) \\ x^p y^q z^r \cdot r \exp(-\zeta(x^2+y^2+z^2)) & ((n-l) \text{ even}). \quad (7.324) \end{cases}$$

We see from equation (7.323) that, if $(n-l)$ is odd, the GTO and all its derivatives have no singularities. This together with the fact that the product of two Gaussians $\exp(-\zeta_1 r_1^2)$ and $\exp(-\zeta_2 r_2^2)$ may be expressed as a single Gaussian on a third centre, permits a simple reduction of all three- and four-centre electron repulsion integrals to two-centre integrals.[39] If on the other hand $(n-l)$ is even the singularity at the origin arising from the factor $r = (x^2+y^2+z^2)^{\frac{1}{2}}$ in equation (7.324) prevents such a reduction. For STOs the factor r in equation (7.322) is unimportant since the factor $e^{-\zeta r}$ is itself singular at the origin.

In most practical applications the set of GTOs (7.320) is further restricted to the set of primitive GTOs

$$(1s)_G, (2p)_G, (3d)_G, \ldots, \tag{7.325}$$

that is, only functions with the smallest permitted value of n are used. The necessary flexibility in the basis set is then obtained by using a wide range of ζ values for each type of orbital. It has been found that this procedure simplifies the integral evaluation without significantly increasing the number of basis functions needed for a given accuracy.

A further step in the direction of simplifying integral evaluation at the expense of an increased number of basis functions has been taken by Whitten.[40] His Gaussian-lobe-function (GLF) bases exploit the fact that all GTOs (7.323) may be obtained from the simplest, namely

$$(1s)_G = \exp(-\zeta((x-X)^2 + (y-Y)^2 + (z-Z)^2)) \tag{7.326}$$

by differentiation with respect to the coordinates X, Y, Z of the expansion centre. For example, we have

$$\begin{aligned}(2p_z)_G &= (z-Z)\exp(-\zeta((x-X)^2 + (y-Y)^2 + (z-Z)^2)) \\ &= (2\zeta)^{-1}\frac{\mathrm{d}}{\mathrm{d}Z}\exp(-\zeta((x-X)^2 + (y-Y)^2 + (z-Z)^2)).\end{aligned} \tag{7.327}$$

If derivatives such as that appearing in equation (7.327) are approximated as ratios of finite differences we obtain approximations to the higher Gaussian functions as differences of closely spaced $1s$ Gaussians (GLFs). Thus for the function (7.327) centred at the origin $X = Y = Z = 0$ we obtain the approximation

$$\begin{aligned}(2p_z)_G \approx (2p_z)_{Gl} = (4\zeta\delta)^{-1}\{&\exp(-\zeta(x^2 + y^2 + (z-\delta)^2)) \\ &-\exp(-\zeta(x^2 + y^2 + (z+\delta)^2))\}.\end{aligned} \tag{7.328}$$

Extending this procedure we obtain

$$\begin{aligned}(3d_{xy})_G \approx (3d_{xy})_{Gl} = (4\zeta\delta)^{-2}\{&\exp(-\zeta((x-\delta)^2 + (y-\delta)^2 + z^2) \\ &-\exp(-\zeta((x+\delta)^2 + (y-\delta)^2 + z^2)) \\ &-\exp(-\zeta((x-\delta)^2 + (y+\delta)^2 + z^2)) \\ &+\exp(-\zeta((x+\delta)^2 + (y+\delta)^2 + z^2))\}\end{aligned} \tag{7.329}$$

and analogous expressions for the other $2p$, $3d$ and higher Gaussians.

Of course, for finite values of δ, the GLFs (7.328) and (7.329) do not have exactly the angular behaviour indicated by the labels p_z, d_{xy}. However, for values of δ used in practical applications ($\leqslant 0.1$ au) departures of the final atomic wave functions from the correct angular behaviour are quite small.

The ease with which multi-centre integrals may be evaluated makes GTO bases extremely flexible. It is quite simple to float the atomic orbitals away from the nuclear centres[41] in the manner required to satisfy the electrostatic theorem for a finite expansion (Chapter 2) or to introduce additional basis functions centred in the bonding region between atoms. These advantages are offset by the slow convergence of GTO expansions, which is largely attributable to their incorrect behaviour near the nuclei. Clearly any finite sum of terms of the form (7.323) fails to reproduce the cusp at the origin needed to cancel the $1/r$-type singularity arising from the nuclear attraction term in the Schrödinger equation (cf. the discussion of single-centre expansions in Chapter 3, especially Fig. 3.8). This especially slow convergence in nuclear regions must be borne in mind for some applications. A Gaussian expansion which gives a good total energy may well lead to poor values for some properties which are dominated by contributions from nuclear regions. Gaussian expansions also show incorrect behaviour at large distances from the nuclei but this is less important both energetically and for the calculation of most molecular properties.

Intermediate in character between GTOs and STOs are the so-called combined or contracted Gaussian type orbitals (CGTOs). Here each basis function is a fixed linear combination of primitive GTOs obtained by a least squares fit to a single STO or in some other way. These CGTO bases are designated by square brackets [ ], to distinguish them from GTOs ( ) and STOs { }. Within the brackets the numbers of functions of $s, p, d \ldots$ types are listed successively. Thus

$$\mathrm{N}[4, 2] = (3, 3, 2, 3; 3, 3) \tag{7.330}$$

denotes a CGTO basis of four s functions and two p functions centred on the N nucleus; the s functions are fixed linear combinations of 3, 3, 2 and 3 primitive GTOs respectively and each p function is a fixed linear combination of 3 primitive GTOs.

Of course, the contracted basis (7.330) is less flexible than the GTO basis N(11, 6) obtained by using all the primitive Gaussian functions independently, and the same primitive integrals must be evaluated in both cases. However, for the contracted basis (7.330), far fewer integrals need be stored and subsequently manipulated in the SCF calculation. For this reason the contraction procedure is used in most molecular SCF calculations based on a large set of GTOs.

In atomic SCF calculations each atomic orbital $1s$, $2p$, $3d$, . . . contains contributions only from basis functions of the same symmetry type. Here it is feasible to fully optimize the non-linear parameters ζ and to obtain convergence to numerical Hartree–Fock results with considerably better than chemical accuracy. Such calculations have established that, for atoms, STO bases are much more efficient than GTO bases. This is illustrated by the results for the ground states of He and the first row atoms Li, . . ., Ne listed in Tables 7.9 and 7.10.

Table 7.9. Total energies (au) for some SCF calculations on the ground state, $(1s)^2$, ^1S of He[a] ($E_{HF} = -2.861680$)

STO bases		GTO bases		
{1} {SZ}	{2} {DZ}	(3) = $(1s, 1s', 1s'')$	(4) = $(1s, 1s', 1s'', 1s''')$	(10)
−2.847656	−2.861670	−2.835680	−2.855160	−2.861669

[a] Values from Ref. 42.

Table 7.10. Orbital errors for some SCF calculations on the ground states of first row atoms (au)[a]

			$\Delta_{orb} \times 10^4$						
			STO bases				GTO bases		GLF basis
Atomic state	Hartree–Fock energy E_{HF}	$-E_{corr} \times 10^4$	{2, 1} {SZ}	{4, 2} {DZ}	{4, 3} {BA}	{5, 4} {BA}	(9, 5) ≈ {DZ}	(10, 6) ≈ {DZ}	(10, 5) ≈ {DZ}
Li ^2S	−7.43273	454	142	0	0	0	4	2	15
Be ^1S	−14.57302	943	163	7	0	0	10	4	27
B ^2P	−24.52906	1248	307	12	1	0	19	8	50
C ^3P	−37.68862	1565	662	19	1	0	34	13	81
N ^4S	−54.40093	1886	1320	30	1	0	56	20	127
O ^3P	−74.80938	2579	2690	52	2	0	91	31	179
F ^2P	−99.40933	3220	4672	81	4	0	137	44	270
Ne ^1S	−128.5471	3896	7348	123	6	0	203	61	368

[a] Values from Ref. 42.

The quantity Δ_{orb} of Table 7.10 is defined by the equation

$$\Delta_{orb} = E_{SCF} - E_{HF}; \tag{7.331}$$

it is usually referred to as the orbital error or expansion error of the basis. Also listed is the correlation error which is inherent in the

Hartree–Fock approximation itself; it is defined by

$$E_{\text{corr}} = E_{\text{exact}} - E_{\text{HF}}. \tag{7.332}$$

where E_{exact} is the appropriate eigenvalue of the full non-relativistic Schrödinger equation. These eigenvalues have been estimated from experimental atomic energies adjusted for small relativistic and other corrections.[43]

For each basis set, E_{SCF} has been minimized with respect to the non-linear parameters ζ. Consequently the Δ_{orb} values of Table 7.10 give the errors in the total energy due to the replacement of accurate Hartree–Fock orbitals by optimal expansions in the various bases listed at the top of the table. It is immediately apparent that STO bases are very much more efficient than GTO bases for atomic SCF calculations. For atoms up to nitrogen even the minimal {SZ} basis leads to an expansion error less than the correlation error inherent in the Hartree–Fock approximation. Although other important quantities such as orbital energies and expectation values for one-electron operators may behave somewhat differently from the total energy, we have the rough working rule that (up to nitrogen) orbital errors from a minimal STO basis are less significant than correlation errors. Beyond nitrogen, however, the expansion error for a minimal STO basis rises rapidly; Table 7.10 shows that by the time Ne is reached it is already almost twice the correlation error and this steep rise continues beyond the atoms listed.[42] The representation of the *d* orbitals of transition metals is especially poor.

A quantitative measure of the relative convergence of STO and GTO expansions of HF orbitals is provided by the ratio

$$\frac{N_{\text{G}}}{N_S} = \frac{n_{\text{G}}(s) + 3n_{\text{G}}(p)}{n_{\text{S}}(s) + 3n_{\text{S}}(p)} \tag{7.333}$$

of the total number of basis functions of the two types required for a given expansion error (Δ_{orb}), allowance being made for the three p-type basis functions (p_1, p_0, p_{-1} or p_x, p_y, p_z) corresponding to each p-type radial function listed in Table 7.10. It is clear from the table that this ratio is a sensitive function of Δ_{orb}. Thus from Table 7.10 the STO set {DZ} = {4, 2} and the GTO sets (9, 5) or (10, 6) are of comparable accuracy and the GLF set $(10, 5)_l$ is only slightly worse. This corresponds to a ratio $N_{\text{G}}/N_{\text{S}}$ which lies between 2 and 3. On the other hand to obtain $\Delta_{\text{orb}} < 10^{-5}$ au for the ground state of He we require at least 10 GTOs as opposed to only 2 STOs (Table 7.9), indicating a ratio $N_{\text{G}}/N_{\text{S}} > 5$. Although no Gaussian calculations of adequate accuracy are currently available a similar, or even larger, ratio $N_{\text{G}}/N_{\text{S}}$ would be required to match the expansion errors of the STO sets {4, 3} and {5, 4} of Table 7.10.

Since the number of many-centred integrals required for a polyatomic molecule increases approximately as N^4 where N is the number of basis functions, it is clear that the relative efficiency of STO and GTO bases is very dependent on the expansion error that can be tolerated. It is difficult to give a precise numerical estimate of the advantage which GTO bases derive from the simpler evaluation of many-centred integrals, but for a molecule containing at least three heavy atoms (i.e. atoms other than H) which are not collinear a decrease by a factor of 100–1000 in the computation time for integral evaluation relative to a STO basis is not unreasonable. On this basis Gaussian functions are more efficient at a STO {DZ} level of accuracy since only between $2^4 = 16$ and $3^4 = 81$ times as many integrals are required. At this level of refinement $\Delta_{orb} \approx 0.002$ au for a typical heavy atom (C), which is only 1–3% of the correlation error and is approaching our criterion of chemical accuracy.

However, if one is aiming at an accuracy comparable with that achieved for atoms with the STO {4, 3} and {5, 4} sets in Table 7.10, that is $\Delta_{orb} < 10^{-5}$ au per heavy atom, the advantage of GTO bases over STO bases is more problematic since, perhaps, 5^4 or more times as many basic integrals would be required. In fact no molecular Hartree–Fock solutions approaching this degree of accuracy are available at present, even for diatomic molecules, and are hardly justifiable economically in view of errors in the Hartree–Fock method itself.

Nevertheless, even at the chemical level of accuracy (0.001 au per heavy atom), extended and highly optimized atomic bases, such as the {4, 3} and {5, 4} STO sets, are extremely valuable for molecular calculations because they are capable of accommodating radial distortions of atomic orbitals through the linear expansion coefficients of the molecular orbitals. The very expensive task of re-optimizing the non-linear ζ parameters in the molecule is thereby eliminated or greatly reduced.

(b) NEAR HARTREE-FOCK WAVE FUNCTIONS FOR N_2 AND NH_3

For N_2 the correlation diagram (Fig. 7.1) predicts a closed-shell ground state

$$N_2\,(1\sigma_g^2\,1\sigma_u^2\,2\sigma_g^2\,2\sigma_u^2\,3\sigma_g^2\,1\pi_u^4,\,X\,^1\Sigma_g^+). \qquad (7.334)$$

One of the most thorough studies of the convergence of a series of SCF-MO calculations towards the Hartree–Fock limit is that of Cade, Sales and Wahl[44] on the state (7.334). These authors carried out two series of calculations. The first started with the basis {SZ} of Table 7.10 on each atom with ζ values optimized for the N_2 molecule. With this basis we have three symmetry orbitals of types σ_g and σ_u, leading to 3 x 3 Fock matrices for the determination of the MOs $1\sigma_g$, $2\sigma_g$, $3\sigma_g$ and $1\sigma_u$, $2\sigma_u$, whilst the MO $1\pi_u$ is completely determined by symmetry. The qualitative form of the resulting MOs is given by Table 7.2 with $|a/b| = |c/d| = |e/f| = |g/h| = |k/l| = 1$. In subsequent calculations of this first series the {SZ} basis on each atom was gradually augmented by additional functions of s, p, d and f type; different ζ values were used for the same basis function in different molecular symmetries and these ζ values were partially optimized at each stage.

The second series of calculations was similar except that it started from a large STO basis for N, ^4S. This "best atom" basis, intermediate between the bases {4, 3} and {5, 4} of Table 7.10 will be denoted {BA}.

Key results of the two series of calculations and the estimated Hartree–Fock limit are shown in Table 7.11. A prime indicates complete or partial optimization of the orbital exponents ζ and the symbol P represents polarization functions, that is basis functions with l values one or more units higher than any valence atomic orbital. The inclusion of such functions leads to an energy decrease of about 0.1 au and is, therefore, essential for chemical accuracy; for N_2 up to three d functions and one f function were included for each symmetry type.

Table 7.11. SCF–MO total energies for N_2, $X\,^1\Sigma_g^+$ (au)[a]

Calculation	Basis	Number of symmetry orbitals			Total energy	
		σ_g	σ_u	π_u	(E_{SCF})	$E_{SCF} - E_{HF}$
1B	{SZ}′	3	3	1	−108.6459	0.3469
1G	{DZ}′	6	6	2	−108.8914	0.1014
1S	{BA + P}′	12	8	6	−108.9888	0.0040
2A	{BA}	8	7	3	−108.8967	0.0961
2B	{BA + P}	12	8	6	−108.9897	0.0031
2D	{BA + P}′	12	8	6	−108.9928	0.0014
	{BA + P}′	12	8	6	−108.9956[b]	0.0014

Estimated limit $E_{HF} = -108.997 \pm 0.002$.
[a] Values from Ref. 44.
[b] At r_e(calc.) = 2.013 au, all other energies at r_e (exp) = 2.068 au.

The basis sets 1S, 2B and 2D are identical apart from ζ values and the superior results with 2B and especially 2D reflect the inefficiency of the procedure of gradually building up the basis with partial optimization at each stage; it is much more efficient to start with large optimal atomic bases and perform limited molecular optimizations with all important basis functions included. Indeed we see from Table 7.11 that the basis {BA + P} without any molecular optimization is already almost adequate for chemical accuracy (0.002 au for N_2).

Two SCF-MO calculations on the ground state

$$NH_3\,(1a_1^2\,2a_1^2\,3a_1^2\,1e^4,\,X\,{}^1A_1) \tag{7.335}$$

of ammonia of accuracy comparable to the above study of N_2, have been carried out by Rauk, Allen and Clementi[(45)] using a very large CGTO basis and by Rajagopal[(46)] using a STO basis. Both basis sets were expressed in local Cartesian coordinate systems (equations (7.321)-(7.324)); the determination of symmetry orbitals for such basis sets has been discussed in Section 4.7 (especially Table 4.9). Total energies from these and simpler calculations are shown in Table 7.12 together with the estimated Hartree–Fock limit.

This estimate is based mainly on the work of Rauk *et al.* and the other Gaussian calculations. Although the final wave functions of Rajagopal's calculation may be of comparable accuracy, the grids used in the numerical integration technique[(28)] are such that the final two decimals quoted in the total energy are not significant. As for N_2, the inclusion of polarization functions, here d functions on nitrogen and p functions on the hydrogens, is essential for near chemical accuracy. Their effect may be estimated as the greater part of $E_{SCF} - E_{HF}$ for the 16 function STO basis N{4, 2} H{2}, that is about 0.05 au.

Table 7.12. SCF-MO total energies for NH_3, $X\,{}^1A_1$ (au)[a]

Basis	Total energy (E_{SCF})	$E_{SCF} - E_{HF}$
8STO, N{2, 1}H{1}	−56.096	0.129
16STO, N{4, 2}H{2}	−56.1678	0.0572
23STO, N{4, 3, 2}H{1}	(−56.2268)	(−0.0018)
45GTO, N(9, 5)H(4, 1)	−56.2015	0.0235
67GTO, N[7, 8, 1]H[3, 2]	−56.2109	0.0141
91GTO, N[10, 5, 2]H[4, 1][b]	−56.2219	0.0031

Estimated limit $E_{HF} = -56.225 \pm 0.002$.
[a] From Refs 47, 48.
[b] Ref. 45.

From these examples and others in the literature (cf. especially, McLean and Yoshimine[(49)]) we may formulate the following requirements for an HFR calculation to converge to the Hartree–Fock limit to within chemical accuracy.

(i) The basis set should include for each atom sufficient functions to approximate the Hartree–Fock limit for the isolated atom to within 0.0005 au or less, that is, considerably better than chemical accuracy. From Table 7.10 we see that, for first row atoms, the STO basis {4, 3} satisfies this requirement comfortably whereas the GTO basis (10, 6) is inadequate beyond C. Something approaching the basis N(15, 8) employed by Rauk *et al.*[45] for NH_3 is required.

(ii) In addition several polarization functions should be included for each atom. The most important of these have l values one greater than the highest l value for the valence electrons. The non-linear parameters (ζ) in the polarization functions should be optimized in a molecular calculation of at least {DZ} quality. In the case of a Slater basis the ζs for the two or three s functions needed on each hydrogen atom should be optimized at this stage since this is not possible for the isolated atom. Limited experience suggests that, for polarization functions, Gaussian bases are almost as efficient as Slater bases. That is, for chemical accuracy, comparable numbers of polarization functions are needed for the two types of basis. With Gaussian bases it is also simple to include additional functions centred at points other than the nuclei; for example, bond functions consisting of 1s Gaussians distributed along chemical bonds. For CH_4 such functions have proved an attractive alternative to the more conventional d functions on carbon and p functions on the hydrogens.[50]

(iii) All basis functions from (i) and (ii) are to be used in a final HFR calculation. If rather small Slater bases are used in step (i) it may be necessary to carry out extensive optimization of many ζs at this stage. A more efficient procedure is to use large atomic bases; little or no optimization of the ζs in the final calculation may be required for chemical accuracy. The Gaussian bases required to satisfy the conditions of step (i) are so extensive that molecular optimization is unnecessary. Indeed it is usually possible to reduce the flexibility of the basis by the contraction procedure and still retain chemical accuracy.

(c) EXTRAPOLATED HARTREE-FOCK TOTAL ENERGIES

The Hartree–Fock total energies listed in Table 7.13 are the results of HFR calculations which satisfy, or come close to satisfying, the above requirements. The best atom Slater bases {BA} are at least as extensive as the {4, 3} basis of Table 7.10. Both the lowest calculated energy and the estimated Hartree–Fock limit are given for each molecule. The estimated limits and their uncertainties are based mainly on estimates of the original authors of the calculations modified in some cases for mutual consistency and to ensure what are intended to be safe estimates. Thus for the linear molecules taken from the work of McLean and Yoshimine,[(49)] their estimate, d, of the "probable" distance from the Hartree–Fock limit was assumed to be a lower limit to this distance with an uncertainty $\pm d$.

For O_2 and C_2 results are given for some excited states as well as the ground states. We note that the observed ground state of C_2 is the $a\ {}^1\Sigma_g^+$ state arising (principally) from the closed-shell configuration $(2\sigma_u)^2(1\pi_u)^4$, whereas the Hartree–Fock calculations predict that the $A\ {}^3\Sigma_g^-$ state from the configuration $(2\sigma_u)^2(3\sigma_g)^2(1\pi_u)^2$ lies almost 3 eV lower than the $a\ {}^1\Sigma_g^+$ state (cf. Table 7.4). This rather large error is understandable from the correlation diagram for homonuclear diatomic molecules (Fig. 7.1) which indicates that for C_2 the MOs $3\sigma_g$ and $1\pi_u$ are almost equal in energy. Thus there is a second ${}^1\Sigma_g^+$ state from $(2\sigma_u)^2(3\sigma_g)^2(1\pi_u)^2$ which, in the single configuration approximation, is almost degenerate with $a\ {}^1\Sigma_g^+$; we have a case of strong first order configuration-interaction (Section 5.6) and the energy of $a\ {}^1\Sigma_g^+$ is substantially lowered. On the other hand, $A\ {}^3\Sigma_g^-$ is the only low-lying state of ${}^3\Sigma_g^-$ symmetry and is, therefore, not affected by strong first-order configuration-interaction. A similar, although less marked, failure of the Hartree–Fock calculations is the substantial overestimation of the $X\ {}^3\Sigma_g^-$–$b\ {}^1\Sigma_g^+$ interval for O_2. This is again attributable to neglect of first-order CI, which lowers the state $b\ {}^1\Sigma_g^+$ relative to $X\ {}^3\Sigma_g^-$. In Ref. 69 we consider the extensions of the theory which are required to eliminate these rather large errors from the Hartree–Fock results.

(d) HARTREE–FOCK BINDING ENERGIES

The Hartree–Fock binding energies $D_e(\text{HF})$ in Table 7.13 were obtained from the extrapolated Hartree–Fock total energies and the Hartree–Fock atomic energies of Table 7.10. Also listed are the observed binding energies and the discrepancies

$$\delta D_e = D_e(\text{obs}) - D_e(\text{HF}). \tag{7.336}$$

Assuming that the observed total energy of an atom or molecule may be obtained from the Hartree–Fock value by adding the correlation energy E_{corr} and a relativistic correction E_{rel} we have

$$\delta D_e = -\Delta E_{corr} - \Delta E_{rel} \tag{7.337}$$

where Δ denotes the change in a quantity in going from the separated atoms to the molecule.

Estimates of relativistic corrections for atoms[43] indicate that contributions from the valence shell are negligible to within chemical accuracy, at least for the first row elements. Appreciable contributions arise only from inner-shell electrons and it is reasonable to assume that these are unchanged on molecular formation. Hence in equation (7.337) we have $\Delta E_{\text{rel}} \approx 0$ and

$$\delta D_{\text{e}} = -\Delta E_{\text{corr}} \tag{7.338}$$

that is, the deficiency in the HF dissociation energy is given by the change in correlation energy on molecular formation.

From Table 7.13 we see that δD_{e} (often referred to as the extra molecular correlation energy or the correlation contribution to binding) is substantial, often amounting to several electron volts even for diatomic molecules. Hartree–Fock estimates of dissociation energies are, therefore, not directly useful. The large changes in correlation energy reflect the extensive electronic reorganization on molecular formation.

The description of the molecules of Table 7.13 in terms of the localized molecular orbitals of Section 7.3 is useful in interpreting correlation energy changes. In terms of this picture the formation of each new bond leads to an additional pair of electrons in a localized MO. Since the LMOs are defined to maximize intra-orbital interactions at the expense of inter-orbital interactions it is to be expected that the formation of these new electron pairs will dominate the correlation energy changes. The values of Table 7.13 bear out this view in a qualitative way, the δD_{e} for closed-shell molecules amounting to some 1.0–2.0 eV per bond. A more thorough analysis along these lines requires the introduction of appropriate valence states for the separated atoms.[69]

This picture in terms of LMOs suggests that the correlation energy should remain almost invariant in any change that preserves the number of electron pairs and, as far as possible, their local spatial relationship to each other. The Hartree–Fock approximation should then provide quite an accurate estimate of the change in total energy.

This property of the correlation energies has been exploited in several ways. One of the simplest changes satisfying the conditions is the formation of a hydride by the extraction of one or more protons from the nucleus of the united atom. Assuming conservation of the electronic correlation energy in this process we obtain improved estimates $D'_{\text{e}}(\text{HF})$ of hydride binding energies from the equation

$$D'_{\text{e}}(\text{HF}) = D_{\text{e}}(\text{HF}) - E_{\text{corr}}(\text{UA}) + E_{\text{corr}}(\text{SA}). \tag{7.339}$$

In applying this equation to the diatomic hydrides of Table 7.13 it is necessary to employ those states of the united atom (UA) and the separated atoms (SA) which correlate with the appropriate hydride

Table 7.13. Hartree–Fock total energies and binding energies for some small molecules

Molecule and state	Basis functions	$-E_{HF}$ (au) Lowest calculated[a]	Estimated limit	D_e (HF) (eV)	δD_e (eV)	D_e(obs)[b] (eV)	D'_e (HF) (eV)
H_2, $X\ ^1\Sigma_g^+$	Elliptic	1.1336	1.1336	3.64	1.11	4.75	4.78
LiH, $X\ ^1\Sigma^+$	$\{BA + P\}'$	7.9873	7.9878 ± 0.0003	1.50	1.02	2.52	2.82
BeH, $X\ ^2\Sigma^+$	$\{BA + P\}'$	15.1531	15.1536 ± 0.0003	2.19	0.41	2.6 ± 0.2	3.01
BH, $X\ ^1\Sigma^+$	$\{BA + P\}'$	25.1315	25.1320 ± 0.0003	2.80	0.78	3.58	3.96
CH, $X\ ^2\Pi$	$\{BA + P\}'$	38.2796	38.2801 ± 0.0003	2.49	1.16	3.65	3.82
NH, $X\ ^3\Sigma^-$	$\{BA + P\}'$	54.9784	54.9789 ± 0.0003	2.12	1.68	3.8 ± 0.2	4.00
OH, $X\ ^2\Pi$	$\{BA + P\}'$	75.4213	75.4218 ± 0.0003	3.06	1.57	4.63	4.80
FH, $X\ ^1\Sigma^+$	$\{BA + P\}'$	100.0708	100.071 ± 0.0003	4.41	1.71	6.12	6.25
Li_2, $X\ ^1\Sigma_g^+$	$\{BA + P\}'$	14.8718	14.873 ± 0.001	0.21	0.84	1.05	—
C_2, $x\ ^1\Sigma_g^+$	$\{BA + P\}'$	75.4062	75.408 ± 0.002	0.84	5.56	6.4 ± 0.2	—
C_2, $A\ ^3\Sigma_g^-$	$\{BA + P\}'$	75.5152	75.517 ± 0.002	3.80	1.80	5.6 ± 0.2	—
N_2, $X\ ^1\Sigma_g^+$	$\{BA + P\}'$	108.9956	108.997 ± 0.002	5.31	4.59	9.90	—
O_2, $X\ ^3\Sigma_g^-$	$\{BA + P\}'$	149.6659	149.670 ± 0.003	1.39	3.79	5.21	—
O_2, $a\ ^1\Delta_g$	$\{BA + P\}'$	149.6172	149.621 ± 0.003	0.06	4.17	4.23	—
O_2, $b\ ^1\Sigma_g^+$	$\{BA + P\}'$	149.5683	149.572 ± 0.003	−2.50	6.07	3.57	—
F_2, $X\ ^1\Sigma_g^+$	$\{BA + P\}'$	198.7761	198.780 ± 0.003	−1.05	2.70	1.65 ± 0.05	—
LiO, $X\ ^2\Pi$	$\{BA + P\}'$	82.3111	82.312 ± 0.001	1.90	1.54	3.5 ± 0.2	—
LiF, $X\ ^1\Sigma^+$	$\{BA + P\}'$	106.9916	106.993 ± 0.001	4.11	1.89	6.0 ± 0.3	—
BeO, $X\ ^1\Sigma^+$	$\{BA + P\}'$	89.4541	89.455 ± 0.001	1.98	2.72	4.7 ± 0.1	—
BF, $X\ ^1\Sigma^+$	$\{BA + P\}'$	124.1671	124.169 ± 0.001	6.28	1.62	7.9 ± 0.2	—
CO, $X\ ^1\Sigma^+$	$\{BA + P\}'$	112.7891	112.791 ± 0.001	7.97	3.25	11.22	—
CF, $X\ ^2\Pi$	$\{BA + P\}'$	137.2259[c]	137.229 ± 0.002	3.57	1.93	5.5 ± 0.2	—

NO, $X\ ^2\Pi$	$\{BA + P\}'$	129.2837	129.286	± 0.002	2.06	4.55	6.61	—
NF, $X\ ^2\Sigma^-$	$\{BA + P\}'$	153.8353[c]	153.838	± 0.002	0.75	3.25	4 ± 1	—
OF, $X\ ^2\Pi$	$\{BA + P\}'$	174.1950[d]	174.199	± 0.003	−0.54	2.94	2.4 ± 0.4	—
HCN, $X\ ^1\Sigma^+$	$\{BA + P\}$	92.9160	92.918	± 0.001	8.94	4.64	13.58	—
CO_2, $X\ ^1\Sigma_g^+$	$\{BA + P\}$	187.7254	187.729	± 0.002	11.47	5.38	16.85	—
NNO, $X\ ^1\Sigma^+$	$\{BA + P\}$	183.7567	183.771	± 0.007	4.35	7.37	11.72	—
FCN, $X\ ^1\Sigma^+$	$\{BA + P\}$	191.7798	191.796	± 0.008	8.08	—	—	—
C_2H_2, $X\ ^1\Sigma_g^+$	$\{BA + P\}$	76.8540	76.858	± 0.002	13.08	4.45	17.53	—
LiCCH, $X\ ^1\Sigma^+$	$\{BA + P\}$	83.7305	83.742	± 0.006	11.76	—	—	—
FCCH, $X\ ^1\Sigma^+$	$\{BA + P\}$	175.7236	175.740	± 0.008	12.34	—	—	—
C_2N_2, $X\ ^1\Sigma_g^+$	$\{BA + P\}$	184.6568	184.677	± 0.009	13.55	8.10	21.65	—
H_2O, $X\ ^1A_1$	O(10, 6, 2), H(4, 2)	76.0596	76.070	± 0.005	7.09	2.99	10.07	10.67
	O{3, 3, 1}, H{2, 1}	76.0047						
NH_3, $X\ ^1A_1$	N[10, 5, 2], H[4, 1]	56.2219[e]	56.225	± 0.002	8.82	4.07	12.89	14.29
	N{4, 3, 2}, H{1}	(56.2268)						
H_2O_2, $X\ ^1A_1$	O(11, 7, 1), H(6, 1)	150.7993	150.850	± 0.009	6.29	5.19	11.48	—
CH_4, $X\ ^1A_1$	C[4, 2, 2], H[2, 1]	40.2045[f]	40.220	± 0.005	14.46	3.72	18.18	20.80
	C{5, 3, 1}, H{2, 1}	40.2045						
H_2CO, $X\ ^1A_1$	C, O[5, 3, 2], H[2, 1]	113.8917	113.932	± 0.009	11.81	4.43	16.24	—
C_2H_6, $X\ ^1A_{1g}$	C(11, 7, 1), H(6, 1)	79.2377[g]	79.270	± 0.008	24.29	6.53	30.82	—
C_2H_4, $X\ ^1A_{1g}$	C[5, 3, 2], H[2, 1]	78.048[h]	78.080	± 0.008	19.12	5.25	24.37	—

[a] Refs 47, 48.
[b] Diatomic molecules Ref. 53; polyatomic molecules Refs 54, 55; zero-point energies Ref. 5.
[c, d] Ref. 56.
[e] Ref. 45.
[f] Ref. 50.
[g] Ref. 68.
[h] Ref. 65.

state. The appropriate correlation rules are given by Herzberg[5] and Hurley[69]. These states and their correlation energies are listed in Table 7.14.

Table 7.14. United and separated atoms for first-row diatomic hydrides[b]

E_{corr} (au)	United atom	Hydride, AH	Separated atom A[a]	E_{corr} (au)
−0.0943	$Be(1s^2 2s^2, {}^1S)$	$LiH(1\sigma^2 2\sigma^2, X^1\Sigma^+)$	$Li(1s^2 2s, {}^2S)$	−0.0454
−0.1248	$B(1s^2 2s^2 2p, {}^2P)$	$BeH(1\sigma^2 2\sigma^2 3\sigma, X^2\Sigma^+)$	$Be(1s^2 2s^2, {}^1S)$	−0.0943
−0.1673	$C(1s^2 2s^2 2p^2, {}^1D)$	$BH(1\sigma^2 2\sigma^2 3\sigma^2, X^1\Sigma^+)$	$B(1s^2 2s^2 2p, {}^2P)$	−0.1248
−0.2057	$N(1s^2 2s^2 2p^3, {}^2D)$	$CH(1\sigma^2 2\sigma^2 3\sigma^2 1\pi, X^2\Pi)$	$C(1s^2 2s^2 2p^2, {}^3P)$	−0.1565
−0.2579	$O(1s^2 2s^2 2p^4, {}^3P)$	$NH(1\sigma^2 2\sigma^2 3\sigma^2 1\pi^2, X^3\Sigma^-)$	$N(1s^2 2s^2 2p^3, {}^4S)$	−0.1886
−0.3220	$F(1s^2 2s^2 2p^5, {}^2P)$	$OH(1\sigma^2 2\sigma^2 3\sigma^2 1\pi^3, X^2\Pi)$	$O(1s^2 2s^2 2p^4, {}^3P)$	−0.2579
−0.3896	$Ne(1s^2 2s^2 3p^6, {}^1S)$	$HF(1\sigma^2 2\sigma^2 3\sigma^2 1\pi^4, X^1\Sigma^+)$	$F(1s^2 2s^2 2p^5, {}^2P)$	−0.3220

[a] Plus $H(1s, {}^2S)$, $E_{corr} = 0$.
[b] Ref. 51.

Equation (7.339) may also be applied to the closed-shell polyhydrides H_2O, NH_3 and CH_4, all of which correlate with ground-state separated atoms and the same united atom $Ne(1s^2 2s^2 2p^6, {}^1S)$. The $D_e'(HF)$ values obtained in this way are listed in the final column of Table 7.13. We see that equation (7.339) works well for H_2 and the diatomic hydrides, the binding energies being consistently overestimated by a small fraction of an electron volt. However, for the ten-electron isoelectronic series

$$Ne \rightarrow HF \rightarrow H_2O \rightarrow NH_3 \rightarrow CH_4 \tag{7.340}$$

the discrepancy increases rapidly and amounts to more than 2.5 eV for CH_4. Recent theories of electronic correlation are consistent with some decrease in the magnitude of the correlation energy in going from Ne to CH_4 in the isoelectronic sequence (7.340), but adequate quantitative calculations are still scarce.[69]

Again the picture in terms of LMOs suggests that there should be little change in correlation energy during an ionic decomposition of the type

$$NaCl \rightarrow Na^+ + Cl^-, \tag{7.341}$$

since the number and local environment of the localized electron pairs are unchanged. This leads to formulae of the type

$$\begin{aligned} D_e(NaCl) = D_e(HF, NaCl) - E_{corr}(Na^+) - E_{corr}(Cl^-) \\ + E_{corr}(Na) + E_{corr}(Cl) \end{aligned} \tag{7.342}$$

for estimating the binding energies of NaCl and other alkali halide molecules; these estimates are of near chemical accuracy. [52]

(e) HARTREE-FOCK ENTHALPIES OF REACTION

We may use the δD_e values of Table 7.13 to compute the change in correlation energy for many chemical reactions involving the listed molecules. When both the reactants and products are closed-shell molecules we find quite small correlation energy changes, very much smaller than the δD_e values themselves.

For example, for the reaction

$$N_2 + 3H_2 \rightarrow 2NH_3 \tag{7.343}$$

the change in correlation energy is given by

$$\begin{aligned}\Delta_{corr}(N_2 + 3H_2 \rightarrow 2NH_3) &= 2E_{corr}(NH_3) - E_{corr}(N_2) - 3E_{corr}(H_2)\\ &= -2\delta\, D_e(NH_3) + \delta D_e(N_2) + 3\delta\, D_e(H_2)\\ &= -0.22 \pm 0.15\ \text{eV}.\end{aligned} \tag{7.344}$$

Here we have used equation (7.338) and values of δD_e from Table 7.13. The uncertainty arises from errors in the experimental binding energies (negligible in this case) and in the extrapolated Hartree–Fock limits for N_2 and NH_3.

This near invariance of the correlation energy for closed-shell reactions is clearly consistent with the conditions enunciated above in terms of LMOs. In such reactions the number of electron pairs is always preserved and changes in their local spatial relationship are quite minor. In this connexion we note that, for N_2, the localization criterion we have adopted leads to three equivalent bent bonds rather than the alternative picture in terms of one σ-bond and two π-bonds (Section 7.3).

Snyder[57] has used Hartree–Fock electronic energies together with estimates of translational, rotational and vibrational energies to calculate enthalpies[54] of reaction ΔH°_{298} for a number of closed-shell molecules. Values accurate to within a few kcal mole^{-1} were obtained. However, uncertainties in the then available Hartree–Fock limits, especially for non-linear molecules made it difficult to separate the error inherent in the Hartree–Fock approximation itself from inadequacies in the HFR calculations.

Here we use a slightly different approach[58] and employ the Hartree–Fock limits of Table 7.13.

Any compilation of enthalpies of formation depends on some definition for the chemical elements. For our purposes the most convenient reference state for the molecules is the standard perfect-gas state at a temperature of absolute zero. In this state both the internal energy U and the enthalpy H are given by

$$H = U = E + v_0 \tag{7.345}$$

where E is the total electronic energy (including nuclear repulsion terms) and v_0 is the zero-point vibrational energy. The term v_0 could, in principle, be calculated to near chemical accuracy using Hartree–Fock total energies for a range of nuclear configurations. However, such calculations are not available for many of the molecules of Table 7.13 and we use spectroscopic values in all cases.

For the chemical elements various choices of reference state are possible. The usual thermodynamic standard state[54] is chosen for experimental convenience and differs for different elements; for example, we have Li (metal), C (graphite), and O_2 (gas). This choice is clearly unsuitable for our purposes since we have no estimate of the Hartree–Fock limit for many of these states.

On the other hand, theoretical discussions are usually based on the binding energies D_e of Table 7.13. The use of these quantities, or more properly $D_0 = D_e - v_0$, corresponds to the choice of a monatomic gas in the ground electronic state as the reference state for each chemical element. The enthalpy of formation of any molecule is then given by $\Delta H_0^\circ = -D_0$. The large values of δD_e in Table 7.13 show that this choice of reference state is a very bad one from the point of view of invariance of the correlation energy, and the above considerations based on the LMO picture suggest that some different choice may lead to much more accurate enthalpies of formation.

The best choice of reference state for each chemical element would seem to be some simple closed-shell molecule which contains one atom of the element and which is easily accessible both experimentally and theoretically. Closed-shell hydrides satisfy these requirements and we have chosen these hydrides in the perfect gas state at absolute zero, as reference states for the chemical elements.

One temporary disadvantage of these reference states is that in two cases (H_2O, CH_4) the best available HFR calculations are hardly adequate to estimate the Hartree–Fock limit to within chemical accuracy and for two others (BeH_2, BH_3) no adequate HFR calculations have been carried out. For these reasons compounds containing Be are omitted from the following tables and the reference state of B is chosen as the diatomic hydride BH($X\,^1\Sigma^+$)

Table 7.15. Standard enthalpies of formation ΔHf_0° for elementary reference states

Element:	H	Li	B	C	N	O	F
Hydride:	H_2	LiH	BH	CH_4	NH_3	H_2O	HF
ΔHf_0° (kcal mole^{-1})	0.0[a]	30.0[b]	106.7[c]	−15.970[c]	−9.34[c]	−57.102[c]	−64.789[c]

[a] By definition.
[b, c] Values from Ref. 55.

rather than BH_3, which is probably more appropriate from correlation energy considerations.

The standard enthalpies of formation ΔHf_0° of these closed-shell hydrides, relative to the thermodynamic standard states of the elements are listed in Table 7.15. These values are used to convert enthalpies of formation from thermodynamic standard states[54] to values relative to the closed-shell hydride reference states and vice versa.

In Table 7.16 enthalpies of formation ΔH_0° are listed for molecules appearing in Table 7.13; all values are relative to the elementary reference states of Table 7.15. The experimental values are based on the data of Gaydon[53] for diatomic molecules and on the data of Lewis and Randall[54] and Wagman *et al.*[55] for polyatomic molecules.

The Hartree–Fock values, $\Delta H_0^\circ(\mathrm{HF})$, were derived from the estimated Hartree–Fock limits of Table 7.13 and spectroscopic values for the zero-point vibrational energies.[5] For each molecule the listed uncertainty in $\Delta H_0^\circ(\mathrm{HF})$ corresponds only to the uncertainty in the Hartree–Fock limit for that molecule; if the molecule contains C, N, or O, additional errors in $\Delta H_0^\circ(\mathrm{HF})$ arise from the uncertainties 3, 1 and 3 kcal mole^{-1}, respectively, in the Hartree–Fock limits of the reference states CH_4, NH_3 and H_2O. For example, in the case of CO_2 the maximum error in the estimate of $\Delta H_0^\circ(\mathrm{HF})$ is $1 + (1 \times 3) + (2 \times 3) = 10$ kcal mole^{-1}. The molecules are arranged in groups with similar types of bonding and the mean deviation $\Delta H_0^\circ(\mathrm{HF}) - \Delta H_0^\circ(\mathrm{exp})$ is listed for each group.

For the closed-shell molecules of Table 7.16a the agreement between $\Delta H_0^\circ(\mathrm{HF})$ and $\Delta H_0^\circ(\mathrm{exp})$ is quite remarkable in view of the large errors in the Hartree–Fock dissociation energies, and vindicates our choice of reference states for the elements. Apart from C_2, $a\ {}^1\Sigma_g^+$ which, as we have seen above (Section 7.4(c)) hardly qualifies as a closed-shell state, the largest discrepancy is for N_2O where $\Delta H_0^\circ(\mathrm{HF}) - \Delta H_0^\circ(\mathrm{exp}) = 16.1 \pm 9$ kcal mole^{-1}. In all other cases the discrepancy is less than 10 kcal mole^{-1} and is usually less than the combined computational and experimental uncertainties. We conclude that, relative to the closed-shell hydrides as reference states for the elements, accurate Hartree–Fock calculations would yield ΔH_0° values for closed-shell molecules correct to within a few kilo-calories per bond, and that existing HFR calculations are hardly adequate to assess the errors of such calculations, even for the molecules of Table 7.16a.

The results for the open-shell diatomic states in Table 7.16b show some interesting trends. As is to be expected, most of the discrepancies are larger than for the closed-shell molecules. However, good agreement with experiment is obtained for the chemically stable species O_2, $X\ {}^3\Sigma_g^-$ and NO, $X\ {}^2\Pi$. This, together with the large positive discrepancies for O_2, $b\ {}^1\Sigma_g^+$ and the formally closed-shell state

Table 7.16. Enthalpies of formation ΔH_0° from closed-shell hydride reference states (kcal mole^{-1})

Compound	ΔH_0°(exp)	ΔH_0°(HF)	ΔH_0°(HF) $-\Delta H_0^\circ$(exp)	ΔH_0°[DZ]	ΔH_0°[DZ] $-\Delta H_0^\circ$(exp)
		a. Closed-shell molecules			
H_2	0	0	–	0	–
LiH	0	0	–	0	–
BH	0	0	–	0	–
CH_4	0	0 ± 3	–	0	–
NH_3	0	0 ± 1	–	0	–
H_2O	0	0 ± 3	–	0	–
HF	0	0	–	0	–
Li_2	–17 ± 1	–16.7 ± 1	0.3	–	–
(C_2, $x\ ^1\Sigma_g^+$	229 ± 5	185.5 ± 1	56.5)	–	–
F_2	130 ± 2	139.4 ± 2	9.4	127.6	–2.4
LiF	–49 ± 8	–42.1 ± 1	6.9	–	–
C_2H_6	15.4	20.1 ± 5	4.7	22.2	6.8
H_2O_2	83.1	90.9 ± 6	7.8	82.8	–0.3
		Mean[a]	5.2	Mean	1.3
C_2H_4	46.5	47.3 ± 5	0.8	55.5	8.2
CH_2O	46.0	45.5 ± 6	–0.5	58.7	13.2
LiCCH	–	76.2 ± 4	–	–	–
FCCH	–	126.0 ± 5	–	–	–
		Mean	0.2	Mean	10.7
N_2	18.7	13.7 ± 1	–5.0	39.7	21.0
BF	–71 ± 4	–63.3 ± 1	7.7	–	–
CO	45.9	43.3 ± 1	–2.6	63.2	17.3
C_2H_2	86.3	94.3 ± 2	8.0	101.5	15.2
HCN	57.7	60.0 ± 1	2.3	71.7	14.0
FCN	–	94.3 ± 5	–	–	–
		Mean	2.1	Mean	16.9
CO_2	36.2	39.7 ± 1	3.5	70.5	34.3
C_2N_2	124.0	132.3 ± 6	8.3	–	–
N_2O	96.0	112.3 ± 4	16.1	–	–
		b. Open-shell states			
C_2, $A\ ^3\Sigma_g^-$	246 ± 5	217.0 ± 1	–29.0	–	–
O_2, $X\ ^3\Sigma_g^-$	114.2	116.1 ± 2	1.9	–	–
O_2, $a\ ^1\Delta_g$	136.7	146.7 ± 2	10.0	–	–
O_2, $b\ ^1\Sigma_g^+$	151.7	177.3 ± 2	25.6	–	–
LiO, $X\ ^2\Pi$	39 ± 5	24.8 ± 1	–14.2	–	–
NO, $X^2\Pi$	88.1	94.6 ± 1	6.5	–	–
OF, $X\ ^2\Pi$	144 ± 10	144.2 ± 2	0.2	–	–
CF, $X\ ^2\Pi$	142 ± 5	127.7 ± 1	–14.3	–	–

[a] Omitting C_2, $x\ ^1\Sigma_g^+$ (see text).

C_2, $x\ ^1\Sigma_g^+$ which are both strongly affected by first-order CI, suggests that the *primary* criterion for the accuracy of ΔH_0°(HF) is the absence of the strong first-order CI which is characteristic of isolated atoms and reactive free radicals. The closed-shell nature of all reactants and products then becomes a *secondary* requirement which improves the invariance of the correlation energy, and hence the accuracy of ΔH_0°(HF), in cases where the first requirement is satisfied. For the ground states of most stable molecules both requirements are satisfied and their priority becomes irrelevant.

It is sometimes convenient to have estimates of Hartree–Fock energies of molecules for which adequate HFR calculations are not available. Such estimates are commonly obtained from Hartree–Fock atomic energies, experimental binding energies D_e (Table 7.13) and empirical estimates of correlation energies.(59) We may obtain much simpler and more accurate estimates by reversing the procedure leading to the ΔH_0°(HF) values of Table 7.16. That is, we use the estimates of Table 7.13 for the Hartree–Fock energies of the closed-shell hydride reference states and experimental enthalpies of formation relative to the thermodynamic standard states.(54) For any closed-shell hydrocarbon $C_m H_n$ this procedure, and the numerical values of Tables 7.13, 7.14 and 7.15 leads to the simple formula (in au)

$$E_{HF}(C_m H_n) + v_0(C_m H_n) = \Delta Hf_0^\circ - 37.9039m - 0.56185n. \tag{7.346}$$

Here v_0 is the zero-point vibrational energy and ΔHf_0° is the standard enthalpy of formation of $C_m H_n$ both expressed in atomic units.

The formula (7.346), which is readily extended to molecules containing other first row atoms, depends on the conservation of correlation energy in reactions involving only closed-shell molecules. From the ΔH_0°(HF) – ΔH_0°(exp) values of Table 7.16 it appears that Hartree–Fock total energies estimated in this way should be accurate to within 0.005 au per carbon atom.

(f) SIMPLER SCF CALCULATIONS

As we have seen, accurate Hartree–Fock total energies provide a good account of the thermochemistry of those (relatively few) chemically stable small molecules for which current SCF calculations are adequate to estimate the Hartree–Fock limit. However, we have also seen that the basis function requirements for such calculations are stringent and difficult to satisfy even for a molecule as small as NH_3. The task of reducing the uncertainties in the Hartree–Fock limits of Table 7.13 and extending the table to include larger molecules is, therefore, difficult and expensive despite the fact that computer programs of sufficient generality are available.(60) It is of interest to see whether simpler SCF calculations with smaller basis sets can produce comparable results.

For linear molecules McLean and Yoshimine(49) showed that the near-Hartree–Fock binding energies D_e from their best calculations ({BA + P} of Table 7.13) could be reproduced to within about 1 kcal mole^{-1} by the consistent use of a smaller basis {DZ + P} despite a substantial rise in all total energies. It seems likely that, if such a {DZ + P} basis were used consistently for all the molecules of Table 7.16, the enthalpies of formation obtained would be close to the Hartree–Fock values, but such calculations are not available for the non-linear molecules. The most comprehensive set of calculations using a comparable basis set for many small molecules seems to be that of Snyder,(61) who treats most of the molecules of Table 7.16 and a number of others using a CGTO basis [DZ] which is comparable in quality to a double-zeta Slater basis. Snyder's results, converted to ΔH_0° values relative to our reference states are shown in the last two columns of Table 7.16.

We see that, for the first group of molecules containing only single bonds the [DZ] results are about as good as those from the estimated Hartree–Fock limits, but that for molecules containing multiple bonds the [DZ] results are consistently too high, the error increasing with both the number of such bonds and their multiplicity.

As a result of these errors hydrogenation reactions such as

$$\begin{aligned} C_2H_4 + H_2 &= C_2H_6 \\ CH_2O + 2H_2 &= H_2O + CH_4 \end{aligned} \tag{7.347}$$

are predicted to be too exothermic by the [DZ] calculations, the average error being 7 kcal mole^{-1} for each hydrogen molecule added. This effect was noted by Snyder,(61) who attributed it principally to changes in electron correlation energy. Since the effect is absent for the ΔH_0°(HF) values of Table 7.15, it seems more likely that it arises from the inadequacy of the [DZ] basis set for multiply-bonded molecules. That is, the introduction of polarization functions would lead to greater energy lowerings for multiply bonded molecules and the errors would be much reduced. However, the small number of cases in Table 7.16 and the uncertainties in the estimated Hartree–Fock limits make this conclusion tentative.

Recent calculations(62) suggest that reliable results can be obtained using quite small CGTO bases (H: (3, 1) = [2], X: (8, 4) = [3, 2] for a first row atom X) provided that the individual basis functions are carefully chosen; the rather large errors ($\sim$0.1 au $\approx$ 60 kcal mole^{-1}) in the total energy of each atom then cancel to high accuracy. With such a small basis it is feasible to extend the calculations to a wide variety of organic molecules.(63)

(g) HARTREE-FOCK POTENTIAL CURVES AND SURFACES

We consider here the variation of the energy of a particular electronic state with the geometrical configuration of the nuclei. For a diatomic molecule this leads to the potential energy curves considered in Chapter 1. Near Hartree–Fock wave functions have been used to calculate a large number of these curves for a variety of diatomic molecules and ions. As typical examples we consider the curves for the ground state

$$N_2: 1\sigma_g^2 1\sigma_u^2 2\sigma_g^2 2\sigma_u^2 1\pi_u^4 3\sigma_g^2, X\,^1\Sigma_g^+$$

of the nitrogen molecule and the three lowest lying states

$$N_2^+: 1\sigma_g^2 1\sigma_u^2 2\sigma_g^2 2\sigma_u^2 1\pi_u^4 3\sigma_g, X\,^2\Sigma_g^+$$

$$N_2^+: 1\sigma_g^2 1\sigma_u^2 2\sigma_g^2 2\sigma_u^2 1\pi_u^3 3\sigma_g^2, A\,^2\Pi_u$$

$$N_2^+: 1\sigma_g^2 1\sigma_u^2 2\sigma_g^2 2\sigma_u 1\pi_u^4 3\sigma_g^2, B\,^2\Sigma_u^+$$

of the N_2^+ ion (cf. Fig. 7.1). Cade, Sales and Wahl[44] have carried out separate HFR calculations on these four states for a range of nuclear separations using a sufficiently large basis of STOs to ensure effective convergence to the Hartree–Fock limit. In Fig. 7.6 their results are compared with accurate potential curves derived from band spectroscopic data by the Rydberg-Klein-Rees method (Section 1.5). Note that, although the ordinate scale is the same in the two halves of Fig. 7.6 and the abscissa scale and range are identical the ordinate ranges of the two halves of the figure are different; they have been adjusted so that the calculated and observed minima of $N_2(X\,^1\Sigma_g^+)$ appear on the same level. The small numbered struts on the RKR curves represent the observed vibrational levels. Hartree–Fock and observed energies of the dissociation limits N (^4S) + N (^4S), for $N_2(X\,^1\Sigma_g^+)$ and $N(^4S) + N^+(^3P)$, for $N_2^+(X\,^2\Sigma_g^+, A\,^2\Pi_u, B\,^2\Sigma_u^+)$ are also shown.

In the previous section we found two necessary conditions for the near invariance of the correlation energy.

(i) Absence of states of the same symmetry and similar energy arising from different configurations; that is absence of strong first-order CI.

(ii) Constancy of the number of electron pairs in the LMO description and, as far as possible, of their local spatial relationship to each other.

For our N_2 and N_2^+ states and indeed for most states of diatomic and polyatomic molecules we expect that condition (ii) will be quite well satisfied for moderate variations of nuclear separation

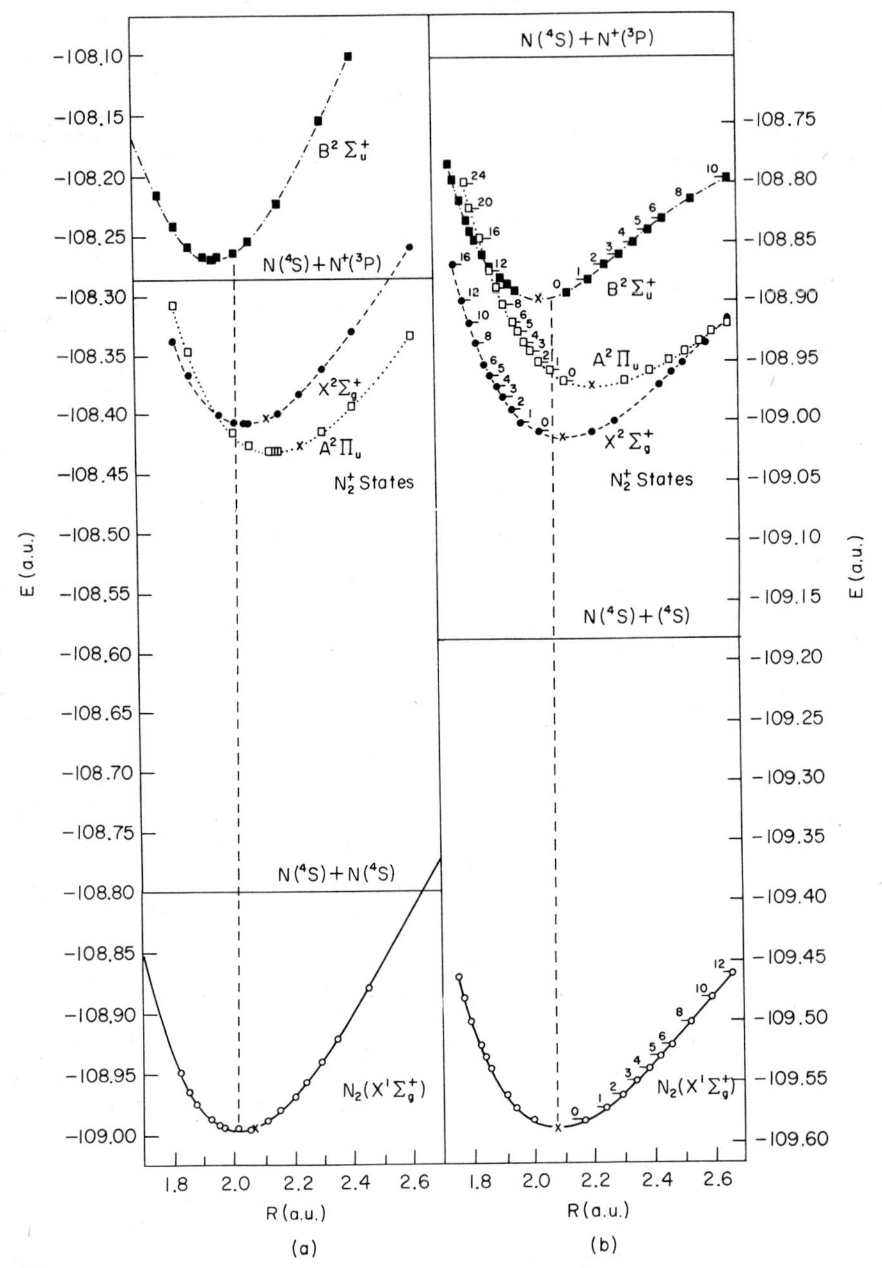

Fig. 7.6. Potential curves, $E(R)$, for $N_2(X\ ^1\Sigma_g^+)$ and $N_2^+(X\ ^2\Sigma_g^+, A\ ^2\Pi_u, B\ ^2\Sigma_u^+)$. (a) Calculated $E_{HF}(R)$ results. (b) Empirical results $E_{RKR}(R)$ derived from band spectroscopic data. Note: the ordinate scales for (a) and (b) are the same but the ranges are different. Calculated and experimental dissociation limits are also shown.

or geometry about equilibrium. From the correlation diagram, Fig. 7.1, we expect that condition (i) will also be well satisfied by our N_2 and N_2^+ states for nuclear separations ranging from well below to slightly above the equilibrium value. For separations appreciably above the equilibrium value, however, the incompletely filled MOs from the $2p$ level of the separated atoms (namely, $3\sigma_g$, $1\pi_u$, $1\pi_g$ and $3\sigma_u$) all start converging to the same energy level. We then have a number of near-degenerate configurations giving rise to a large number of states, many of which will have the same symmetries as the states of Fig. 7.6. There is, therefore, strong first-order CI and a rapid rise in the magnitude of the correlation energy. At still larger nuclear separations the single configuration restricted Hartree–Fock approximation breaks down entirely, just as for the H_2 molecule (Chapter 5), and the Hartree–Fock curve tends to quite the wrong dissociation limit.

These expectations are well borne out by the curves of Fig. 7.6. From the smallest R values up to just beyond the minimum the Hartree–Fock curves parallel the RKR curves quite closely, but for larger R values all four Hartree–Fock curves start rising much too rapidly. At $R = 2.7$ au the Hartree–Fock curve for N_2, $X\ {}^1\Sigma_g^+$, has already risen above the Hartree–Fock energy of the dissociation products $N(^4S) + N(^4S)$.

These deficiencies in the Hartree–Fock potential curves have the expected effect on the spectroscopic constants of the various states, if these constants are determined by a Dunham analysis (Section 1.4). The upward distortion of the curve for large R leads, in general, to slightly too small a value for R_e, too large a value for ω_e and quite unreliable values for the higher spectroscopic constants (except for hydrides). These trends are illustrated in Table 7.17 which shows results for a variety of diatomic states. We note that changes in R_e and ω_e values on excitation or ionization, which reflect the bonding character of the MOs, are quite accurately reproduced.

For closed-shell polyatomic molecules, too, we expect that the correlation energy should be little changed by small displacements of the nuclei from their equilibrium positions or by certain larger changes, such as the inversion of NH_3, the internal rotation of ethane and the change of a cyclohexane ring from chair to boat form, which preserve the LMO pairs and their local spatial relationships. This implies that Hartree–Fock calculations should yield quite accurate molecular geometries, vibration frequencies, and energy barriers to internal rotation. For those few non-linear polyatomic molecules where the Hartree–Fock limit has been approached to near chemical accuracy (Table 7.13) these hopes have been realized.

In some cases, indeed, quite simple MO wavefunctions of LCAO form have given qualitatively correct results. Thus the first *ab initio*

Table 7.17. Hartree–Fock spectroscopic constants for some diatomic states[a]

Molecular state	R_e (Å)		ω_e		$\omega_e x_e$		α_e	
	Calc.	Expt	Calc.	Expt	Calc.	Expt	Calc.	Expt
LiH($X\ ^1\Sigma^+$)	1.605	1.595	1433	1405.6	23.26	23.20	0.1945	0.213
CH($X^2\Pi$)	1.104	1.124	3053	2868.5	55.50	64.4	0.4712	0.530
OH($X^2\Pi$)	0.950		4062	3735.2	74.54	82.81	0.6501	0.714
FH($X\ ^1\Sigma^+$)	0.897	0.917	4469	4139.0	80.34	90.44	0.7693	0.797
AlH($X\ ^1\Sigma^+$)	1.647	1.648	1741	1682.6	26.86	29.09	0.1719	0.186
HCl($X\ ^1\Sigma^+$)	1.264	1.275	3181	2789.7	47.76	52.05	0.2682	0.302
Li_2 ($X\ ^1\Sigma_g^+$)	2.78	2.67	326	351	—	—	—	—
N_2 ($X\ ^1\Sigma_g^+$)	1.065	1.097	2729.6	2358.1	8.378	14.188	0.0135	0.0178
N_2^+($X\ ^2\Sigma_g^+$)	1.080	1.118	2570.5	2207.1	9.809	16.14	0.0148	0.020
($A\ ^2\Pi_u$)	1.129	1.176	2312.5	1902.8	6.082	14.91	0.0155	0.018
($B\ ^2\Sigma_u^+$)	1.024	1.075	3101.8	2419.8	19.88	23.19	0.0128	0.020
F_2 ($X\ ^1\Sigma_g^+$)	1.33	1.409	1257	923.1	9.85	16.04	0.0108	0.022
BeO($X\ ^1\Sigma_g^+$)	1.300	1.331	1732	1487	—	11.8	—	0.019
($a\ ^2\Pi$)	1.436	—	1372	—	—	—	—	—
($A\ ^1\Pi$)	1.441	1.463	1399	1144	—	8.4	—	0.016
($b\ ^3\Sigma^+$)	1.403	—	1409	—	—	—	—	—
($B\ ^1\Sigma^+$)	1.356	1.362	1554	1371	—	7.7	—	0.015

[a] Refs.47, 48.

calculation on ethane using a single-zeta Slater basis and assuming rigid rotation of the two CH_3 groups gave a barrier to internal rotation of 3.3 kcal mole^{-1} in fair agreement with the experimental barrier of 2.928 kcal mole.$^{-1}$ This calculation has since been improved in two respects; a large Gaussian basis has been used to approach the Hartree–Fock limit to within near chemical accuracy (Table 7.13) and the CC bond length and HCH bond angle have been varied to minimize the energy of both eclipsed and staggered conformations. The geometry and rotational barriers predicted by this calculation are shown in Table 7.18.

We see that both improvements have a small but appreciable effect on the calculated barrier and that the final results, both for the height of the barrier and for the staggered geometry are very satisfactory.

The qualitative success of the much simpler, single-zeta, STO calculation is, perhaps, surprising in view of the fact that it leads to a total energy some 0.7 au ($\approx$ 400 kcal mole^{-1}) above the Hartree–Fock limit. Since this is over two orders of magnitude larger than the barrier itself, the success of the single-zeta calculation implies almost

Table 7.18. Geometry and rotational barriers predicted by near Hartree–Fock wave functions for ethane (C_2H_6)

	C–C (Å)	HĈH (degrees)
Experimental staggered geometry	1.543	109.3
Calculated staggered geometry	1.551	107.31
Calculated eclipsed geometry	1.570	106.97

Calculated barrier to internal rotation (kcal mole^{-1})	
(i) Assuming experimental staggered geometry	3.654
(ii) Optimizing C–C	3.468
(iii) Optimizing C–C and HĈH	3.071
Experimental barrier	2.928

complete cancellation of the energy improvements for the two conformations as the basis set is extended to approach the Hartree–Fock limit. An analysis of the various terms contributing to the barrier leads us to expect this very favourable cancellation for similar situations with a high degree of symmetry and, indeed, good results for rotational barriers in CH_3NH_2, N_2H_4, CH_3OH and NH_2OH have been obtained using Gaussian wave functions of approximately {DZ} quality. By contrast the less symmetrical double barrier in hydrogen peroxide (H_2O_2) and the inversion of NH_3 require near Hartree–Fock wave functions and careful geometrical optimization for any adequate treatment.(45, 47, 48)

There are also important regions of potential energy curves and surfaces where the restricted Hartree–Fock theory based on a single configuration breaks down completely. The dissociative behaviour of nearly all diatomic states falls in this category and so does the potential surface for any chemical reaction involving the breaking or formation of a covalent bond. In these situations the predictions of single-determinant MO–SCF calculations are qualitatively incorrect even at the Hartree–Fock limit, and this constitutes, perhaps, the most serious limitation of this theory. The extensions of the theory which are needed to overcome these difficulties are considered by Hurley.(69)

(h) IONIZATION AND EXCITATION ENERGIES

In assessing Hartree–Fock ionization potentials we should distinguish between an adiabatic ionization potential, I_a, and a vertical value, I_v. The former corresponds to the energy difference between the zeroth vibrational levels of the neutral and ionized states whereas the latter gives the energy separation between the potential energy curves (or surfaces) at the equilibrium nuclear configuration of the neutral molecule (cf. Fig. 7.6). The difference between I_a and I_v is small (~0.1 eV) unless there is a substantial change in equilibrium geometry following ionization. Experimentally the most accurately determined quantity is usually I_a, but I_v is simpler to calculate. According to Koopmans' theorem (Section 7.2(a), (b)) the orbital energies for a closed-shell ground state provide fairly good estimates of vertical ionization potentials. Better estimates of I_v may be obtained from separate Hartree–Fock calculations on the neutral and ionized species as in Fig. 7.6. Any theoretical estimate of an adiabatic ionization potential necessarily involves two separate molecular calculations for different nuclear configurations and, for high accuracy, an estimate of the zero-point vibrational energies.

Table 7.19 compares the magnitudes ($-\epsilon_\mu$) of Hartree–Fock orbital energies with experimental ionization potentials both vertical and adiabatic. In some cases I_v and I_a values obtained from the appropriate differences between calculated Hartree–Fock total energies are also shown. In these calculations vibrational contributions have been ignored as have any differences between calculated and experimental geometry. Uncertainties from these two sources (~0.1 eV) are much less than the errors caused by changes in correlation energy on ionization.

The calculated values of Table 7.19 show that each of the steps $-\epsilon_\mu \rightarrow I_v$ (electronic relaxation) and $I_v \rightarrow I_a$ (nuclear relaxation) leads to a lowering in the ionization potential, the total decrease being as much as 2 eV in some cases. The directly calculated values are usually in better agreement with experiment than are values of $-\epsilon_\mu$, although in a few cases, notably N_2^+, $A\,^2\Pi_u$ and BH^+, $X\,^2\Sigma^+$ there is a fortuitous cancellation between the relaxation energy and the change in electronic correlation energy.

For polyatomic molecules separate Hartree–Fock calculations on the ions are not usually available; the orbital energies then provide the only estimate of ionization potentials. From Table 7.19 we see that these estimates are typically 1–2 eV too high, reflecting the decrease in the magnitude of the correlation energy on ionization. The order and general pattern of the ionic states is usually correct and such calculations (and simpler SCF–MO calculations) have proved useful in assigning and interpreting the many accurate ionization potentials that have recently been obtained from molecular photoelectron spectroscopy. However, errors in the ordering of the ionic states (or neutral molecule states) do sometimes occur as for N_2^+ where the orbital energies predict the $A\,^2\Pi_u$ state to lie below the experimental ground state $X\,^2\Sigma_g^+$. This error persists when allowance is made for electronic relaxation (I_v) and nuclear relaxation (I_a); it arises from differences in the electronic correlation energies of the two states.

Table 7.19. Hartree–Fock and experimental ionization potentials (eV)[a]

Molecular state	Orbital ionized (μ)	$-\epsilon_\mu \approx I_v$	Difference of HF energies I_v	I_a	Experiment I_v	I_a
$N_2, X\ ^1\Sigma_g^+$	$3\sigma_g$	17.36	16.01	15.99	15.60	15.57
	$1\pi_u$	17.10	15.67	15.34	16.98	16.69
	$2\sigma_u$	20.92	19.93	19.74	18.78	18.75
$H_2, X\ ^1\Sigma_g^+$	$1\sigma_g$	16.18			15.88	15.45
$CO, X\ ^1\Sigma^+$	5σ	15.08		13.51	14.01	14.01
	1π	17.40		15.17	16.91	16.53
	4σ	21.87			19.72	19.68
$F_2, X\ ^1\Sigma_g^+$	$1\pi_g$	17.71	16.27		15.83	15.70
	$3\sigma_g$	21.29	19.60			17.35
	$1\pi_u$	22.54	21.19		18.80	18.39
	$2\sigma_u$	40.17	38.79			
$LiH, X\ ^1\Sigma^+$	2σ	8.21	7.02		6.5 ± 0.5	
$BH, X\ ^1\Sigma^+$	3σ	9.48		8.45		9.73
$CH, X\ ^2\Pi$	1π	(11.29)[b]		10.08		10.64
$NH, X\ ^3\Sigma^-$	1π	(14.63)[b]		12.82		13.10
$OH, X\ ^2\Pi$	1π	(15.57)[b]		11.44		13.36
$HF, X\ ^1\Sigma^+$	1π	17.69	14.54		15.8 ± 0.2	
$CO_2, X\ ^1\Sigma_g^+$	$1\pi_g$	14.85			13.78	13.78
	$1\pi_u$	19.77			17.6	17.32
	$3\sigma_u$	20.34			18.1	18.08
	$4\sigma_g$	21.68			19.4	19.40
$H_2O, X\ ^1A_1$	$1b_1$	13.79			12.6	12.61
	$3a_1$	15.84			14.8	13.7
	$1b_2$	19.56			18.3	17.22
$NH_3, X\ ^1A_1$	$3a_1$	11.64			10.85	10.16
	$1e$	17.28			15.8	14.8
$H_2CO, X\ ^1A_1$	$2b_2$	11.98			10.9	10.88
	$1b_1$	14.53			14.38	14.09
	$3a_1$	17.70			16.0	15.85

[a] Data from Refs 47, 48.
[b] Koopmans' theorem is not strictly valid for these open-shell ionizations.

(i) MOLECULAR PROPERTIES AND CHARGE DENSITY FUNCTION

Once the electronic wavefunction for a molecule has been determined it is a relatively simple task to compute a number of useful molecular properties. Many important properties including dipole, quadrupole and higher moments of the charge distribution, diamagnetic susceptibilities and shielding factors, electric fields and field gradients at the nuclei are obtained as expectation values of one-electron operators.[47, 48] For a single determinant wave function constructed from orthonormal spin orbitals each of these expectation values appears as a sum of contributions from the individual spin orbitals (equation (6.40) with $l = k$, $\Delta_i = 1$).

Furthermore from Brillouin's theorem and its corollaries (Section 7.2(a), (b)) we expect quite accurate values for these one-electron operators if near HF wavefunctions are employed. This has been found in practice.[47, 48] Tables 7.20 and 7.21 give a representative sample of results for dipole moments, electric field gradients, and forces on the nuclei from the electrostatic formula (Section 2.3). For comparison, values from simple LCAO SCF–MO calculations (that is, {SZ} calculations) are also given. We see that these values differ widely from those obtained with near HF wavefunctions; at best they have a qualitative significance.

Table 7.20. Computed and experimental dipole moments (Debyes)[a]

Molecule AB	SCF–MO calculations: Minimal basis {SZ}[b]	SCF–MO calculations: Near HF {BA + P}′	Near HF +CI[c]	Experiment A^+B^-
LiH	6.48	6.00	5.86	5.83
BH	−1.03	−1.73	—	—
HF	0.88	1.94	—	—
CO	−0.59	0.27	−0.12	−0.112 (±0.005)
CS	—	−1.56	−2.03	−1.97

[a] Refs 47, 48, 1 au = 2.54158 Debye.
[b] With "best atom" exponents.
[c] Ref. 64.

From Tables 7.20 and 7.21 we see that the best available Hartree–Fock results still differ significantly from experiment. How much of this error is attributable to inadequacies in the basis sets used in the HFR calculations (orbital error) and how much to deficiencies in the Hartree–Fock approximation itself (correlation error)? For a general one-electron operator this is a difficult question but in two particular cases we can give a definite answer.

For the dipole moment it has been shown[64] that the errors in the near HF values of Table 7.20 may be reduced by an order of magnitude using a wave function of the form

$$\Psi = \Psi_0 + \sum_i \sum_a \Psi_i^a C_i^a + \sum_{i<j} \sum_{a<b} \Psi_{ij}^{ab} C_{ij}^{ab}$$

where Ψ_0 is the near HF determinant and $\Psi_i^a(\Psi_{ij}^{ab})$ are determinants obtained by exciting one (two) spin orbitals from Ψ_0 (cf. equation (7.105)). The techniques employed in these configuration interaction (CI) calculations are discussed by Hurley.[(69)] The results are shown in the fourth column of Table 7.20. We see that some 90% of the error in the near-HF dipole moments is attributable to correlation effects leaving at most 10% for orbital error.

The situation is quite the reverse for the forces on the nuclei shown in the last four lines of Table 7.21. Accurate Hartree–Fock functions, both restricted and unrestricted, clearly satisfy the conditions established in Section 2.4, for the validity of the electrostatic theorem. In each case the accurate HF function is the best (lowest energy) function in a well-defined class of functions and the specification of this class (e.g. as a single Slater determinant) is quite independent of the value of any parameter in the Hamiltonian. Consequently for an accurate Hartree–Fock function at the calculated equilibrium geometry the electrostatic force on each nucleus must vanish; furthermore for *any* nuclear geometry the vector sum of the forces on all the nuclei must vanish. The quite large deviations from this behaviour in Table 7.21 must be ascribed entirely to orbital error. In particular we see that:

(i) the forces calculated using simple LCAO functions (that is, {SZ} functions) are quite ridiculous, despite the fact that such functions predict quite reasonable geometries in a conventional calculation;[(62, 63)]

(ii) for CO there are quite large differences between forces calculated using the large contracted Gaussian basis [5, 3, 2] and the much more accurate Slater basis $\{BA + P\}'$; neither basis succeeds in predicting the correct zero results, but the basis $\{BA + P\}'$ does best. The good agreement between the two bases for the other properties of CO gives considerable support to the values obtained for H_2CO with the corresponding Gaussian basis [5, 3, 2 | 2, 1].

We may draw two conclusions from this discussion. Firstly, although the electrostatic point of view based on the Hellmann–Feynman theorem has led to valuable qualitative insights into the nature of the chemical bond and the charge shifts associated with bond formation,[(11, 66, 67)] it is very difficult to obtain quantitative results; even for diatomic molecules the best available wavefunctions are hardly adequate for this purpose.

Secondly we conclude that even for a diatomic molecule such as CO the best available HFR calculation is rather far from an accurate solution of the Fock equations. Although the HFR total energy has probably converged to the Hartree–Fock limit to within chemical

Table 7.21. Near Hartree–Fock ground state properties of CO and H_2CO (au)[a]

		CO				H_2CO		
	Basis set	C, O{2, 1} {SZ}	C, O(10, 5, 2) [5, 3, 2]	{BA + P}′ ≈ HF		C, O{2, 1}H{1} {SZ}	C, O(10, 5, 2)H(4, 1) [5, 3, 2∣2, 1]	
$E - E_{HF}$		0.400	0.029	0.003		0.482	0.040	
Property[b]					*Expt*			*Expt*
Dipole moment	μ_z	−0.183	0.096	0.108	−0.044	0.396	1.110	0.917
Electric[c]	$q_{zz}(C)$	−1.02	−1.135	−1.180			−0.347	
field	$q_{yy}(C)$	0.51	0.567	0.590			−0.301	
gradients	$q_{zz}(O)$	−1.16	−0.697	−0.679	−0.794	0.837	0.404	0.335
at nuclei	$q_{yy}(O)$	0.58	0.349	0.340	0.397	−2.929	−2.270	−2.194
Forces on	$F_z(C)$	−1.16	−0.153	0.146	(0.0)	−0.251	0.012	(0.0)
nuclei from				0.071[d]	0.0			
electrostatic	$F_z(O)$	2.69	0.361	−0.138	(0.0)	2.648	0.416	(0.0)
formula				−0.088[d]	0.0			
	$F_z(H)$	—	—	—		−0.063	0.005	(0.0)
	$F_z(tot)$	1.53	0.208	0.008	0.0	2.334	0.438	0.0
				−0.016[d]	0.0			

[a] Ref. 65.
[b] z axis from C to O, y axis in molecular plane.
[c] $q_{xx} + q_{yy} + q_{zz} = 0$; for CO, $q_{xx} = q_{yy}$.
[d] At the calculated equilibrium geometry. All other values at the experimental equilibrium geometry.

accuracy (Table 7.13), the HFR orbital functions must show appreciable deviations from accurate Hartree–Fock orbitals; these errors are much larger than for atomic Hartree–Fock functions.

References

1. F. Hund (1927), *Z. Physik* **40**, 742.
2. R. S. Mulliken (1928), *Phys. Rev.* **32**, 186.
3. J. E. Lennard-Jones (1929), *Trans. Faraday Soc.* **25**, 668.
4. R. S. Mulliken (1930), *Rev. Mod. Phys.* **2**, 60, 506; ibid. (1931) **3**, 90; ibid. (1932) **4**, 1.
5. G. Herzberg, "Molecular Spectra and Molecular Structure. Vol. I. Spectra of Diatomic Molecules (1950). Vol. II. Infrared and Raman Spectra of Polyatomic Molecules (1945). Vol. III. Electronic Spectra and Electronic Structure of Polyatomic Molecules (1966)", D. Van Nostrand, New York.
6. J. C. Slater (1960), "Quantum Theory of Atomic Structure", Vols I and II. McGraw-Hill, New York.
7. R. K. Nesbet (1958), *Annals of Physics* **3**, 397.
8. M. Kotani (1937), *Proc. Phys. Math. Soc. Japan* **19**, 460; see also J. S. Lomont (1959). "Applications of finite groups", Academic Press, London and New York.
9. A. D. Walsh (1953), *J. Chem. Soc.* p. 2260.
10. S. D. Peyerimhoff, R. J. Buenker and L. C. Allen (1966), *J. Chem. Phys.* **45**, 734.
11. A. C. Hurley (1964), *In* "Molecular Orbitals in Chemistry, Physics and Biology, a tribute to R. S. Mulliken" (Eds P.-O. Löwdin and B. Pullman) p. 161, Academic Press, London and New York; A. C. Hurley and V. W. Maslen (1961), *J. Chem. Phys.* **34**, 1919.
12. L. Brillouin (1933), *Actualités sci. et ind.* **71**; ibid. (1934) **159**.
13. T. A. Koopmans (1933), *Physica* **1**, 104.
14. P. O. Löwdin (1955), *Phys. Rev.* **97**, 1509.
15. V. Fock (1930), *Z. Physik* **61**, 126.
16. D. R. Hartree (1957), "The Calculation of Atomic Structures", Wiley Interscience, New York.
17. R. K. Nesbet (1955), *Proc. Roy. Soc.* **A230**, 312.
18. C. C. J. Roothaan (1960), *Rev. Mod. Phys.* **32**, 179.
19. G. G. Hall (1951), *Proc. Roy. Soc.* **A205**, 541.
20. C. C. J. Roothaan (1951), *Rev. Mod. Phys.* **23**, 69.
21. C. C. J. Roothaan and P. S. Bagus (1963), *Methods in Computational Physics* **2**, 47.
22. D. J. Thouless (1961), "The Quantum Mechanics of Many-Body Systems", Academic Press, New York and London.
23. H. P. Kelly (1963), *Phys. Rev.* **131**, 684.
24. C. Froese (1963), *Can. J. Phys.* **41**, 1895.
25. R. Lefebvre (1957), *J. Chim. phys.* **54**, 168.
26. W. J. Hunt, W. A. Goddard and T. H. Dunning (1969), *Chem. Phys. Letts.* **3**, 606.
27. W. J. Hunt, W. A. Goddard and T. H. Dunning (1970), *Chem. Phys. Letts.* **6**, 147.

28. S. F. Boys and P. Rajagopal (1965), *Advan. Quantum Chem.* **2**, 1.
29. C. A. Coulson (1937), *Trans. Faraday Soc.* **33**, 388.
30. J. E. Lennard-Jones (1949), *Proc. Roy. Soc.* **A198**, 14.
31. C. Edmiston and K. Ruedenberg (1963), *Rev. Mod. Phys.* **35**, 457.
32. W. England, L. S. Salmon and K. Ruedenberg (1971), Fortschritte der Chemischen Forschung, Topics in Current Chemistry, Band 23, 31.
33. G. G. Hall (1950), *Proc. Roy. Soc.* **A202**, 336.
34. W. England and K. Ruedenberg (1973), *J. Amer. Chem. Soc.* **95**, 8769.
35. J. M. Foster and S. F. Boys (1960), *Rev. Mod. Phys.* **32**, 303.
36. T. L. Gilbert (1964), *In* "Molecular Orbitals in Chemistry, Physics and Biology, a Tribute to R. S. Mulliken", (Eds P.-O. Löwdin and B. Pullman) p. 405, Academic Press, London and New York.
37. J. C. Slater (1930), *Phys. Rev.* **36**, 57.
38. Quantum Chemistry Program Exchange, Catalog and Procedures, Vol. X (1974), Chemistry Department, Indiana University.
39. S. F. Boys (1950), *Proc. Roy. Soc.* **A200**, 542.
40. J. L. Whitten (1966), *J. Chem. Phys.* **44**, 359.
41. A. A. Frost (1967), *J. Chem. Phys.* **47**, 3707.
42. S. Huzinaga, D. McWilliams and B. Domsky (1971), *J. Chem. Phys.* **54**, 2283.
43. A. Veillard and E. Clementi (1968), *J. Chem. Phys.* **49**, 2415.
44. P. E. Cade, K. D. Sales and A. C. Wahl (1966), *J. Chem. Phys.* **44**, 1973.
45. A. Rauk, L. C. Allen and E. Clementi (1970), *J. Chem. Phys.* **52**, 4133.
46. P. Rajagopal (1965), *Z. Naturforschung,* **A20**, 1557.
47. M. Krauss (1967), *Nat. Bur. Stand. (U.S.) Tech. Note* **438**.
48. W. G. Richards, T. E. H. Walker and R. K. Hinkley (1971), "A Bibliography of *ab initio* Molecular Wave Functions", Oxford University Press (Clarendon), London and New York.
49. A. D. McLean and M. Yoshimine (1967), *Int. J. Quantum Chem.* **1S**, 313; *IBM J. Res. Develop.* **11**, suppl.
50. S. Rothenberg and H. F. Schaefer (1971), *J. Chem. Phys.* **54**, 2764.
51. P. E. Cade and W. M. Huo (1967), *J. Chem. Phys.* **44**, 1973.
52. R. L. Matcha (1968), *J. Chem. Phys.* **48**, 335.
53. A. G. Gaydon (1968), "Dissociation Energies", Chapman and Hall, London.
54. G. N. Lewis and M. Randall (1961), "Thermodynamics", McGraw-Hill, New York.
55. D. D. Wagman, W. H. Evans, V. B. Parker, I. Harlow, S. M. Bailey and R. H. Schumm (1968), *Nat. Bur. Stand. (U.S.), Tech. Note.* **270–3**.
56. P. A. G. O'Hare and A. C. Wahl (1970), *J. Chem. Phys.* **57**, 2469; ibid. (1971) **55**, 666.
57. L. C. Snyder (1967), *J. Chem. Phys.* **46**, 3602; (1969) *J. Amer. Chem. Soc.* **91**, 2189.
58. A. C. Hurley (1973), *Advan. Quantum Chem.* **7**, 315.
59. C. Hollister and O. Sinanoglu (1966), *J. Amer. Chem. Soc.* **88**, 13.
60. E. Clementi (1969), *Int. J. Quantum Chem.* **3S**, 179.
61. L. C. Snyder (1969), *J. Amer. Chem. Soc.* **91**, 2189.
62. W. J. Hehre, R. Ditchfield and J. A. Pople (1972), *J. Chem. Phys.* **56**, 2257.
63. L. Radom, W. J. Hehre and J. A. Pople (1971), *J. Amer. Chem. Soc.* 93, 289.
64. S. Green (1971), *J. Chem. Phys.* **54**, 827; ibid. (1971) **54**, 3051.
65. D. B. Neumann and J. W. Moskowitz (1969), *J. Chem. Phys.* **50**, 2216.
66. T. Berlin (1951), *J. Chem. Phys.* **19**, 208.

67. R. F. W. Bader, W. H. Henneker and P. E. Cade (1967), *J. Chem. Phys.* **46**, 3341.
68. A. Veillard (1969), *Chem. Phys. Letts.* **3**, 128.
69. A. C. Hurley (1976), "Electron Correlation in Small Molecules", Academic Press, London and New York.

APPENDIX 1

Orthonormalization

Given a basic set of functions,

$$\psi_i \qquad (i = 1 \ldots n), \tag{A1.1}$$

which may be orbitals, spin-orbitals or Slater determinants, there are various procedures for constructing an orthonormal set which spans the same function space as the set $\{\psi_i\}$.

Initially, we assume that the functions (A1.1) are linearly independent. A necessary and sufficient condition for this is that the determinant constructed from the overlap integrals

$$\Delta_{ij} = (\psi_i \mid \psi_j) \tag{A1.2}$$

does not vanish. That is

$$\det(\Delta_{ij}) \neq 0. \tag{A1.3}$$

In this case any of the possible orthonormal sets will contain the same number (n) of functions as the original set (A1.1).

A1.1. The Schmidt process

Here the members of the orthonormal set

$$\{\psi_j^0\} \qquad (j = 1 \ldots n) \tag{A1.4}$$

are constructed one at a time. At each stage the components of ψ_j parallel to earlier members ($i < j$) of the set (A1.4) are subtracted out; the resulting function is then normalized. In this way we obtain the expressions

$$\psi_1^0 = \frac{\psi_1}{(\psi_1 \mid \psi_1)^{\frac{1}{2}}} \tag{A1.5}$$

$$\psi_2^0 = \frac{\psi_2 - \psi_1^0(\psi_1^0 \mid \psi_2)}{\{(\psi_2 \mid \psi_2) - (\psi_2 \mid \psi_1^0)(\psi_1^0 \mid \psi_2)\}^{\frac{1}{2}}} \tag{A1.6}$$

$$\ldots \tag{A1.7}$$

$$\psi_j^0 = \frac{\psi_j - \sum_{i=1}^{j-1} \psi_i^0(\psi_i^0 \mid \psi_j)}{\{(\psi_j \mid \psi_j) - \sum_{i=1}^{j-1} (\psi_j \mid \psi_i^0)(\psi_i^0 \mid \psi_j)\}^{\frac{1}{2}}} \tag{A1.8}$$

$$\cdots$$

Successive substitution of these equations ((A1.5) into (A1.6), (A1.6) into (A1.7), . . .) leads to explicit formulae

$$\psi_j^0 = \sum_{i=1}^{n} \psi_i r_{ij} \tag{A1.9}$$

for the members of the orthonormal set (A1.4) in terms of the original functions (A1.1), the coefficients r_{ij} being determined by the overlap integrals (A1.2). The transformation (A1.9) and its inverse may be expressed in the matrix forms

$$\boldsymbol{\psi}^0 = \boldsymbol{\psi}\mathbf{r} \tag{A1.10}$$

and

$$\boldsymbol{\psi} = \boldsymbol{\psi}^0\mathbf{t} \tag{A1.11}$$

with

$$\mathbf{t} = \mathbf{r}^{-1}. \tag{A1.12}$$

From the structure of equations (A1.5)-(A1.8) it is clear that the matrices r and t are both triangular

$$r_{ij} = t_{ij} = 0 \qquad (i > j). \tag{A1.13}$$

This property of the Schmidt transformation is very convenient when we are concerned with a basis of Slater determinants, or spin-eigenfunctions, each of which contains a closed-shell core constructed from, say c, of the spin orbitals (A1.1). In such a case we index the spin orbitals so that these core orbitals ($\psi_1, \ldots, \psi_c$) appear first and are succeeded by the valence spin orbitals ($\psi_{c+1}, \ldots, \psi_n$) which show variable occupancy in the Slater determinants.

The relations (A1.13) then ensure that, when the spin orbitals are orthonormalized by the Schmidt process, the closed-shell structure remains intact in the new basis of orthonormal Slater determinants. This may lead to great simplifications; thus, for an N-electron system we have $\binom{n}{N}$ possible Slater determinants of which $\binom{n-c}{N-c}$ preserve the closed-shell structure. If the Schmidt process is employed the orthogonalized basis contains only $\binom{n-c}{N-c}$ determinants, whereas, if some other scheme is used all $\binom{n}{N}$ determinants may appear (cf. Section 6.3).

A1.2. Symmetric orthogonalization[1]

Here a basis

$$\{\phi_i^0\} \qquad (i = 1 \ldots n) \tag{A1.14}$$

is defined by the matrix equation

$$\boldsymbol{\phi}^0 = \boldsymbol{\psi}\boldsymbol{\Delta}^{-\frac{1}{2}} \tag{A1.15}$$

where $\boldsymbol{\Delta}$ is the matrix of overlap integrals (A1.2).

Since $\boldsymbol{\Delta}$, and hence $\boldsymbol{\Delta}^{-\frac{1}{2}}$, are Hermitian it is readily verified that the basis (A1.14) is orthonormal. Thus we have

$$\begin{aligned}(\phi_j^0 \mid \phi_k^0) &= \left(\sum_i \psi_i(\boldsymbol{\Delta}^{-\frac{1}{2}})_{ij} \mid \sum_l \psi_l(\boldsymbol{\Delta}^{-\frac{1}{2}})_{lk}\right) \\ &= \sum_i \sum_l (\boldsymbol{\Delta}^{-\frac{1}{2}})^*_{ij} \Delta_{il} (\boldsymbol{\Delta}^{-\frac{1}{2}})_{lk} \\ &= \sum_i \sum_l (\boldsymbol{\Delta}^{-\frac{1}{2}})_{ji} (\boldsymbol{\Delta})_{il} (\boldsymbol{\Delta}^{-\frac{1}{2}})_{lk} \\ &= (\boldsymbol{\Delta}^{-\frac{1}{2}}\, \boldsymbol{\Delta}\boldsymbol{\Delta}^{-\frac{1}{2}})_{jk} \\ &= \delta_{jk},\end{aligned} \tag{A1.16}$$

since the matrices $\boldsymbol{\Delta}$, $\boldsymbol{\Delta}^{-\frac{1}{2}}$ commute.

Several techniques have been used for evaluating the matrix $\boldsymbol{\Delta}^{-\frac{1}{2}}$ of equation (A1.15). If the functions of the original basis (A1.1) are normalized

$$\Delta_{ii} = (\psi_i \mid \psi_i) = 1 \tag{A1.17}$$

and weakly overlapping

$$\Delta_{ij} = (\psi_i \mid \psi_j) \equiv S_{ij}, \qquad |S_{ij}| \ll 1 \qquad (i \neq j) \tag{A1.18}$$

then we have

$$\boldsymbol{\Delta} = \mathbf{1} + \mathbf{S} \tag{A1.19}$$

where $\mathbf{1}$ is the $(n \times n)$ unit matrix, $S_{ii} = 0$ $(i = 1 \ldots n)$ and all off-diagonal elements of the matrix $\mathbf{S}$ are small. Under these circumstances the binomial expansion

$$\boldsymbol{\Delta}^{-\frac{1}{2}} = (\mathbf{1} + \mathbf{S})^{-\frac{1}{2}} = \mathbf{1} - \tfrac{1}{2}\mathbf{S} + \frac{(-\frac{1}{2})(-3/2)}{2!}\mathbf{S}^2 + \cdots$$

$$= \mathbf{1} - \tfrac{1}{2}\mathbf{S} + \tfrac{3}{8}\mathbf{S}^2 - \tfrac{5}{16}\mathbf{S}^3 + \cdots \quad \text{(A1.20)}$$

is rapidly convergent and the first few terms provide an efficient technique for calculating $\boldsymbol{\Delta}^{-\frac{1}{2}}$. This approach was used by Löwdin[1] in a calculation of the cohesive energies of alkali halide crystals.

In the general case of large overlaps we may transform the Hermitian matrix $\boldsymbol{\Delta}$ to diagonal form by a unitary transformation

$$\mathbf{U}^{\dagger}\boldsymbol{\Delta}\mathbf{U} = \boldsymbol{\mu} = \text{diag}\,(\mu_1, \mu_2, \ldots, \mu_n). \quad \text{(A1.21)}$$

The condition (A1.3) implies that $\boldsymbol{\Delta}$ is positive definite, so that all the diagonal elements μ_i are real and positive; we may choose $\mathbf{U}$ so that they occur in decreasing order

$$\mu_1 \geqslant \mu_2 \geqslant \ldots \geqslant \mu_n > 0. \quad \text{(A1.22)}$$

The orthonormalizing matrix $\boldsymbol{\Delta}^{-\frac{1}{2}}$ of equation (A1.15) is now defined by

$$\mathbf{U}^{\dagger}\,\boldsymbol{\Delta}^{-\frac{1}{2}}\mathbf{U} = \boldsymbol{\mu}^{-\frac{1}{2}} = \text{diag}\,(\mu_1^{-\frac{1}{2}}, \mu_2^{-\frac{1}{2}}, \ldots, \mu_n^{-\frac{1}{2}}) \quad \text{(A1.23)}$$

that is, since $\mathbf{U}$ is unitary

$$\boldsymbol{\Delta}^{-\frac{1}{2}} = \mathbf{U}\boldsymbol{\mu}^{-\frac{1}{2}}\mathbf{U}^{\dagger}. \quad \text{(A1.24)}$$

As its name implies the symmetric orthogonalization procedure treats all members of the original basis on the same footing. This has the effect of preserving symmetry relations among the basis functions. For example, the six π-type atomic orbitals of benzene form the basis for a reducible representation of the symmetry group D_{6h}; the symmetrically orthogonalized basis transforms in exactly the same way, whereas Schmidt orthogonalized bases do not. Similar remarks apply to the construction of Bloch orbitals for crystals.

It is sometimes convenient to combine Schmidt orthogonalization and symmetric orthogonalization. Thus given a set of core orbitals and a set of valence orbitals, we first orthogonalize the core orbitals to each other and the valence orbitals to each other using symmetrical transformations (A1.15); the resulting set of orbitals is then Schmidt-orthogonalized with the core orbitals preceding the valence orbitals.

A1.3. Canonical orthogonalization[2-4]

Both the Schmidt and symmetrical orthogonalization schemes may encounter difficulties if there is exact or approximate linear dependence among the basis functions (A1.1). In such cases the canonical orthornormal orbitals defined by

$$\boldsymbol{\chi}^0 = \boldsymbol{\psi}\mathbf{U}\boldsymbol{\mu}^{-\frac{1}{2}} \tag{A1.25}$$

or

$$\chi_k^0 = \frac{1}{\sqrt{\mu_k}} \sum_j \psi_j U_{jk}, \tag{A1.26}$$

may be used. As is readily verified, the kth column of the matrix $\mathbf{U}$ which appears in equation (A1.26), is an eigenvector of $\mathbf{\Delta}$ with eigenvalue μ_k. Exact or approximate linear dependence among the ψ_j corresponds to zero or near zero eigenvalues μ_k. The corresponding functions χ_k^0 are then simply omitted from the orthonormalized basis which is now of order $n - q$, where q is the number of (near) zero eigenvalues.

References

1. P.-O. Löwdin (1950), *J. Chem. Phys.* **18**, 365.
2. P.-O. Löwdin (1970), *Advan. Quantum Chem.* **5**, 185.
3. P.-O. Löwdin (1967), *Int. J. Quantum Chem.* **1S**, 811.
4. K. R. Roby (1971), *Chem. Phys. Letts.* **11**, 6; ibid. (1972), **12**, 579.

APPENDIX 2

Character Tables and Basis Functions for the Crystallographic Point Groups

A2.1. Symmetry elements

(a) PURE ROTATIONS

E = identity.

C_n = rotation through $2\pi/n$ $\quad (n = 2, 3, 4, 6)$.

(b) IMPROPER ROTATIONS

i = inversion.

$\sigma = iC_2 = C_2 i$ = reflection in plane perpendicular to axis of C_2.

σ_h = reflection in a plane perpendicular to a principal axis of symmetry.

σ_v = reflection in a plane containing a principal axis of symmetry (e.g. σ_a, σ_b, σ_c for C_{3v}, cf. Section 4.1).

σ_d = reflection in a plane containing a principal axis of symmetry which bisects the angle between two two-fold rotation axes perpendicular to the principal axis (e.g. σ_{ab}, σ_{ac}, σ_{ad}, σ_{bc}, σ_{bd}, σ_{cd} for T_d, cf. Section 4.1).

$S_n = \sigma_h C_n$, $\qquad S_2 = \sigma_h C_2 = i$.

Basis functions for the various representations are built up from the functions x, y and z and quantities R_x, R_y and R_z which transform like the functions x, y and z, respectively, under pure rotations but are invariant under inversion.

$$i(x, y, z) = (-x, -y, -z), \qquad i(R_x, R_y, R_z) = (R_x, R_y, R_z).$$

A2.2. Direct product groups

For some groups (say P) with a centre of inversion, character tables are not given explicitly; they can be derived from those of the corresponding groups (Q) without inversion by replacing each symmetry species by two species, one even (g) the other odd (u) under inversion. This relationship between the groups is expressed by

$$P \equiv Q \times C_i.$$

A2.3. Characters and basis functions

C_2			E	$C_2(z)$	Bases for C_2	Bases for C_{1h}	Bases for C_i
	C_{1h}		E	$\sigma(xy)$			
		$C_i = S_2$	E	i			
A	A'	A_g	1	1	z, R_z	x, y, R_z	R_x, R_y, R_z
B	A''	A_u	1	−1	x, y, R_x, R_y	z, R_x, R_y	x, y, z

$$C_{2h} \equiv C_2 \times C_i$$

D_2		E	$C_2(z)$	$C_2(y)$	$C_2(x)$	Bases for D_2	Bases for C_{2v}
	C_{2v}	E	$C_2(z)$	$\sigma_v(xz)$	$\sigma_v(yz)$		
A	A_1	1	1	1	1		z
B_1	A_2	1	1	−1	−1	z, R_z	R_z
B_2	B_1	1	−1	1	−1	y, R_y	x, R_y
B_3	B_2	1	−1	−1	1	x, R_x	y, R_x

$$D_{2h} \equiv D_2 \times C_i$$

					Bases	Bases
S_4	E	$S_4(z)$	C_2	S_4^3	for S_4	
C_4	E	$C_4(z)$	C_2	C_4^3		for C_4
A	1	1	1	1	R_z	z, R_z
B	1	−1	1	−1	z, xy	xy
E	$\{1$	i	−1	$-i$	$x - iy, R_x + iR_y$	$x - iy, R_x - iR_y$
	$\{1$	$-i$	−1	i	$x + iy, R_x - iR_y$	$x + iy, R_x + iR_y$

$$C_{4h} \equiv C_4 \times C_i$$

D_4	E	$2C_4(z)$	$C_4^2 \equiv C_2''$	$2C_2$	$2C_2'$			
C_{4v}	E	$2C_4(z)$	$C_4^2 \equiv C_2''$	$2\sigma_v$	$2\sigma_d$	Bases	Bases	Bases
D_{2d}	E	$2S_4(z)$	$S_4^2 \equiv C_2''$	$2C_2$	$2\sigma_d$	for D_4	for C_{4v}	for D_{2d}
A_1	1	1	1	1	1		z	
A_2	1	1	1	−1	−1	z, R_z	R_z	R_z
B_1	1	−1	1	1	−1	$x^2 - y^2$	$x^2 - y^2$	$x^2 - y^2$
B_2	1	−1	1	−1	1	xy	xy	xy, z
E	2	0	−2	0	0	$(x, y), (R_x, R_y)$	$(x, y), (R_x, R_y)$	$(x, y), (R_x, R_y)$

$D_{4h} \equiv D_4 \times C_i$

C_3	E	C_3	C_3^2	Bases
A	1	1	1	z, R_z
E	1	ω	ω^2	$x - iy, R_x - iR_y$
	1	ω^2	ω	$x + iy, R_x + iR_y$

$\omega = \exp \frac{2\pi i}{3}$

$C_{3i} \equiv S_6 \equiv C_3 \times C_i$

D_3	E	$2C_3$	$3C_2$	Bases	Bases
C_{3v}	E	$2C_3$	$3\sigma_v$	for D_3	for C_{3v}
A_1	1	1	1		z
A_2	1	1	−1	z, R_z	R_z
E	2	−1	0	$(x, y), (R_x, R_y)$	$(x, y), (R_x, R_y)$

$D_{3d} \equiv D_3 \times C_i$

C_6		E	C_6	C_3	C_2	C_3^2	C_6^5	Bases	Bases
	C_{3h}	E	S_3	$S_3^2 = C_3^2$	$S^3 = \sigma_h$	$S_3^4 = C_3$	$S_3^5 = \sigma_h C_3^2$	for C_6	for C_{3h}
A	A'	1	1	1	1	1	1	z, R_z	R_z
B	A''	1	−1	1	−1	1	−1	$(x + iy)^3$	z
E_1	E''	1	ω	ω^2	−1	$-\omega$	$-\omega^2$	$x - iy$	$R_x + iR_y$
		1	$-\omega^2$	$-\omega$	−1	ω^2	ω	$x + iy$	$R_x - iR_y$
E_2	E'	1	ω^2	$-\omega$	1	ω^2	$-\omega$	$(x - iy)^2$	$x - iy$
		1	$-\omega$	ω^2	1	$-\omega$	ω^2	$(x + iy)^2$	$x + iy$

$$\omega = \exp\left(\frac{\pi i}{3}\right)$$

$C_{6h} \equiv C_6 \times C_i$

	D_6	E	$2C_6(z)$	$2C_6^2 \equiv 2C_3$	$C_6^3 \equiv C_2''$	$3C_2$	$3C_2'$			
	C_{6v}	E	$2C_6(z)$	$2C_6^2 \equiv 2C_3$	$C_6^3 \equiv C_2''$	$3\sigma_v$	$3\sigma_d$	Bases	Bases	Bases
D_{3h}		E	$2S_3(z)$	$2C_3(z)$	σ_h	$3C_2$	$3\sigma_v$	for D_6	for C_{6v}	for D_{3h}
A_1'	A_1	1	1	1	1	1	1		z	
A_2'	A_2	1	1	1	1	−1	−1	z	R_z	R_z
A_1''	B_1	1	−1	1	−1	1	−1	$y^3 - 3x^2y$	$x^3 - 3xy^2$	zR_z
A_2''	B_2	1	−1	1	−1	−1	1	$x^3 - 3xy^2$	$y^3 - 3x^2y$	z
E''	E_1	2	1	−1	−2	0	0	(x, y)	(x, y)	(R_x, R_y)
E'	E_2	2	−1	−1	2	0	0	$(x^2 - y^2, xy)$	$(x^2 - y^2, xy)$	(x, y)

$$D_{6h} \equiv D_6 \times C_i$$

T	E	$3C_2$	$4C_3$	$4C_3^2$	Bases
A	1	1	1	1	
E	$\{1$	1	ω	ω^2	$2z^2 - x^2 - y^2 - i\sqrt{3}(x^2 - y^2)$
	$\{1$	1	ω^2	ω	$2z^2 - x^2 - y^2 + i\sqrt{3}(x^2 - y^2)$
T	3	−1	0	0	$(x, y, z), (R_x, R_y, R_z)$

$$\omega = \exp \frac{2\pi i}{3}$$

$$T_h \equiv T \times C_i$$

O T_d	E E	$8C_3$ $8C_3$	$6C_2$ $6\sigma_d$	$6C_4$ $6S_4$	$3C_4^2 \equiv 3C_2''$ $3S_4^2 \equiv 3C_2$	Bases for O	Bases for T_d
A_1	1	1	1	1	1		
A_2	1	1	−1	−1	1	xyz	$R_xR_yR_z$
E	2	−1	0	0	2	$[2z^2 - x^2 - y^2, \sqrt{3}(x^2 - y^2)]$	$[2z^2 - x^2 - y^2, \sqrt{3}(x^2 - y^2)]$
T_1	3	0	−1	1	−1	$(x, y, z), (R_x, R_y, R_z)$	(R_x, R_y, R_z)
T_2	3	0	1	−1	−1	(yz, xz, xy)	(x, y, z)

$$O_h \equiv O \times C_i$$

Character tables and basis functions for the infinite groups $C_{\infty v}$ and $D_{\infty h}$ are given in Chapter 4 (Tables 4.6 and 4.7).

Author Index

The numbers in italics refer to the reference lists where the references are listed in full.

A

Ahlrichs, R., 165, *170*
Allen, L. C., 224, 284, 285, 289, 301, *308*
Alliluev, S. P., 47, *65*
Amemiya, A., 172, *197*

B

Bader, R. F. W., 305, *309*
Bagus, P. S., 253, 259, 261, *307*
Bailey, S. M., 292, 293, *308*
Bates, D. R., 44, 56, *65*
Battino, R., 149, *169*
Berlin, T., 305, *308*
Bingel, W., 165, *170*
Born, M., 1, 2, *15*
Boys, S. F., 264, 275, 277, *307*, *308*
Brandas, E., 127, *169*
Brillouin, L., 231, *307*
Bruner, B. L., 168, *170*
Buenker, R. J., 224, *307*
Byers-Brown, W., 145, *169*

C

Cade, P. E., 282, 283, 290, 297, 305, *308*, *309*
Clementi, E., 281, 284, 285, 287, 289, 296, 301, *308*
Cohen, E. R., 15, *16*
Cohen, M., 57, 59, *65*
Condon, E. U., 61, *65*
Conroy, H., 61, *65*, 168, *170*
Cooley, J. W., 149, *170*
Coolidge, A. S., 146, *169*
Cooper, I. L., 193, *197*
Coulson, C. A., 50, 57, *65*, 101, *103*, 114, *169*, 267, *308*

D

Das, G., 137, 142, 159, 160, 163, *169*
Davidson, E. R., 137, 138, 139, 157, 158, 159, 160, 161, *169*, *170*
de Heer, J., *197*
Dickinson, B. N., 54, *65*
Dirac, P. A. M., 17, *41*, 61, *65*, 194, *197*
Ditchfield, R., 296, 305, *308*
Domsky, B., 280, 281, *308*
DuMond, J. W. M., 15, *16*
Dunham, J. L., 9, *16*
Dunning, T. H., 262, 263, 264, *307*

E

Ebbing, D. D., 53, 55, *65*, 110, *169*
Edmiston, C., 270, *308*
Eliason, M. A., 111, *169*
England, W., 270, 274, *308*
Epstein, S. T., 31, *41*
Evans, W. H., 292, 293, *308*
Eyring, H., 4, *15*, 23, *41*, 193, *197*

F

Feynman, R. P., 20, *41*
Finkelstein, B. N., 51, *65*
Fischer, I., 101, *103*, 114, *169*
Fock, V., 236, *307*
Foster, J. M., 275, *308*
Froese, C., 248, *307*
Frost, A. A., 279, *308*

G

Gaydon, A. G., 293, *308*
Gianinetti, E., 189, *197*
Gilbert, T. L., 275, *308*

Goddard, W. A., *103*, 137, *169*, 262, 263, 264, *307*
Green, S., 304, *308*
Guillemin, V., 55, *65*
Gurney, E. F., 110, *169*

H

Hagstrom, S., 143, 159, 160, *169*
Hall, G. G., 66, *103*, 248, 273, *307*, *308*
Hamermesh, M., 66, 76, *103*
Handler, G. S., 57, 58, *65*
Harlow, I., 292, 293, *308*
Harris, F. E., *197*
Hartree, D. R., 129, *169*, 236, 248, *307*
Hehre, W. J., 296, 305, *308*
Heitler, W., 106, *169*
Hellmann, H., 20, *41*
Henneker, W. H., 305, *309*
Herzberg, G., 6, *16*, 123, 150, 153, *169*, *170*, 198, 219, 224, 225, 293, *307*
Hinkley, R. K., 284, 289, 300, 301, 303, 304, *308*
Hirschfelder, J. O., 7, 14, *16*, 111, *169*
Hollister, C., 295, *308*
Horowitz, G. E., 51, *65*
Houser, T. J., 57, 58, *65*
Howe, L. L., 150, *170*
Howell, K. M., 57, *65*
Huang, K., 2, *15*
Hulburt, H. M., 7, 14, *16*
Hund, F., 117, *169*, 198, *307*
Hunt, W. J., 262, 263, *307*
Huo, W. M., 290, *308*
Hurley, A. C., 11, *16*, 31, 33, *41*, 50, 53, 55, 61, *65*, 87, *103*, 109, 110, 127, 155, 165, 168, *169*, *170*, 220, 233, 236, 287, 290, 291, 301, 305, *307*, *308*, *309*
Huzinaga, S., 124, *169*, 280, 281, *308*

I

Ishiguro, E., 172, *197*

J

James, H. M., 54, *65*, 146, *169*
Jarmain, W. R., 11, *16*
Jones, L. L., 137, 138, 139, 157, 158, 159, 160, 161, *169*, *170*
Joy, H. W., 57, 58, *65*
Jucys, A. P., *103*

K

Kato, T., 145, *169*
Kayama, K., 45, 46, 47, *65*, 66, *103*
Kelly, H. P., 240, *307*
Kimball, G. E., 4, *15*, 23, *41*, 193, 197
Kimura, T., 172, *197*
Klein, O., 11, 12, 14, *16*
Kolos, W., 131, 146, 147, 151, 152, 153, 159, *169*, *170*
Koopmans, T. A., 233, *307*
Kotani, M., 45, 46, 47, *65*, 66, *103*, 172, *197*, 213, *307*
Krauss, M., 284, 289, 300, 301, 303, 304, *308*
Kutzelnigg, W., 165, *170*

L

Landau, L. D., 9, *16*
Ledsham, K., 44, 56, *65*
Lefebvre, R., 262, 263, 264, *307*
Lennard-Jones, J. E., 117, 134, *169*, 198, 267, *307*, *308*
Lewis, G. N., 291, 292, 293, 295, *308*
Lifshitz, E. M., 9, *16*
Lippincott, E. R., 9, 14, *16*
London, F., 106, *169*
Löwdin, P. O., *103*, 116, 121, 126, 131, 153, 156, *169*, *170*, 174, 177, 182, 196, *197*, 236, *307*, 312, 313, *314*
Lykos, P. G., 57, 58, *65*

Mc

MacDonald, J. K. L., 20, *41*
McLean, A. D., 142, 165, 168, *169*, 284, 286, 296, *308*
McWeeny, R., 66, 76, *103*, 193, 194, *197*
McWilliams, D., 280, 281, *308*

M

Magee, J. L., 110, *169*
Maslen, V. W., 220, *307*
Matcha, R. L., 290, *308*

Matsen, F. A., 56, *65*
Matveenko, A. V., 47, *65*
Mehler, E. L., 57, 58, *65*
Monfils, A., 150, *170*
Morse, P. M., 7, 14, *16*
Moskowitz, J. W., 289, 306, *308*
Mulliken, R. S., 117, 124, 125, *169*, 198, 208, *307*

N

Nesbet, R. K., 102, *103*, 215, 243, *307*
Neumann, D. B., 289, 306, *308*

O

O'Hare, P. A. G., 289, *308*
Ohno, K., 45, 46, 47, *65*, 66, *103*
Oppenheimer, J. R., 1, *15*

P

Pack, R. T., 145, *169*
Parker, V. B., 292, 293, *308*
Parr, R. G., 25, 31, *41*, 119, *169*
Pauling, L., 193, *197*
Pauncz, R., *197*
Peyerimhoff, S. D., 224, *307*
Phillipson, P. E., 124, 125, *169*
Pople, J. A., 134, *169*, 296, 305, *308*
Preuss, H., 126, *169*

R

Radom, L., 296, 305, *308*
Rajagopal, P., 264, 275, *307*, *308*
Randall, M., 291, 292, 293, 295, *308*
Rauk, A., 284, 285, 289, 301, *308*
Rees, A. L. G., 11, 14, *16*
Reeves, C., 193, *197*
Richards, W. G., 284, 289, 300, 301, 303, 304, *308*
Roby, K. R., 313, *314*
Roothaan, C. C. J., 131, 146, 152, *169*, 244, 248, 251, 253, 259, 261, 266, *307*
Rose, M. E., 61, *65*
Rothenberg, S., 285, 289, *308*
Ruedenberg, K., 270, 274, *308*
Rydberg, R., 11, 14, *16*

S

Sales, P. E., 282, 283, 297, *308*
Salmon, L. S., 270, *308*
Sandeman, I., 10, *16*
Schaefer, H. F., 285, 289, *308*
Schumm, R. H., 292, 293, *308*
Shortley, G. H., 61, *65*
Shull, H., 53, 55, 57, *65*, 110, 126, 143, 156, 159, 160, *169*, *170*
Simonetta, M., 189, *197*
Sinanoglu, O., 295, *308*
Slater, J. C., 23, *41*, 116, *169*, 172, 174, *197*, 203, 204, 214, 276, *307*, *308*
Snyder, L. C., 291, 296, *308*
Steele, D. S., 9, 14, *16*
Stewart, A. L., 44, 56, *65*
Stoicheff, B. P., 151, *170*
Sugiura, Y., 107, *169*
Sutcliffe, B. T., 193, 194, *197*

T

Teller, E., 56, *65*
ter Haar, D., 7, *16*
Thouless, D. J., 231, *307*

V

Vanderslice, J. T., 9, 14, *16*, 149, *169*
Vandoni, I., 189, *197*
Veillard, A., 281, 287, 289, *308*, *309*

W

Wagman, D. D., 292, 293, *308*
Wahl, A. C., 137, 142, 159, 160, 163, *169*, 282, 283, 289, 297, *308*
Walker, T. E. H., 284, 289, 300, 301, 303, 304, *308*
Walsh, A. D., 224, *307*
Walter, J., 4, *15*, 23, *41*
Wang, S. C., 110, *169*
Weinbaum, S., 112, *169*
Weiss, A., 142, 165, 168, *169*
Weissman, S., 149, *169*

Whitten, J. L., 278, *308*
Wigner, E. P., 66, 76, *103*
Wilkinson, P. G., 6, *16*
Wolniewicz, L., 147, 149, 151, 152, 153, 159, *169*, *170*
Wyatt, R. E., 31, *41*

Y

Yoshimine, M., 142, 165, 168, *169*, 284, 286, 296, *308*

Z

Zener, C., 55, *65*

Subject Index

A

Accidental degeneracy, 76
Accumulative accuracy in molecular Hartree–Fock calculations, 264–266
Adiabatic approximation, 1–3
Angular correlation, 140, 162
  momentum operators, 61–65
Antibonding molecular orbitals, 46
Antisymmetry principle, 171
Atomic orbitals, 48, 124
  units, 14, 15
Axial angular momentum, 43

B

BA, Best Atom, 280, 283
BA + P, Best Atom plus Polarization functions, 283
Barriers to internal rotation, 301
Basis functions, 129, 253, 276–282
  for representations of point groups, 315–319
Binding energy (D_e), 5
Bond eigenfunctions, 189–192
Bonding molecular orbitals, 47
Born–Oppenheimer approximation, 1, 3
Branching diagram for spin eigenfunctions, 187, 192
Brillouin's conditions, 226, 231
  Theorem, 231

C

Canonical Hartree–Fock (spin) orbitals, 229
  orthogonalization, 313
CGTO, Contracted Gaussian Type Orbital, 279
Chains, 193
Character of a representation, 84
  for molecular orbital spin multiplets, 212, 213, 214, 215
  tables for 32 crystallographic point groups, 315–319
Charge density function, 22, 50
Chemical accuracy, 266, 275
CI, Configuration Interaction, 122, 164, 166
Class (of conjugate elements), 67
Closed shell, 185, 199
  Hartree–Fock theory, 237
CMO, Canonical Molecular Orbital, 267
Coefficients for open shell Hartree–Fock theory, 253
Commutation relations of angular momentum operators, 61
Complete (discrete) set
  of functions, 57
  of ordered configurations, 182
Configuration interaction, 122–127, 166
Constrained variational calculations, 33–41
Continuous compact groups, 83
Contracted Gaussian-type-orbital (CGTO), 279
Correlating orbitals, 162, 164
Correlation diagrams, 203–210, 221, 222, 223
  for bent XH_2 molecules, 222
  for heteronuclear diatomic molecules, 209
  for homonuclear diatomic molecules, 207
  for linear XH_2 molecules, 221
Correlation energy, 131, 286, 287
Coulomb
  integral, 119, 242
  operator, 128, 227
  supermatrix, 261

Coulson–Fischer function, 101, 114–117
Covalent-ionic resonance, 113, 166
Cusps of wave functions
  at nuclei, 58, 59, 277–279
  at $r_{12} = 0$, 144, 145

D

Density matrices in Hartree–Fock theory, 257, 261
Determinantal method, 171–197
Determinantal wave functions, 172–174
  for open-shell states, 215–217
Diatomic molecule molecular orbitals (approximate), 210
Different orbitals for different spins, 243
Direct product representation, 87
Dissociation energy (D_0), 5
Dunham's method, 9–11
DZ, Double Zeta, 276

E

E chains, 194
Electronic Schrödinger equation, 2, 15
  formal solution by determinantal method, 181, 182
Electronic states
  of diatomic molecules and ions, 218, 219
  of XH_2 molecules, 225
Electrostatic
  formula, 22
  theorem, 26, 54, 60, 166
Elimination of off-diagonal Lagrange multipliers, 35, 247, 257
Elliptical coordinates, 42, 54
Empirical potential curves, 7–9
Energy
  curves for H_2, 108, 109, 121, 149
  units, conversion of, 15
Equilibrium nuclear separation, 4
Equivalent
  electrons, 211
  representations, 74
Exchange
  integral, 119, 242
  operator, 133, 227
  supermatrices, 261
Excited state orbitals, 164
Exclusive orbitals, 275

F

First-order
  configuration interaction (CI), 164
  energies of molecular orbital states, 202
Floating (wave-) function, 31, 53–54, 110, 111, 166
Fock operator, 128, 227

G

Gaussian–Lobe-function (GLF), 278
Gaussian-type-orbital (GTO), 276–282
Generalized valence-bond function (Heitler–London function with optimum orbitals), 136, 137, 167
GLF, Gaussian Lobe Function, 278
Group
  $C_{3v}(NH_3)$, 67–69, 77–79, 91–95, 317
  $C_{\infty v}, D_{\infty h}$ (linear molecules), 70, 80–82
  projection operators, 90–97
  T_d (CH_4), 69, 78–80, 319
GTO, Gaussian Type Orbital, 277

H

Hartree–Fock
  binding energies, 286–290
  charge density function and related properties, 303–307
  conditions, 227, 228
  dipole moments, 304
  dipole moments, forces and field gradients for CO, H_2CO, 306
  enthalpies of formation from hydrides, 294
  enthalpies of reaction, 291–295
  equations, 229
  ionization and excitation energies, 301–303
  method, 198–307
  orbitals, 127–130, 166
  orbitals by the expansion method, 253–261
  potential curves and surfaces, 297–301
  spectroscopic constants, 300
  theory, 127–130, 225–266
  total energies and binding energies for small molecules, 288, 289
  wave functions for N_2 and NH_3, 282–285

Heitler–London
function with optimum orbitals (generalized valence-bond), 136, 137, 166
theory of H_2, 106, 117, 166
Hellmann–Feynman
formulae, 20–26
theorems, 26–33
Hermitian conjugate of a matrix, 37
HF, Hartree–Fock, 225
HFR, Hartree–Fock–Roothaan, 275
H_2 ground state, summary of calculations, 166, 167
Hulburt–Hirschfelder curve, 8
Hybridized LCAO function, 54
Hydrogen-like orbitals, 55
Hydrogen molecule ion, 42–65
calculated equilibrium constants, 55

I

In-out correlation, 138, 162
Integral Hellmann–Feynman formula, 25, 167
theorem, 32, 60, 167
Ionic-covalent resonance, 113, 166
Irreducible representations, 76
Islands, 193

J

James–Coolidge
method, 144–153, 166
type calculations, 166, 168
J–C, James–Coolidge, 159

K

Kinetic Energy Operators, 1
Koopmans' theorem, 233, 302, 303

L

Lagrange multipliers, 33, 229, 243, 247
LCAO, Linear Combination of Atomic Orbitals, 48
LMO, Localized Molecular Orbitals, 26
Left–right correlation, 138, 162
Linear
combination of atomic orbitals (LCAO) 48–54, 119–122, 166
independence, 310
variation method, 19, 20
Localization criterion for orbitals, 269, 275
Localized molecular orbitals, 266–275
for NH_3, 270–273

M

Matrix Elements between Slater determinants, 174–181
formulae for general spin-orbitals, 177
formulae for orthonormal spin-orbitals, 178, 179
Matrix elements between
spin eigenfunctions, 192–197
symmetry adapted functions, 96
symmetry orbitals, 96
Matrix notation, 37
Minimum energy principle, 17–19
MO, Molecular Orbital, 117
Molecular electronic states from equivalent electrons, 214
Molecular orbital (MO) method, 117–137, 166, 198–307
Morse curve, 7, 8
MSO, Molecular spin orbital, 200
Multi-configuration self consistent field (MC–SCF) function, 142, 143, 156, 160, 166
Multiplets (sets of angular momentum eigenfunctions), 61, 183

N

Natural orbital, 154, 166
expansion (NOE), 153–165, 166
Nesbet's method of symmetry and equivalence restrictions, 243
NO, Natural Orbital, 156
NOE, Natural Orbital Expansion, 166
Non-crossing rule, 205, 206
NSO, Natural spin-orbital, 155
Nuclear wave equation for diatomic molecules, 3
Numerical integration in molecular Hartree–Fock theory, 264–266

O

O chains, 193
ODC, Optimum Double Configurations, 132–137
One-centre expansions, 55–61
Open shell, 200, 210, 211

Open shell Hartree–Fock theory, 242–253
  coefficients for simple configurations, 253
  range of applicability of simple version, 252
Optimized valence configurations, 143
Optimum double configurations (ODC) 132–137
Optimum orbital calculations, 166, 168
Orbital angular momentum operators, 62
Orbitals, 44
Ordered configurations, 182
Orthogonality relations
  for characters, 85
  for matrix representation of a group, 82–85
Orthogonalized
  atomic orbitals, 116
  basis, 39–41, 262–264
Orthonormalization, 310–314
OVC, Optimized Valence Configuration, 143

P

Parity under inversion, 44
Parr's
  formula, 25
  theorem, 26, 32
Pauli
  (exclusion) principle, 104
  matrices, 63
Permutation group, 172
Phase convention for surface harmonics, 62
Potential energy curves and surfaces 1–16
Projection operator, 34, 248, 249
Projected unrestricted
  Hartree–Fock theory, 236
  variational calculation 99–103
PUHF, Projected unrestricted Hartree–Fock, 236

R

Reducible representations, 76
Reduction
  of a representation, 85–89
  of direct products (C_{3v}, T_d, $C_{\infty v}$, $D_{\infty h}$), 88, 89
Representations of (symmetry) group, 73
Restricted
  Hartree–Fock theory (for open shell states), 242–253
  variational calculation, 99–103
Results of molecular orbital calculations, 275–307
RHF, Restricted Hartree–Fock, 243
RKR, Rydberg-Klein-Rees method, 11
Rumer diagram, 189–192

S

SA, Separated Atom, 46, 287
SCF, Self Consistent Field, 130, 275
Schmidt orthogonalization, 180, 181, 209–210, 310
Second order configuration interaction (CI), 164–165
Secular equation, 20, 122, 182
Separated atom
  designation, 46, 207
  states, 290
Shell of (molecular) orbitals, 199
Shift operators (step-up and step-down operators), 62, 186
σ-Limit wave function, 138–140, 166
Singularities of wave function
  at nuclei, 58, 59, 277–279
  at $r_{12} = 0$, 144, 145
Slater determinant, 172–174
Slater-type-orbitals (STO), 142, 276–282
Spherical harmonics, 55
Spin
  eigenfunctions, 182–185
  functions, 61–65
  multiplet, 183–189
  orbitals, 64, 172
  projected Slater determinants, 197
  projection operators, 194–197
Step-up and step-down operators (shift operators), 62, 186
STO, Slater Type Orbital, 98, 276
Sub-group, 67
Supermatrices in Hartree–Fock theory, 261
Superposition
  of configurations: *see* second order configuration interaction (CI)
  patterns, 193

Surface harmonics, 55
Symmetric
orthogonalization, 312
permutation group, 213
Symmetry
adapted basis, 96, 259
dilemma, 103
groups of molecules, 66–70
of molecular electronic states, 66–103
of molecular orbital states, 210–217
orbitals, 91–96
paradox, 97–103
SZ, Single Zeta, 276

T

Transformation
of (electronic wave) functions, 70–72
to orthonormal basis, 179, 310
Turn over rule, 195

U

UA, United Atom, 44, 287
UHF, Unrestricted Hartree–Fock, 225
United atom states, 44, 290
Unrestricted Hartree–Fock
theory, 225–237, 243
(UHF) Hamiltonian, 227
Unrestricted variational calculation, 99–103

V

Valence atomic orbital (VAO) calculations, 166, 168
Valence bond (VB) functions, 123, 166, 189–194, 197
VAO, Valence Atomic Orbital, 166
Variational
calculations with linear constraints, 33–41
criterion for Hellmann–Feynman theorem, 29
method, 17–41, 171
method for excited states, 19
wave functions, 17–41
VB, Valence Bond, 123
Vector notation, 37
Vibrational term values, 6
Virial
formula, 23
theorem, 26, 51, 60, 167
Virtual orbitals, 130, 164

W

Walsh diagram for XH_2 molecules, 223
WKBJ, Wigner–Kramers–Brillouin–Jeffreys approximation, 8

Z

Zeroth approximation of molecular orbital method, 198–203